USING ARTIFICIAL INTELLIGENCE IN CHEMISTRY AND BIOLOGY

A Practical Guide

USING ARTIFICIAL INTELLIGENCE IN CHEMISTRY AND BIOLOGY

A Practical Guide

Hugh Cartwright

Chapter 10, *Evolvable Developmental Systems*, contributed by

Nawwaf Kharma

Associate Professor, Department of Electrical and Computer Engineering,
Concordia University, Montreal, Canada

CRC Press
Taylor & Francis Group
Boca Raton London New York

CRC Press is an imprint of the
Taylor & Francis Group, an **informa** business

CRC Press
Taylor & Francis Group
6000 Broken Sound Parkway NW, Suite 300
Boca Raton, FL 33487-2742

© 2008 by Taylor & Francis Group, LLC
CRC Press is an imprint of Taylor & Francis Group, an Informa business

---

Library of Congress Cataloging-in-Publication Data

Cartwright, Hugh M., 1948-
    Using artificial intelligence in chemistry and biology : a practical guide /
Hugh Cartwright ; chapter 10, Evolvable developmental systems, contributed by
Nawwaf Kharma.
        p. cm.
    Includes bibliographical references and index.
    ISBN 978-0-8493-8412-7 (acid-free paper) 1. Chemistry--Data processing. 2.
Biology--Data processing. 3. Artificial intelligence. I. Kharma, Nawwaf. II.
Title.

QD39.3.E46C375 2008
540.285'63--dc22   2007047462

Visit the Taylor & Francis Web site at
http://www.taylorandfrancis.com

and the CRC Press Web site at
http://www.crcpress.com

Contents

Preface

The use of Artificial Intelligence within science is growing at a remarkable rate. In the early 1990s, little was known about how Artificial Intelligence could be applied in a practical way to the physical and life sciences. At that stage, few experimental scientists showed any interest in the area. Now, hundreds of research papers are published each year and the numbers are rising rapidly. The change has been dramatic and yet, despite the growth, the field is still very young.

The upsurge of interest owes much to an increasing understanding of the scientific potential of Artificial Intelligence. This book explains in a lucid and straightforward way how these methods are used by scientists and what we can accomplish with them. Recognizing that not all experimental scientists are computer experts, the approach adopted here assumes no prior knowledge of Artificial Intelligence and no unusual skills in computer science or programming. It does, however, presume some scientific background. Each chapter is designed to take the reader quickly to the point at which meaningful scientific applications can be investigated.

Computer scientists may use this book to gain a clearer picture of how experimental scientists use Artificial Intelligence tools. Chemists, biochemists, physicists, and others in the experimental sciences who have data to analyze or simulations to run will find tools within these pages that may speed up their work or make it more effective. For both groups, the aim of this book is to encourage a broader application of these methods.

Many people have contributed to the production of this book. The final chapter was written by Nawwaf Kharma, following several months on sabbatical at Oxford University. His perceptive and challenging chapter gives a glimpse of the future of Artificial Intelligence in science. EJS, a software tool for the construction of simulations in Java, has been used to perform a number of calculations for this book. A complete version of EJS is included on the accompanying CD with the permission of Francisco Esquembre, to whom I am most grateful. Most of the figures in this text have been prepared by John Freeman, adding both clarity and humor to it. The editorial staff at Taylor & Francis Group, particularly Lance Wobus, Russ Heaps, and Pat Roberson, have been a regular source of encouragement and expert advice. Numerous members of my research group, past and present, have contributed to the ideas that are crystallized here, notably "the two Alexes,"whose comments about the interface between Artificial Intelligence and science are often thought-provoking. Finally, my wife, Susie, tolerating my long and unsociable working hours, and my daughter, Jenny, a distant but engaging e-mail voice, have provided less tangible but invaluable contributions. I am grateful to them all.

Errors are almost unavoidable in a text of this sort and I shall be grateful to be notified of any that may be spotted by readers.

Hugh Cartwright

Chemistry Department, Oxford University
St Anne's College and Oriel College, Oxford, United Kingdom

The Author

Hugh M. Cartwright is a member of the Chemistry Department at Oxford University in the United Kingdom, where he is Laboratory Officer. He is also a Supernumerary Fellow at St. Anne's College, Oxford, and a Lecturer in Physical Chemistry at Oriel College, Oxford. His research interests lie in the broad field of the application of computational methods from Artificial Intelligence to problems within science. Past and current research, carried out by some seventy students, has covered a wide variety of areas. Typical of these are studies of biodegradation, the prediction of the properties of polymers, the analysis of oil spill data, automatic elucidation of scientific rules, the design of organic synthetic routes, the optimization of solid-state phosphor composition, quantitative structure-activity relationships (QSARs), unraveling mass spectrometric data from crude oils, and the analysis of large biomedical datasets. The work of his group focuses primarily on the use of genetic algorithms, neural networks, self-organizing maps, and growing cell structures.

He has written, edited, and provided invited chapters in a number of textbooks on the use of Artificial Intelligence in science and continues to run an active research group in the area. He has taught at universities in Canada, the United States, Ireland, and the United Kingdom, and is a board member of the CoLoS (Conceptual Learning of Science) International Consortium.

1

Artificial Intelligence

Computers don't understand. At least, they don't understand in the way that we do. Of course, there are some things that are almost beyond comprehension: Why do women collect shoes? What are the rules of cricket? Why do Americans enjoy baseball? And why are lawyers paid so much?

These are tough questions. As humans, we can attempt to answer them, but (at least for the next few years) computers cannot help us out. Humans possess far more "intelligence" than computers, so it is to be expected that if a human struggles to understand shoe collecting, a computer will be totally baffled by it. This makes it all the more surprising that computers that use "artificial intelligence" can solve a variety of scientific tasks much more effectively than a human.

Artificial Intelligence (AI) tools are problem-solving algorithms. This book provides an introduction to the wide range of methods within this area that are being developed as scientific tools. The application of AI methods in science is a young field, hardly out of diapers in fact. However, its potential is huge. As a result, the popularity of these methods among physical and life scientists is increasing rapidly.

In the conventional image of AI, we might expect to see scientists building laboratory robots or developing programs to translate scientific papers from Russian to English or from English to Russian, but this does not accurately reflect the current use of AI in science. The creation of autonomous robots and the perfection of automatic translators are among the more important areas of research among computer scientists, but the areas in which most experimental scientists are involved are very different.

Science is dominated by problems; indeed, without them there would be hardly any science. When tackling a new problem, scientists are generally less concerned about *how* they solve it, provided that they get a solution, than they are about the quality of the solution that is obtained. Thus, the pool of mathematical and logical algorithms that scientists use for problem solving continues to grow. That pool is now being augmented by AI methods.

Until the second half of the 1990s, experimental scientists, by and large, knew little about AI, suspecting (without much evidence) that it might be largely irrelevant in practical science. This turned out to be an unduly pessimistic view. AI algorithms are in reality widely applicable and the power of these methods, allied to their simplicity, makes them an attractive proposition in the analysis of scientific data and for scientific simulation. The purpose of this book is to introduce these methods to those who need to solve scientific problems, but do not (yet) know enough about AI to take advantage of it. The text includes sufficient detail that those who have some modest programming skills, but no prior contact with AI, can rapidly set about using the techniques.

1.1 What Is Artificial Intelligence?

Computer scientists usually define AI in terms of what it can accomplish. They are especially interested in the creation of intelligent software that can reproduce the kind of behavior that humans would recognize as intelligent, such as understanding language or conducting a conversation.

Some of this work on human-like behavior spills over into the scientific laboratory where work progresses on producing intelligent robotic systems for chemical or biological laboratories. Robotic sample handlers are commonplace in science, and already incorporate enough rudimentary intelligence to be able to cope with the demands of the job. However, a robotic sampler is really pretty dumb and, in truth, that is how it should be. A sampler whose role is to feed samples into a mass spectrometer should do no more than unquestioningly follow predefined rules of priority. If it had ideas above its station and started making its own decisions about whose sample would be processed next, based on the color of the users' hair or the number of syllables in their name perhaps, it would soon be headed for the great sampler scrap heap in the sky.

Experimental scientists tend not to be interested in robotic samplers, even those with attitude; instead they want software that will solve problems. For them, AI tools may be of considerable value as problem solvers. In this book, we explore such uses.

For most scientific purposes, a reasonable working definition of AI is: "A computer program that can learn."

Although computer scientists might quibble at this definition, it describes most AI methods rather well. Methods that depend on learning, such as artificial neural networks and genetic algorithms, are already proving their worth in science, and there are others, such as growing cell structure networks, whose potential is just starting to be recognized.

The learning that AI algorithms depend on can be achieved in several different ways. An expert system (Chapter 7) is tutored by a human expert and starts off resembling an empty box awaiting information. This requires that a scientist spend long periods feeding relevant information into the system. By contrast, artificial neural networks (Chapter 2) inspect examples of what they are required to understand and learn by extracting examples from a database or by interacting with a simulation. Classifier Systems (Chapter 9) talk directly with their environment, attempting to control it, receiving feedback from it, and learning from any mistakes. The self-organizing map (Chapter 3) makes its deductions by just looking at data without having to be told what it is expected to learn.

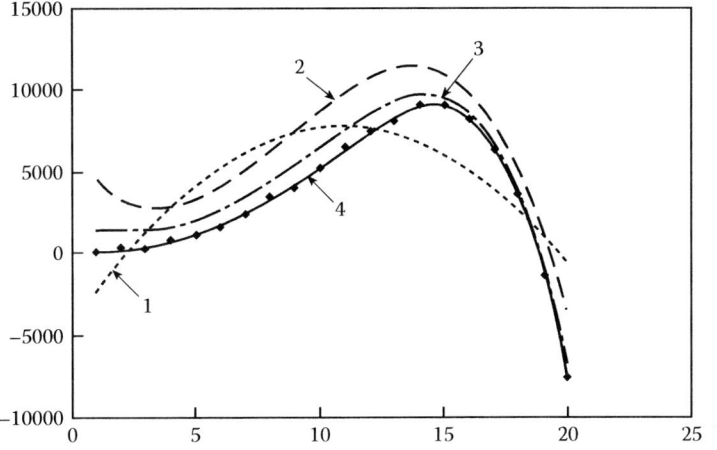

FIGURE 1.1
Progressive improvements in the attempts by a genetic algorithm to find a curve that fits a set of data points well (lines 1 to 3). The solid line is fitted using standard least squares.

But, you may have spotted a problem here. If an AI program has to learn, does that not suggest that its performance might be fairly hopeless when we first ask it to solve a problem? Indeed it does, as Figure 1.1 indicates.

The figure shows a polynomial fit to a dataset calculated according to a standard least squares algorithm (solid line); this is compared with a series of attempts to find a fit to the same data using a genetic algorithm.

The genetic algorithm is unimpressive, at least to begin with (line 1). The quality of the fit that it finds does improve (lines 2 and 3) and eventually it will reach a solution that matches the least squares fit, but the algorithm takes far longer to find the solution than standard methods do, and the fit is no better.

1.2 Why Do We Need Artificial Intelligence and What Can We Do with It?

So why bother? If established and reliable methods exist to solve scientific problems, surely there is no point in using methods from AI that may be slower or less effective? The key here is the phrase: "if established and reliable methods exist." In many situations, there is no need to turn to AI methods, but not every scientific problem is as straightforward as finding the line of best fit. When more conventional methods are unavailable, or are of insufficient power, an AI method may be just what is needed.

There are several reasons why it may be difficult to find an adequate solution to a scientific problem; the two most common are scale and complexity.

In many types of problems, the number of possible solutions can be huge. The folding of a protein is a typical example. Proteins fold to take up a conformation of minimum energy, although the conformation that the protein adopts in a living organism may not be that of the global minimum energy. The number of ways in which a large molecule, such as a protein, can be folded so that its energy is at least locally minimized and all bond lengths and angles are reasonable is so large that it is impossible to investigate all of them computationally. As it is not feasible to check every conformation, some search method is needed that makes intelligent decisions about which conformations to assess and which to ignore.

A problem of similar scale is the identification of useful drugs from among the huge number that could, in principle, be prepared. The number of molecules that might potentially act as drugs can only be estimated, but it is probably of the order of 10^{40}. Of this huge number of possibilities only a tiny fraction will ever be synthesized, and of those only a handful will make it through the process of high throughput screening to human trials. Any computational procedure that is devised to screen molecules for possible value as drugs must be able, somehow, to search through this huge number of candidates to identify promising structures without actually inspecting every molecule (Figure 1.2).

The chemical flowshop, which we shall meet in Chapter 5, is similarly challenging to solve. Many different chemicals are synthesized within a single line of processing units in a defined sequence. The efficiency of the flowshop, which is determined by the length of time required to synthesize the complete group of chemicals, is strongly correlated with the order in which chemicals are made. Therefore, determining a near-optimum order is important in the operation of flowshops if the process is to be financially viable. For n chemicals, the number of possible sequences is $n!$, a number

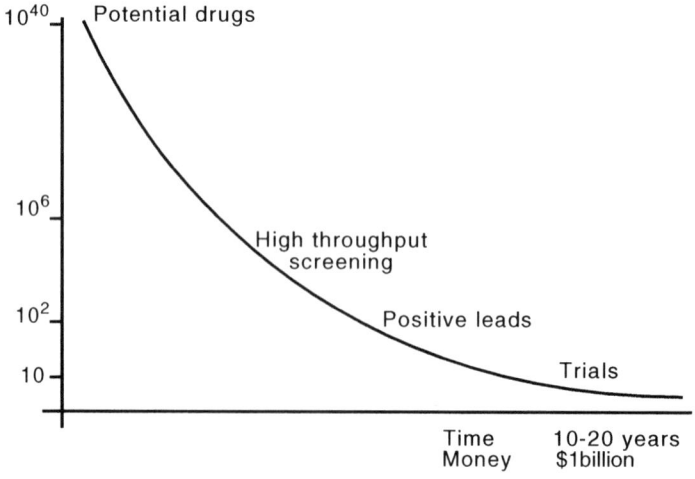

FIGURE 1.2
The drug development process.

that, for a typical shop producing twenty chemicals, is much too large to allow every sequence to be investigated individually.

A second reason why AI is of value to scientists is that it offers powerful tools to cope with complexity. In favorable circumstances, the solutions to problems can be expressed by rules or by a well-defined, possibly trivial, model. If we want to know whether a compound contains a carbonyl group, we could record its infrared spectrum and check for a peak near 1760 cm^{-1}. The spectrum, paired with the rule that ketones generally show an absorption in this region, is all that we need. But other correlations are more difficult to express by rules or parametrically. What makes a good wine? We may (or may not) be able to recognize a superior wine by its taste, but would have considerable difficulty in determining whether a wine is good, or even if it is palatable, if all we had to go on was a list of the chemicals of which it is comprised.

We shall meet a number of examples in this book of the sort of scientific problems in which AI can be valuable, but it will be helpful first to view a few typical examples.

1.2.1 Classification

Scientists need to classify and organize complex data, such as that yielded by medical tests or analysis via GC-MS (gas chromatography-mass spectrometry). The data may be multifaceted and difficult to interpret, as different tests may conflict or yield inconclusive results. Growing cell structures may be used to assess medical data for example, such as that obtained from patient biopsies, and determine whether the test results are consistent with a diagnosis of breast cancer.[1]

1.2.2 Prediction

The prediction of stable structures that can be formed by groups of a few dozen atoms is computationally expensive because of the time required to determine the energy of each structure quantum mechanically, but such studies are increasingly valuable because of the need in nanochemistry to understand the properties of these small structures. The genetic algorithm is now widely used to help predict the stability of small atomic clusters.[2]

1.2.3 Correlation

The relationships between the molecular structure of environmental pollutants, such as polychlorinated biphenyls (PCBs), and their rate of biodegradation are still not well understood, though some empirical relationships have been established. Self-organizing maps (SOMs) have been used to rationalize the resistance of PCBs to biodegradation and to predict the susceptibility to degradation of those compounds for which experimental data are lacking.[3] The same technique has been used to analyze the behavior of lipid bilayers, following a molecular dynamics simulation, to learn more about how these layers behave both in natural systems and when used as one component in biosensors.[4]

1.2.4 Model Creation

Neural networks are extensively used to develop nonparametric models and are now the method of choice when electronic noses are used to analyze complex mixtures, such as wines and oils.[5] Judgments made by the neural network cannot rely on a parametric model that the user has supplied because no model is available that correlates chemical composition of a wine to the wine's taste. Fortunately, the network can build its own model from scratch, and such models often outperform humans in determining the composition of oils, perfumes, and wines.

1.3 The Practical Use of Artificial Intelligence Methods

AI methods may be used in various ways. The models may be used as a standalone application, e.g., in recent work on the design of microwave absorbers using particle swarm optimization (PSO).[6] Alternatively, a computational tool, such as a finite element analysis or a quantum mechanical calculation, may be combined with an AI technique, such as an evolutionary algorithm.

The relationships between the principal AI methods of value in science are shown in Figure 1.3.

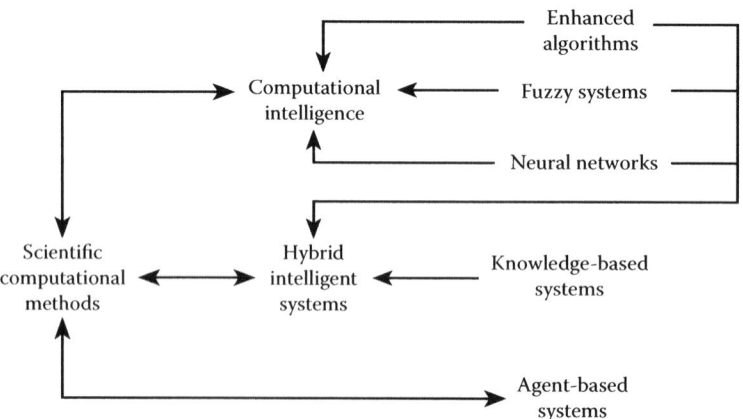

FIGURE 1.3
Relationships between the AI methods of the most current value in science.

1.4 Organization of the Text

Each of the remaining chapters in this book, other than the final one, addresses a different area of AI. The methods are described in a nontechnical way, but in sufficient detail, and with mathematical details introduced as needed that scientists of most backgrounds should be able to follow the description and determine how they might use the methods in their own work. Although the approach assumes no prior knowledge of the field, there is enough depth that a scientist with some understanding of computer programming should be able to rapidly write working programs to solve scientific problems.

Each chapter on technique ends with the short section called: "Where Do I Go from Here?" This includes suggestions for those who wish to investigate a topic further. It is not possible for a book of this nature to cover all issues of implementation for each technique. We assume that those who want to pursue a topic further will have access to library facilities and to the Internet, so no attempt has been made to provide a comprehensive reading list. Regular journal articles provide overviews of research in each area, so the section on applications toward the end of each chapter similarly provides examples of how each method is used rather than a comprehensive review.

The chapters close with a few problems whose purpose is to encourage a deeper investigation of the chapter material, often with suggestions for computational exercises.

In the final chapter, Nawwaf Kharma looks ahead to some of the methods that experimental scientists may be using in the coming years. While the main aim of the book is to provide an introduction to AI methods, Kharma delves more deeply into some challenging new areas, opening a window on how a computer scientist views the use of AI to solve practical problems.

The CD that accompanies this text contains a full version of the EJS (Easy Java Simulations) software. Written by Francisco Esquembre (University of Murcia, Spain) and collaborators, this Java tool can be used to investigate many of the techniques described in this book. It is of particular value in providing a variety of plotting tools that can be invoked in a simple way. Sample programs and data are provided, details of which can be found on the ReadMe file on the CD.

Finally, a word on terminology. Many AI methods learn by inspecting examples, which come in a variety of forms. They might comprise a set of infrared spectra of different samples, the abstracts from a large number of scientific articles, a set of solid materials defined by their composition and their emission spectrum at high temperature, or the results from a series of medical tests. In this text, we refer to these examples, no matter what their nature, as "sample patterns."

It is time to turn to our first AI method: Chapter 2, Artificial Neural Networks.

References

1. Walker, A.K., Cross, S.S., and Harrison, R.F., Visualisation of biomedical data-sets by use of growing cell structure networks: A novel diagnostic classification technique, *Lancet*, 354, 1518, 1999.
2. Johnston, R.L., et al., Application of genetic algorithms in nanoscience: Cluster geometry optimisation. *Applications of Evolutionary Computing, Proceedings, Lecture Notes in Computer Science*, Springer, Berlin, 2279, 92, 2002.
3. Cartwright, H.M., Investigation of structure-biodegradability relationships in polychlorinated biphenyls using self-organising maps, *Neural Comput. Apps.*, 11, 30, 2002.
4. Murtola, T., et al., Conformational analysis of lipid molecules by self-organizing maps, *J. Chem. Phys.*, 125, 054707, 2007.
5. Escuderos, M.E., et al., Instrumental technique evolution for olive oil sensory analysis, *Eur. J. Lipid Sci. Tech.*, 109, 536, 2007.
6. Goudos, S.K. and Sahalos, J.N., Microwave absorber optimal design using multi-objective particle swarm optimisation, *Microwave Optic. Tech. Letts.*, 48, 1553, 2006.

2

Artificial Neural Networks

The role of an artificial neural network is to discover the relationships that link patterns of input data to associated output data. Suppose that a database contains information on the structure of many potential drug molecules (the input) and their effectiveness in treating some specific disease (the output). Since the clinical value of a drug must in some way be related to its molecular structure, correlations certainly exist between structure and effectiveness, but those relationships may be very subtle and deeply buried.

Although the ways in which the effectiveness of a drug is linked to its molecular shape, the dipole moment and other molecular features (so-called *molecular descriptors*) are incompletely understood, this in no way diminishes the importance of uncovering these relationships. If pharmaceutical companies were to neglect computer studies before starting on synthetic work, the development of new drugs would return to the time-consuming and expensive synthesize-and-test methods that were the only feasible means of developing new drugs until late in the twentieth century.

The links between the molecular descriptors for a large group of drugs and their efficacy comprise a pattern of relationships, so discovering these links is a pattern-matching problem. "Pattern matching" might at first seem a field of rather restricted interest to an experimental scientist, related more to image processing than to the physical or life sciences. In fact, pattern matching is a very broad subject, in which links between data items of any type are sought. Numerous types of problems in science can be cast in the form of a pattern-matching problem, and artificial neural networks are among the most effective methods within machine learning for revealing the links. Once trained, artificial neural networks are fast in operation and are particularly valuable in circumstances where the volume of raw data that is available is large, but not enough is known about the relationships within the data to allow the scientist to build an analytical model.

2.1 Introduction

Pattern matching may sound dull or esoteric, but we all use an excellent, free pattern-matching tool every day—the human brain excels at just this sort of task. Babies are not taught how to recognize their mother, yet without instruction quickly learn to do so. By the time we are adults, we have become

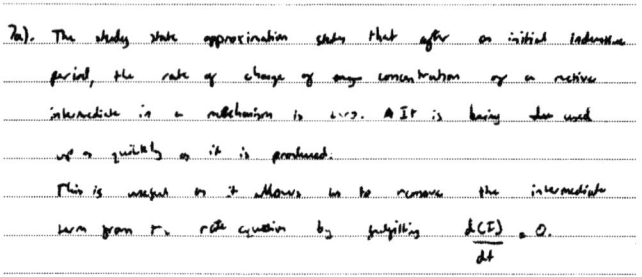

FIGURE 2.1
The handwriting of a typical chemist is (just) readable.

so accomplished at reading handwriting that we can read words written in a
style that we have never encountered before and in which individual letters
may be barely distinguishable (Figure 2.1).

The brain's remarkable ability to learn through a process of pattern recognition suggests that, if we wish to develop a software tool to detect patterns
in scientific or, indeed, any other kind of data, the structure of the brain
could be a productive starting point. This view led to the development of
artificial neural networks (ANNs). The several methods that are gathered
under the ANN umbrella constitute some of the most widely used applications of Artificial Intelligence in science. Typical areas in which ANNs are of
value include:

- Analysis of infrared spectra, in which molecules that share structural features, such as a common functional group, give rise to
  spectra with spectral similarities. ANNs can be used to link spectral features, such as strong absorption at a particular energy, with
  molecular structure and, thus, help in the interpretation of spectra.

- The prediction of the properties of molecules from a knowledge
  of their structure (quantitative structure-property relationships
  [QSPRs] or quantitative structure-activity relationships [QSARs]).
  ANNs can be used to determine QSPRs or QSARs from experimental data and, hence, predict the properties of a molecule, such as its
  toxicity in humans, from its structure.

- Industrial process control, when the interrelation of process variables cannot be expressed analytically. ANNs can be used to find
  empirical relationships between process control parameters and the
  performance of an industrial plant when these cannot be readily
  derived from theory.

- The optimization of the formulation of fuels. ANNs can be used to
  develop relationships between the chemical composition of fuels
  and their performance, thus allowing the prediction of the properties of fuels that have yet to be formulated in the laboratory or tested
  in the field.

- "Electronic noses." ANNs can take the electronic signals provided by sensors and use these signals to recognize complex mixtures, such as oils, wines, and fragrances.

Artificial neural networks are as common outside science as they are within it, particularly in financial applications, such as credit scoring and share selection. They have even been used in such eccentric (but, perhaps, financially rewarding) activities as trying to predict the box office success of motion pictures.[1]

2.2 Human Learning

In the human brain, very large numbers of neurons are connected in a three-dimensional mesh of great complexity (Figure 2.2). Each neuron is a simple computational unit which, on its own, can accomplish little. When many of them are joined in a large-scale structure, however, the ensemble is capable of impressive feats of computation (such as making sense of the words on this page).

Because of the scale and complexity of the brain, it will be some years before the 10^{11} neurons in a typical human brain and the 10^{15} connections between them can be fully modeled in software. Not only is the brain very complex, it is fast; we can recognize the face of a friend within a few hundred milliseconds by comparing it against the many thousands of faces that we have seen in the past. It is evident, therefore, that spreading a demanding computational task across large numbers of simple units can be very efficient.

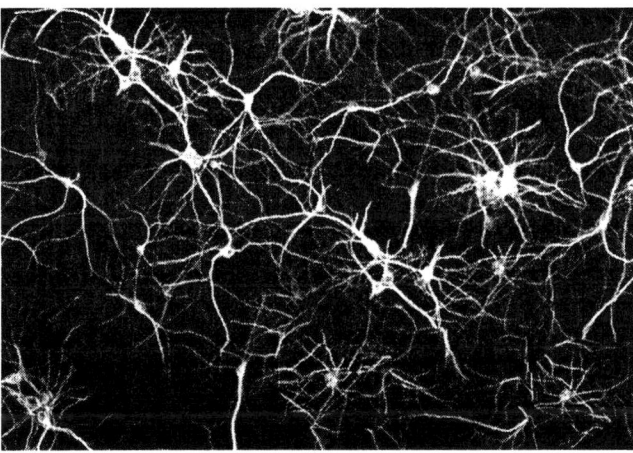

FIGURE 2.2
The complex pattern of interneuron connections within the brain.

FIGURE 2.3
A novice user's view of the operation of an artificial neural network.

Any software model of the brain that we could create using a current PC could emulate only an infinitesimal amount of it, so it might seem that our expectations of what an ANN could accomplish should be set at a thoroughly modest level. Happily, the reality is different. Though computer-based neural networks are minute compared with the brain, they can still outperform the brain in solving many types of problems.

The first contact that many scientific users have with neural networks is through commercially available software. The manuals and Help systems in off-the-shelf neural network software may not offer much on the principles of the method, so, on a first encounter, ANNs might appear to be black boxes whose inner workings are both mysterious and baffling (Figure 2.3).

This cautious and distant view of the innards of the box is no disadvantage to the user if the network is already primed for use in a well-defined domain and merely waiting for the user to feed it suitable data. Scientists, though, will want to employ ANNs in innovative ways, taking full advantage of their potential and flexibility; this demands an appreciation of what lies within the box. It is the purpose of this chapter to lift up the lid.

2.3 Computer Learning

The design of ANNs drew initial inspiration from the structure of the brain. It was presumed that a software model based around an understanding of how the brain was constructed and how individual neurons function would be best able to emulate the brain's abilities in pattern recognition (Figure 2.4). However, as the field has developed, computer scientists have come to the view that, though the human brain is adept at pattern recognition, there is no reason to believe that the method that it uses is the only way, or even the best way among several, of recognizing and making sense of patterns.

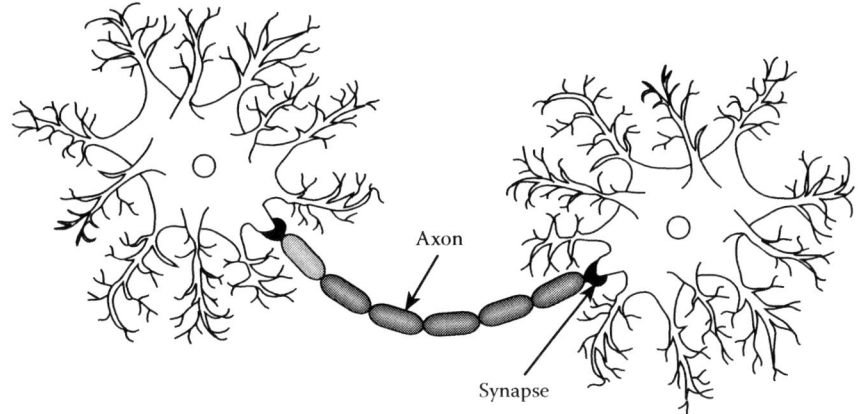

FIGURE 2.4
A schematic view of two neurons in the brain.

Consequently, the direction of development of artificial neural networks is no longer influenced by the desire to prepare an accurate model in silicon of the brain in the hope and expectation that this must form the ideal route to a problem-solving tool. Instead, computer scientists take a pragmatic view, tweaking, expanding, and molding neural network models in any way that might improve performance. Inevitably, the models that have been developed in recent years have tended to diverge from models of the way that the brain is believed to operate, but this is not undesirable, provided that the new methods can at least match nature-based models in speed and power. Although little attention is now paid to whether any particular modification to an ANN model brings it more or less into line with the operation of the brain, the resulting computer models still show intriguing areas of correspondence with their biological counterparts.

In the human brain, it is the combined efforts of many neurons acting in concert that creates complex behavior; this is mirrored in the structure of an ANN, in which many simple software processing units work cooperatively. It is not just these artificial units that are fundamental to the operation of ANNs; so, too, are the connections between them. Consequently, artificial neural networks are often referred to as *connectionist models*.

While ANNs cannot do anything that would be impossible to accomplish using an alternative algorithmic method, they can execute tasks that would otherwise be very difficult, such as forming a model from sensory data or from data that is extracted from a continuous industrial process for which no comprehensive theoretical model exists. Their principal limitation is that the numerical model that they create, while open to inspection, is difficult to interpret. This is in marked contrast to the model used by expert systems, in which the knowledge of the system is expressed in readily interpreted statements of fact and logical expressions. Nevertheless, the power of the ANN is so considerable that, even though the model that it generates cannot always be translated into plain English rules and equations, its development is still justified by superior performance.

2.4 The Components of an Artificial Neural Network

Four components are needed to form a conventional artificial neural network:

1. Some *nodes*.
2. The *connections* between them.
3. A *weight* associated with each connection.
4. A recipe that defines how the output from a node is determined from its input.

2.4.1 Nodes

ANNs are built by linking together a number of discrete nodes (Figure 2.5). Each node receives and integrates one or more *input signals*, performs some simple computations on the sum using an *activation function*, then outputs the result of its work. Some nodes take their input directly from the outside world; others may have access only to data generated internally within the network, so each node works only on its local data. This parallels the operation of the brain, in which some neurons may receive sensory data directly from nerves, while others, deeper within the brain, receive data only from other neurons.

2.4.2 Connections

Small networks that contain just one node can manage simple tasks (we shall see an example shortly). As the name implies, most networks contain several nodes; between half a dozen and fifty nodes is typical, but even much larger networks are tiny compared with the brain. As we shall see in this chapter and the next, the topology adopted by a network of nodes has a profound influence on the use to which the network can be put.

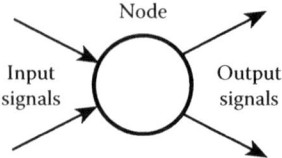

FIGURE 2.5
A node, the basic component in an artificial neural network.

Nodes are joined by virtual highways along which messages pass from one node to another. As well as possessing internal connections, the network must be able to accept data from the outside world, so some network connections act as inputs, providing a pipeline through which data arrive. The input to the network can take many forms. It might consist of numerical data drawn from a database, streams of numbers generated by a piece of simulation software, the RGB (red, green, blue) values of pixels from a digital image, or the real-time signal from a sensor, such as a pH electrode or a temperature probe.

A network that computed, but kept the result of its deliberations to itself, would be valueless, thus in every network at least one output connection is present that feeds data back to the user. Like the input data, this output can be of several different types; it could be numerical, such as a predicted boiling point or an LD_{50} value. Alternatively, the network might generate a Boolean value to indicate the presence or absence of some condition, such as whether the operating parameters in a reactor were within designated limits. The network might output a string to be interpreted as an instruction to turn on a piece of equipment, such as a heater. Networks may have more than one output connection and then are not limited to a single type of output. A Boolean output which, if set, would sound an alarm to alert the user to a drop in the temperature of a reaction cell, could be combined with a real-valued output that indicated the level to which a heater in the cell should be set if this event occurred.

2.4.3 Connection Weights

Associated with each internode connection is a connection weight. The connection weight is a scaling factor that multiplies the signal traveling along the connection. Collectively, these weights comprise the memory of the network. Their values are determined during a training process, so for each application to which the ANN is applied the weights are different; finding their values is the most important step in using a neural network.

When first set to work on a new problem, the network knows nothing and this is reflected in the connection weights, all of which are given random values. A process of *supervised learning* (also known as *supervised training*) then takes place, during which the network is shown a large number of examples of the sort of information that it is required to learn, and, in response, adjusts its connection weights in order to give meaningful behavior. A network in which all the weights have been determined is fully trained and ready for use. Section 2.5 describes how this training is performed.

2.4.4 Determining the Output from a Node

There are two stages to the calculation of the output from a node. The first is to calculate the total input to the node; that input is subsequently fed into a function that determines what signal the node will output (Figure 2.6).

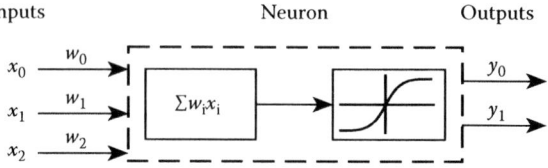

FIGURE 2.6
The calculations performed by an artificial neuron.

2.4.4.1 The Input Signal

Any node that is not an input node (we will learn more about input nodes in section 2.7) receives signals from other nodes and carries out a computation on those signals; this computation is straightforward. The node multiplies each of the signals that arrive through the incoming connections by the weight on that connection. It sums the results to give the total incoming signal, *net*:

$$net = \sum_{i=0}^{n} w_i x_i \tag{2.1}$$

The summation in equation (2.1) is over all $n + 1$ inputs to the node. x_i is the signal traveling into the node along connection i and w_i is the weight on that connection. Most networks contain more than one node, so equation (2.1) can be written in the slightly more general form:

$$net_j = \sum_{i=0}^{n} w_{ij} x_{ij} \tag{2.2}$$

in which w_{ij} is the weight on the connection between nodes i and j, and the signal travels down that connection in the direction $i \rightarrow j$.

Example 1: A Simple Network

The simple network shown in Figure 2.7 contains just one active node. There are three inputs; in this application we choose to feed in the Cartesian coordinates X and Y of a data point through two of the inputs; the third input is provided by a *bias node*, which produces a constant signal of 1.0. The values shown beside each connection are the connection weights.

As we shall see in the next section, the output of a node is computed from its total input; the bias provides a threshold in this computation. Suppose that a node follows a rule that instructs it to output a signal of +1 if its input is greater than or equal to zero, but to output zero otherwise. If the input signal from the bias node, after multiplication by the connection weight, was +0.1, the remaining inputs to the node would together have to sum to a value no smaller than –0.1 in order to trigger a

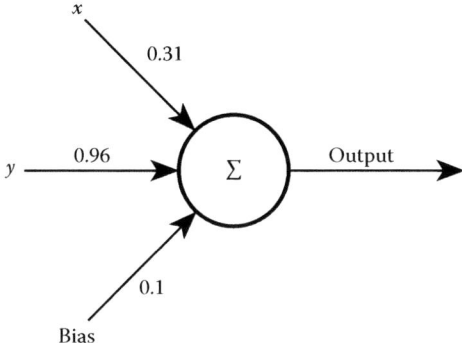

FIGURE 2.7
A simple artificial neural network.

nonzero output from the node. It is in this sense that input from the bias node can be regarded as providing a threshold signal.

For the node shown, if $X = 3$ and $Y = 7$, the total input signal at the node is

$$net = 3 \times 0.31 + 7 \times 0.96 + 1.0 \times 0.1 = 7.75 \qquad (2.3)$$

2.4.4.2 The Activation Function

The node uses the total input to calculate its *output signal* or *activation, y_j*. This is the signal that the node sends onwards to other nodes or to the outside world. To calculate its output, the node feeds net_j into an *activation function* or *squashing function*:*

$$y_j = f_j(net_j) \qquad (2.4)$$

Several types of activation functions have been used in ANNs; the *step* or *binary threshold activation function* is one of the simplest (Figure 2.8).** A node that employs the binary activation function first integrates the incoming signals as shown in equation (2.1). If the total input, including that coming in on the connection from the bias node, is below a threshold level, T, the node's output is a fixed small value Φ, which is often chosen to be zero. If the summed input is equal to or greater than this threshold, the output is a fixed, larger value, usually set equal to one.

---

* The curious name "squashing function" reflects the action of the function. Activation functions can take as input any real number between $-\infty$ and $+\infty$; that value is then squashed down for output into a narrow range, such as {0, +1} or {–1, +1}.
** The step function is also known as a Heaviside function.

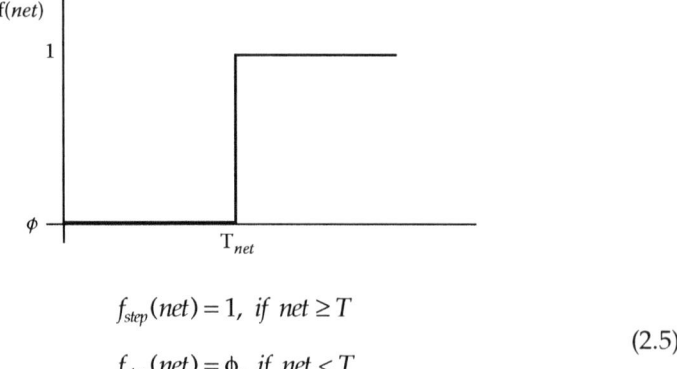

$$f_{step}(net) = 1, \ if \ net \geq T$$
$$f_{step}(net) = \phi, \ if \ net < T$$

(2.5)

FIGURE 2.8
A binary or step activation function.

Nodes that employ a step activation function are sometimes known as threshold logic units (TLUs).

The *sign function* is a second activation function with an abrupt threshold T, this time set at zero input (Figure 2.9). The output from a node that uses this function is +1 if the input is zero or greater; otherwise it is –1.

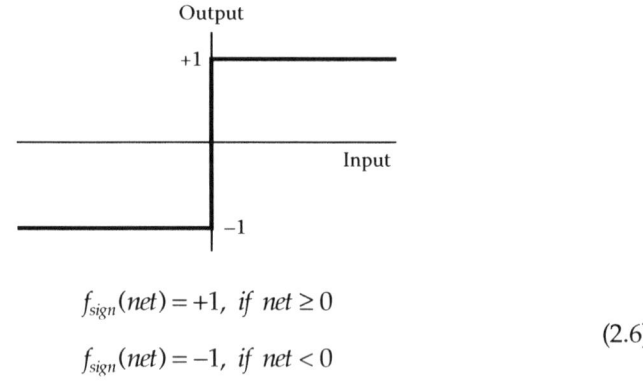

$$f_{sign}(net) = +1, \ if \ net \geq 0$$
$$f_{sign}(net) = -1, \ if \ net < 0$$

(2.6)

FIGURE 2.9
The sign function.

The binary threshold activation function was used in early attempts to create ANNs because of a perceived parallel between this function and the way that neurons operate in the brain. Neurons require a certain level of activation before they will "fire," otherwise they are quiescent. A TLU functions in the same way.

Although a TLU has features in common with a neuron, it is incapable of acting as the building block of a versatile computer-based learning network. The reason is that the output from it is particularly uninformative. The output signal tells us only that the sum of the weighted input signals did, or did not, exceed the threshold T. As the response from the unit is the same whether the input is fractionally above the threshold or well beyond it, almost all information about the size of the input signal is destroyed at the node. A network of TLUs has few applications in science unless the input signals are themselves binary, when the use of a binary activation function does not unavoidably cause loss of information.

If the input data are not binary, an activation function should be chosen that allows the node to generate an output signal that is related in some way to the size of the input signal. This can be accomplished through several types of activation function.

The *linear activation function* passes the summed input signal directly through to the output, possibly after multiplication by a scaling factor.

$$y_j = f_j(net_j) = k \times net_j = k \times \sum_{i=0}^{n} w_{ij} x_{ij} \tag{2.7}$$

In equation (2.7), k is a scaling factor and usually $0 < k \le 1.0$. The simplest linear activation function is the *identity function,* in which k is 1.0, thus the output from a node that uses the identity function is equal to its input.

$$y_j = f_j(net_j) = net_j = \sum_{i=0}^{n} w_{ij} x_{ij} \tag{2.8}$$

The output from a linear function may be capped so that input and output are linearly related within a restricted range. Beyond these limits, either positive or negative, the output does not vary with the size of the incoming signal, but has a fixed value (Figure 2.10). Capping is used to prevent signals from becoming

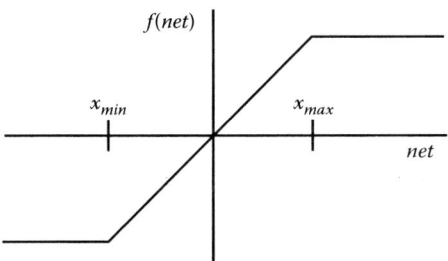

$$y = k \times net, \ x_{min} \le x \le x_{max}$$
$$y = -\Phi, \ x < x_{min} \tag{2.9}$$
$$y = +\Phi, \ x > x_{max}$$

FIGURE 2.10
A capped linear scaling activation function.

unreasonably large as they pass through the network, which may happen in a large network if the connection weights are much greater than one.

Example 2: A Simple Classification Task

In our first working network, shown in Figure 2.11, the single node uses the identity function to determine its output. This little network can be used to perform a simple classification of two-dimensional data points. Suppose that a group of data points whose Cartesian coordinates are $\{X_n, Y_n\}$ can be divided by a straight line into two sets: set 0 comprising points that lie below the line and set 1, those points that lie above the line (Figure 2.12). Given the coordinates of an arbitrary point $[X_i, Y_i]$, the network can determine the group within which the point lies.

To classify a point, its Cartesian coordinates are fed into the network. If the output from the network is less than zero, this indicates that the point lies in group 1, while if the output is greater than zero, the point is a member of group 0. Suppose that we feed in a point whose coordinates are $[X = 3, Y = 7]$; the input to the node is easily calculated from equation (2.1) to be –7.8. Since the node uses the identity activation function, the output from the node is also –7.8, which is less than zero, so the point [3, 7] is correctly allocated to group 1.

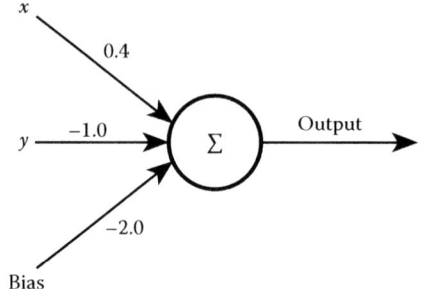

FIGURE 2.11
A one-node network that can tackle a classification task.

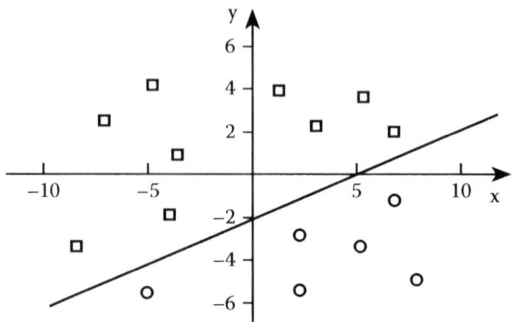

FIGURE 2.12
A set of data points that can be separated into two groups by a single straight line.

How did the network do this? Encoded in the network's connection weights is the equation of the line that separates the two groups of points; this is

$$Y = 0.4 \ X - 2.0 \tag{2.10}$$

In the network in Figure 2.11, the connection weight on the X input signal and the bias node are equal to the coefficients of X^1 and X^0 in equation (2.10), while the connection weight on the Y input is -1.0. When the node calculates the sum of the weighted inputs, it is computing:

$$net_i = (0.4X - 2.0) - Y \tag{2.11}$$

The term in brackets is the value of Y calculated using equation (2.10). The actual value of Y for the point is subtracted from this, and, if the result is negative, the point lies above the line that divides the two groups and, therefore, is in group 1, while, if the value is greater than zero, the point lies below the line and is in group 0. If the value is zero, the point lies exactly on the line.

2.5 Training

The network in Figure 2.11 can allocate any two-dimensional data point to the correct group, but it is only able to do this because it was prepared in advance with suitable connection weights. This classification problem is so simple that the connection weights can be found "by inspection," either by calculating them directly or by testing a few straight lines until one is found that correctly separates the points. Not every problem that we meet will be as straightforward as this.

Indeed, if the problem is simple enough that the connection weights can be found by a few moments work with pencil and paper, there are other computational tools that would be more appropriate than neural networks. It is in more complex problems, in which the relationships that exist between data points are unknown so that it is not possible to determine the connection weights by hand, that an ANN comes into its own. The ANN must then discover the connection weights for itself through a process of supervised learning.

The ability of an ANN to learn is its greatest asset. When, as is usually the case, we cannot determine the connection weights by hand, the neural network can do the job itself. In an iterative process, the network is shown a sample pattern, such as the X, Y coordinates of a point, and uses the pattern to calculate its output; it then compares its own output with the correct output for the sample pattern, and, unless its output is perfect, makes small adjustments to the connection weights to improve its performance. The training process is shown in Figure 2.13.

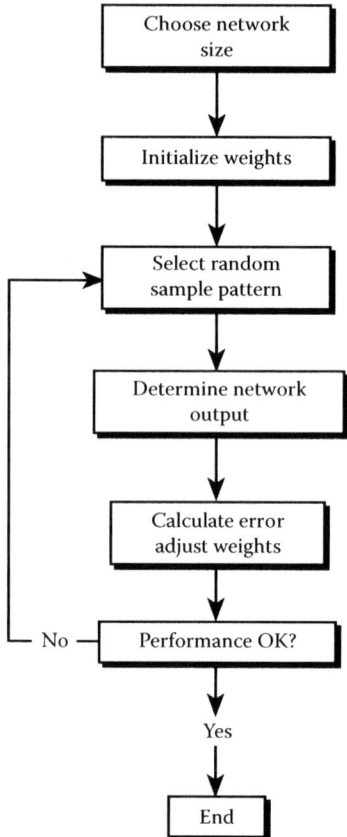

FIGURE 2.13
The steps in training an artificial neural network.

The first step in the preparation from scratch of a working ANN is to decide how many nodes it should contain and to set the weights of the connections between them to random values. Once this has been done, a sample pattern is drawn from the database, together with the output (known as the *target response)* that the network will give if it is doing its job perfectly. If the network was being trained for the classification problem, the input might be $X = 3$, $Y = 7$ and the desired output -7.8. The pattern is fed through the network and the output from the network is compared with the target response. The difference between the target response and the output of the network is a measure of how well the network is performing.

To improve performance, the connection weights are now adjusted by an amount that is related to how good the match was between target response and network output. This is measured by the error, δ, which, if the network contains just a single output node, is the difference between the output of the node, y, and the target response, t.

$$\delta = t - y \tag{2.12}$$

If the actual output and the target output are identical, the network has worked perfectly for this sample, therefore, no learning need take place. Another sample is drawn from the database and fed into the network and the process is continued. If the match is not perfect, the network needs to learn to do better. This is accomplished by adjusting the connection weights so that the next time the network is shown the same sample it will provide an output that is closer to the target response.

The size of the adjustment, Δw, that is made to the weight on a connection into the node is proportional both to the input to the node, x, and to the size of the error:

$$\Delta w = \eta \delta x \tag{2.13}$$

The input to the node enters into this expression because a connection that has sent a large signal into the node bears a greater responsibility for any error in the output from the node than a connection that has provided only a small signal. In equation (2.13), η is the learning rate, which determines whether changes to the connection weights should be large or small in comparison with the weight itself; its value is chosen by the user and typically is 0.1 or less. The connection weight is then updated:

$$w_{new} = w_{old} + \Delta w \tag{2.14}$$

This process is known as the *delta rule*, or the *Widrow–Hoff rule* and is a type of gradient descent because the size of the change made to the connection weights is proportional to the difference between the actual output and the target output. Once the weights on all connections into the node have been adjusted, another pattern is taken from the database and the process is repeated. Multiple passes through the database are made, every sample pattern being included once in each cycle or *epoch*, until the error in the network's predictions becomes negligible, or until further training produces no perceptible improvement in performance.

Example 3: Training

Suppose that Figure 2.7 shows the initial connection weights for a network that we wish to train. The first sample taken from the database is $X = 0.16$, $Y = 0.23$, with a target response of 0.27. The node uses the identity function to determine its output, which is therefore:

$$0.16 \times 0.31 + 0.96 \times 0.23 + 0.1 \times 1.0 = 0.3704. \tag{2.15}$$

As the target response is 0.27, the error is −0.1004. If the learning rate is 0.05, the adjustment to the weight on the x connection is

$$0.05 \times (-0.1004) \times 0.31 = -0.00016$$

so the new weight is

$$0.31 - 0.00016 = 0.3084.$$

2.6 More Complex Problems

The separation of points into two groups by a single straight line is an example of a *linearly separable problem*. The Boolean AND function is linearly separable, as Figure 2.14 illustrates. In this function $Y(x_1, x_2)$, Y is equal to 1 only if x_1 and x_2 are both equal to 1. The region of the figure in which $Y = 1$ can be divided from the region in which $Y = 0$ by a single straight line.

Problems that are not linearly separable are easy to generate. The classic example is the XOR function, Figure 2.15, in which $Y(x_1, x_2)$ equals one if just one of x_1 and x_2 equals 1, but is zero if x_1 and x_2 have the same value. In this

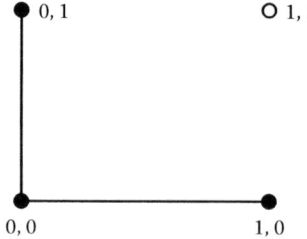

FIGURE 2.14
The Boolean AND function. A single straight line can separate the open and filled points.

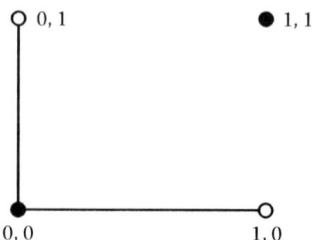

FIGURE 2.15
The Boolean XOR function. Two straight lines are required to divide the area into regions that contain only one type of point.

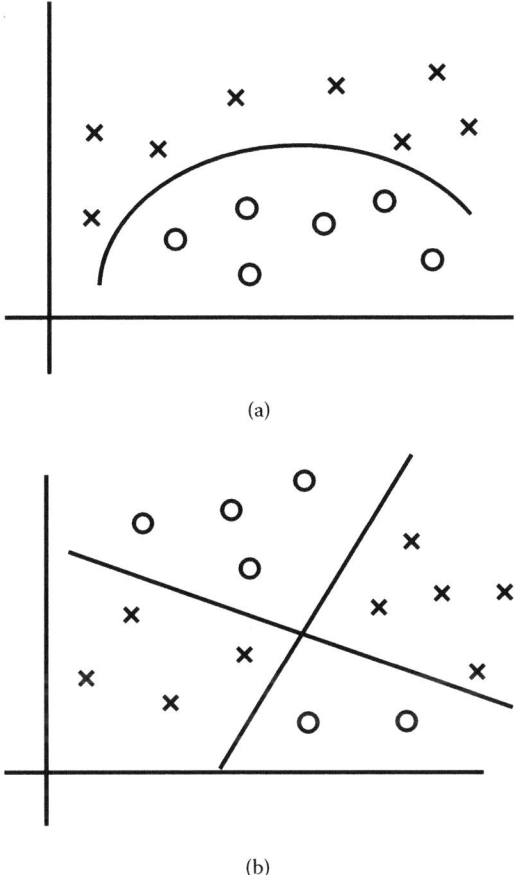

FIGURE 2.16
Scientific data points usually require curves, several lines, or both to divide them into homogeneous groups.

instance, two straight lines are required to separate the space into regions in which Y has the same value.

Most scientific problems are nonlinear, requiring (if the data points are two-dimensional) either a curve (Figure 2.16a) or more than one line (Figure 2.16b) to separate the classes.

Although some problems in more than two dimensions are linearly separable (in three dimensions, the requirement for linear separability is that the points are separated by a single plane, Figure 2.17), almost all problems of scientific interest are not linearly separable and, therefore, cannot be solved by a one-node network; thus more sophistication is needed. The necessary additional power in the network is gained by making two enhancements: (1) the number of nodes is increased and (2) each node is permitted to use a more flexible activation function.

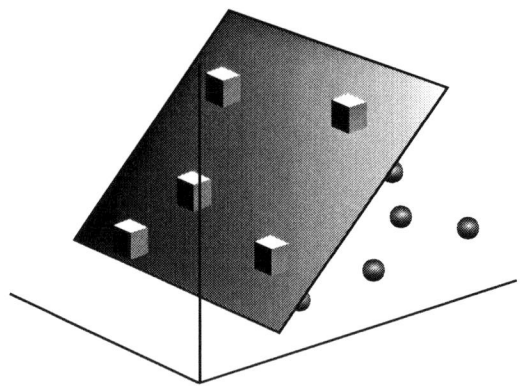

FIGURE 2.17
Linear separability in three dimensions.

2.7 Layered Networks

The connections within a "network" that consists of just one node can only link it with its environment, but as soon as the number of nodes in a network is increased, choices exist in the way that the nodes are connected. Two nodes could be arranged in parallel, so that both accept input data and both provide output (Figure 2.18). Alternatively, nodes could be placed in a serial configuration, so that the output of one node becomes the input of the second (Figure 2.19).

When the network contains more than two nodes, the options increase even further. Nodes might be arranged in a chain, as a single layer, or some combination of the two; they might all receive input signals from the outside world or just some of them may; the bias node might be connected to every node or just to a selection of them; there may be *recursive* links (Figure 2.20); and so on, so a rich variety of geometries is possible.

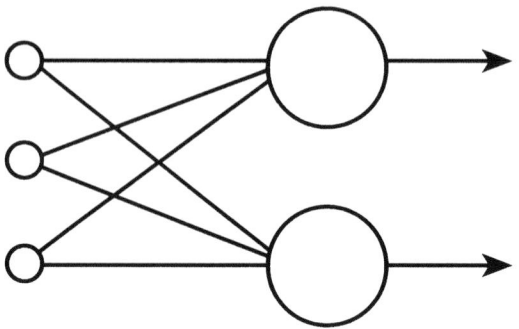

FIGURE 2.18
A parallel arrangement of nodes.

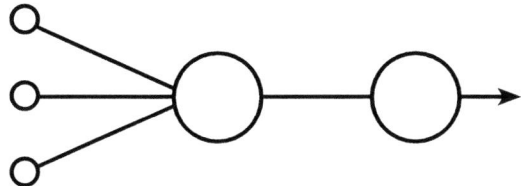

FIGURE 2.19
A serial arrangement of nodes.

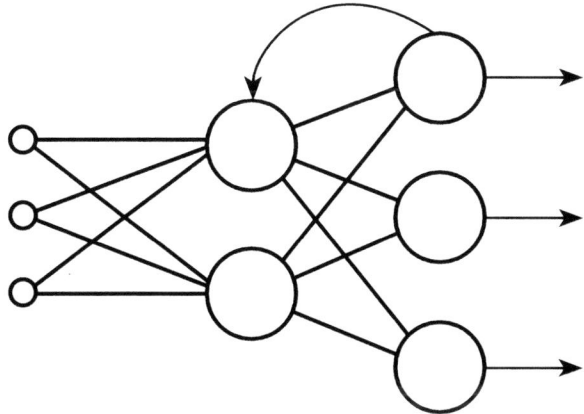

FIGURE 2.20
A network that contains a recursive (backward) link.

In the brain, connections between neurons are quite numerous and apparently random, at least at the local level. The impressive logical power of the brain might suggest that we should aim for a network designed along similarly random lines, but this type of disorganized network is hard to use: Where do we feed data in? Where are the outputs? What is to stop the data from just going round and round in an endless loop? Even worse, such a network is very difficult to train. Instead, the most common type of artificial neural network is neither entirely random nor completely uniform; the nodes are arranged in layers (Figure 2.21). A network of this structure is simpler to train than networks that contain only random links, but is still able to tackle challenging problems.

One layer of input nodes and another of output nodes form the bookends to one or more layers of *hidden nodes.** Signals flow from the input layer to the hidden nodes, where they are processed, and then on to the output nodes, which feed the response of the network out to the user. There are no recursive links in the network that could feed signals from a "later" node to an "earlier" one or return the output from a node to itself. Because the messages in this type of layered network move only in the forward direction when input data are processed, this is known as a *feedforward network*.

---

* So-called because nodes in the hidden layer have no direct contact with the outside world.

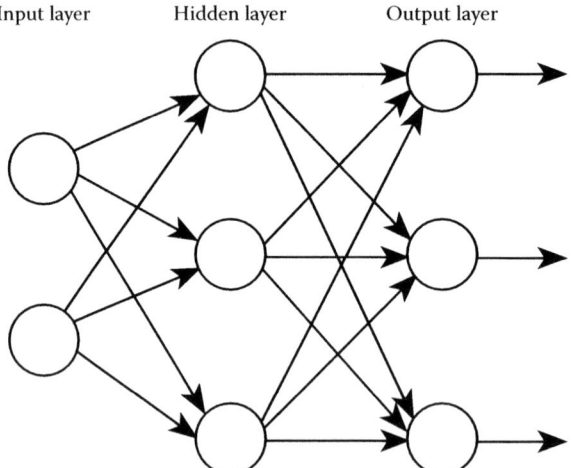

FIGURE 2.21
A feedforward neural network.

Feedforward networks that employ a suitable activation function are very powerful. A network that contains a sufficient number of nodes in just one hidden layer can reproduce any continuous function. With the addition of a second hidden layer, noncontinuous functions can also be modeled.

There is some disagreement in the published literature about whether the type of network shown in Figure 2.21 should be referred to as a two-layer network or a three-layer network. The three-layer fan club argues that it is self-evident that the network contains three layers—just count them. Those supporting the two-layer terminology point out that the first layer actually does no processing, but merely distributes the input signals to the hidden nodes, therefore, there are only two active layers. We shall refer to this type of network as a three-layer model, but there are no right answers in this argument and one should remain aware of the ambiguity.

2.7.1 More Flexible Activation Functions

The creation of a layered structure that includes one or more hidden layers of nodes as well as input and output layers takes us halfway along the road to a more versatile network; the journey is completed by the introduction of a more responsive activation function.

Although the linear activation function passes more information from the input to a node to its output than a binary function does, it is of limited value in layered networks as two nodes in succession that both use a linear activation function are equivalent to a single node that employs the same function, thus adding an extra layer of nodes does not add to the power of the network. This limitation is removed by the use of curved activation functions.

2.7.1.1 Sigmoidal Activation Functions

Sigmoidal activation functions generate an output signal that is related uniquely to the size of the input signal, so, unlike both binary and capped linear activation functions, they provide an unambiguous transfer of information from the input side of a node to its output.

A commonly used sigmoidal function is the logistic function (Figure 2.22).

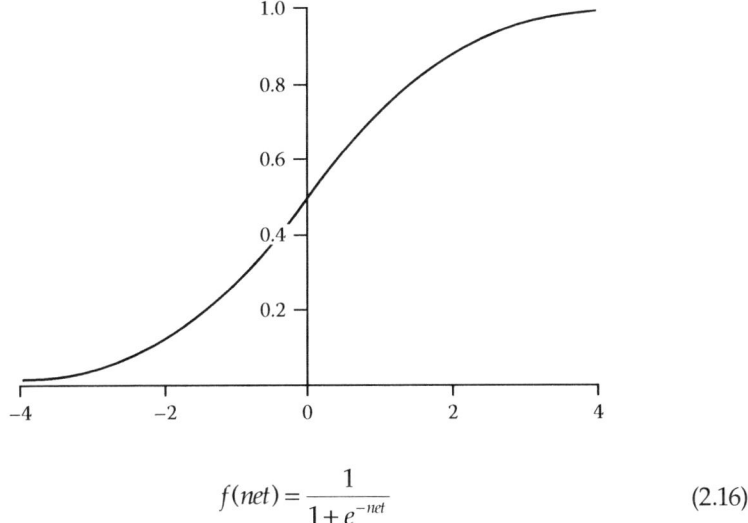

$$f(net) = \frac{1}{1 + e^{-net}} \tag{2.16}$$

FIGURE 2.22
The logistic activation function.

From any real input, this provides an output in the range {0, 1}. The Tanh function (Figure 2.23) has a similar shape, but an output that covers the range {–1, +1}. Both functions are differentiable in closed form, which confers a speed advantage during training, a topic we turn to now.

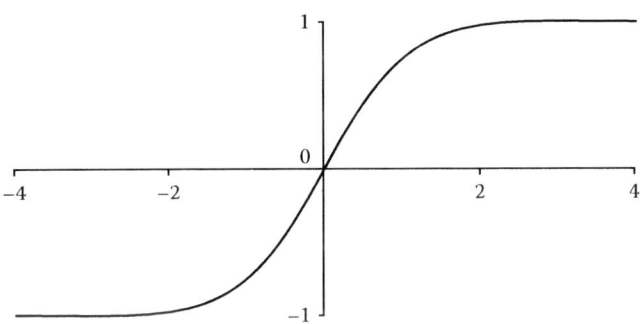

FIGURE 2.23
The Tanh activation function.

2.8 Training a Layered Network: Backpropagation

Once the network geometry and the type of activation function have been chosen, the network will be ready to use as soon as the connection weights have been determined, which requires a period of training.

Training a network is a much slower process than using it once the training is complete. While the network is learning, every time a pattern is fed through it the connection weights must be adjusted, and this process is repeated many times for large numbers of patterns. The speed at which suitable connection weights are discovered depends on the values of adjustable parameters, such as the learning rate and on the procedure used for training. Several training rules exist, which prescribe how the network weights should be adjusted so that the network learns rapidly and productively.

When the network contains no hidden layer, training is simple. The error at an output node can be found immediately, since both the actual network output and the target output are known; this error can then be used to modify the weights on connections to the output nodes through gradient descent. Iterative adjustment of connection weights gradually reduces the difference between the output of the network and the target value for each sample pattern so that eventually an optimum set of weights is obtained.

With the introduction of a hidden layer, training becomes trickier because, although the target responses for the output nodes are still available from the database of sample patterns, there are no target values in the database for hidden nodes. Unless we know what output a hidden node should be generating, it is not possible to adjust the weights of the connections into it in order to reduce the difference between the required output and that which is actually delivered.

The lack of a recipe for adjusting the weights of connections into hidden nodes brought research in neural networks to a virtual standstill until the publication by Rumelhart, Hinton, and Williams[2] of a technique now known as *backpropagation* (BP). This offered a way out of the difficulty.

Backpropagation is a generalized version of the delta rule, extended to multiple layers. The central assumption of BP is that when the target output and actual output at a node differ, the responsibility for the error can be divided between:

1. The weights of connections into the node.
2. The output of hidden nodes in the immediately preceeding layer that generated the input signals into the node.

A hidden node that sends a large signal to an output node is more responsible for any error at that node than a hidden node that sends a small signal, so changes to the connection weights into the former hidden node should be larger.

Backpropagation has two phases. In the first, an input pattern is presented to the network and signals move through the entire network from its inputs to its outputs so the network can calculate its output. In the second phase, the error signal, which is a measure of the difference between the target response and actual response, is fed backward through the network, from the outputs to the inputs and, as this is done, the connection weights are updated.

The recipe for performing backpropagation is given in Box 1 and illustrated in Figure 2.24 and Figure 2.25. Figure 2.24 shows the first stage of the process, the calculation of the network output; this is the *forward pass*. Input signals of 0.8 and 0.3 are fed into the network and pass through it, being multiplied by connection weights before being summed and fed through a sigmoidal activation function at each node.

Box 1: The Backpropagation Algorithm

1. Set a learning rate η.
2. Set all connection weights to random values.
3. Select a random input pattern, with its corresponding target output.
4. Assign to each node in the input layer the appropriate value in the input vector. Feed this input to all nodes in the first hidden layer.
5. For each node in the first hidden layer, find the total input,

$$net_j = \sum_{i=0}^{n} x_{ij} w_{ij},$$

   then use this as input to the activation function on the node to determine the output to the next layer of nodes $f(net_j)$.
6. If other hidden layers exist, repeat step 5 as required.
7. Repeat for the output layer.
8. Calculate the error at each output node,

$$\delta_j = (t_j - o_j)o_j(1 - o_j)$$

9. Calculate the error for each node in the final hidden layer

$$\delta_j = o_j(1 - o_j)\sum_{k} \delta_k w_{kj}$$

10. Repeat step 9 for any other hidden layers.
11. Update the weights for all layers $\Delta w_{ij}(\eta + 1) = \eta(\delta_j o_i) + \alpha \Delta w_{ij}(\eta)$.
12. Continue at step 3.

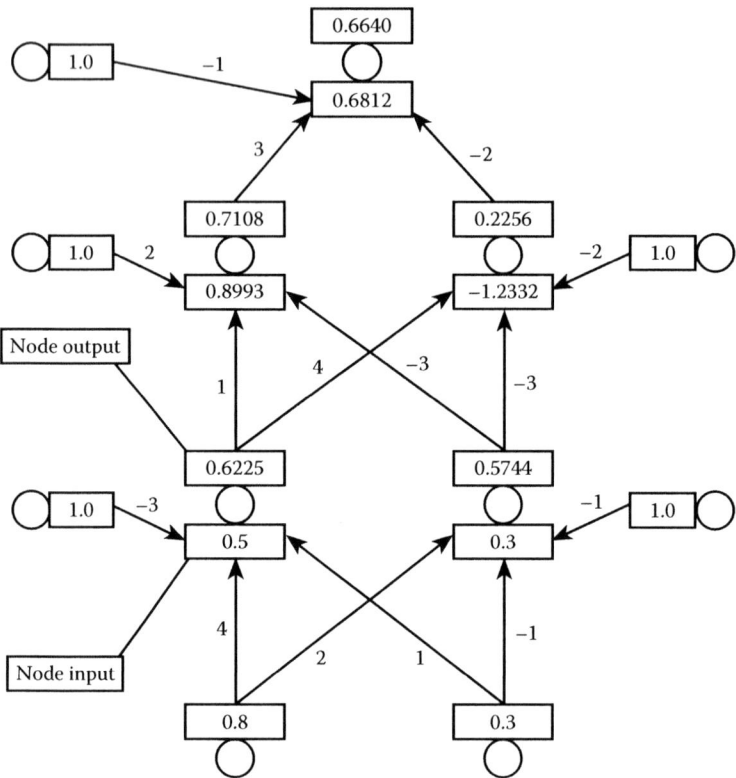

FIGURE 2.24
Backpropagation: The forward pass.

The output generated by the network is 0.6640. Figure 2.25 shows the reverse pass in which this output is compared with the target response of 1.4 and the weights subsequently adjusted to reduce the difference.

2.8.1 The Mathematical Basis of Backpropagation

The mathematical interlude that follows is a justification of the formulae in Box 1. If you are interested only in using neural networks, not the background mathematics, you may want to skip this section.

We write the error signal for unit j as:

$$\delta_j = -\frac{\partial E}{\partial net_j} \tag{2.17}$$

and the gradient of this error with respect to weight w_{ij} as:

$$\Delta w_{ij} = \frac{\partial E}{\partial w_{ij}} \tag{2.18}$$

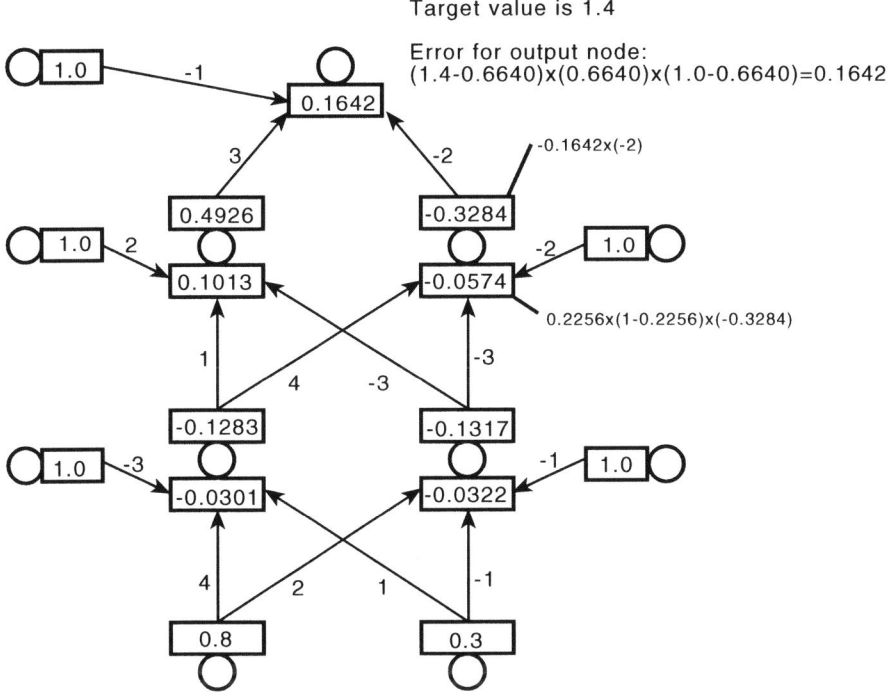

FIGURE 2.25
Backpropagation: Updating the weights.

The gradient of the weights can be expanded using the chain rule:

$$\Delta w_{ij} = -\frac{\partial E}{\partial net_i} \frac{\partial net_i}{\partial w_{ij}} \tag{2.19}$$

The first term is the error in unit i, while the second can be written as:

$$\frac{\partial net_i}{\partial w_{ij}} = \frac{\partial}{\partial w_{ij}} \sum_{k \in A_i} w_{ik} y_k = y_j \tag{2.20}$$

Putting these together we get:

$$\Delta w_{ij} = \delta_i y_j \tag{2.21}$$

To find this term, we need to calculate the activity and error for all relevant network nodes. For input nodes, this activity is merely the input signal x. For all other nodes, the activity is propagated forward:

$$y_i = f_i(\sum_{j \in A_i} w_{ij} y_j) \tag{2.22}$$

Since the activity of unit *i* depends on the activity of all nodes closer to the input, we need to work through the layers one at a time, from input to output. As feedforward networks contain no loops that feed the output of one node back to a node earlier in the network, there is no ambiguity in doing this.

The error output is calculated in the normal fashion, summing the contributions across all the outputs from the node:

$$E = \frac{1}{2} \sum_o (t_o - y_o)^2 \tag{2.23}$$

while the error for unit o is simply:

$$\delta_o = t_o - y_o \tag{2.24}$$

To determine the error at the hidden nodes, we backpropagate the error from the output nodes. We can expand this error in terms of the posterior nodes:

$$\delta_j = -\sum_{i \in P_j} \frac{\partial E}{\partial net_i} \frac{\partial net_i}{\partial y_j} \frac{\partial y_j}{\partial net_j} \tag{2.25}$$

The first factor is as before, just the error of node *i*. The second is

$$\frac{\partial net_i}{\partial y_j} = \frac{\partial}{\partial y_j} \sum_{k \in A_i} w_{ik} y_k = w_{ij} \tag{2.26}$$

while the third is the derivative of the activation function for node *j*:

$$\frac{\partial y_i}{\partial net_j} = \frac{\partial f_j(net_j)}{\partial net_j} = f'_j(net_j) \tag{2.27}$$

Equation (2.27) reveals why some activation functions are more convenient computationally than others. In order to apply BP, the derivative of the activation function must be determined. If no closed form expression for this derivative exists, it must be calculated numerically, which will slow the algorithm. Since training is in any case a slow process because of the number of samples that must be inspected, the advantage of using an activation function whose derivative is quick to calculate is considerable.

When the logistic function is used we can make use of the identity that:

$$f'(net_j) = f(net_j) \times [1 - f(net_j)] \tag{2.28}$$

so that, combining all the bits, we get:

$$\delta_j = f'_j(net_j) \sum_{i \in P_j} \delta_i w_{ij} \tag{2.29}$$

To calculate the error for unit j, the error for all nodes that lie in later layers must previously have been calculated. The updating of weights, therefore, must begin at the nodes in the output layer and work backwards.

2.9 Learning Rate

The aim of training an ANN is to diminish the difference between the target response and actual network output, averaged over a large number of input patterns. For a given pattern, the error could be reduced to zero in a single step by making a suitable change to each connection weight. However, the weights are required to store information about all sample patterns, not just one, and this information builds up over time, as training proceeds. If a large and abrupt change is made to the weights, knowledge gained previously about these other patterns will be damaged. Consequently, when weights are modified, they are only pushed a little in the direction that helps them to represent the current sample pattern, ensuring minimal loss of previously stored knowledge.

At the start of training, as the connection weights have random values, there is no stored knowledge and these weights will be far from their optimum values. At this stage one can be bolder with the size of the changes in connection weights because there is little or no knowledge to destroy and much to learn. The value of the learning rate η, which is chosen by the user, must not be too large, though, or the updating of the weights that follows the presentation of each sample to the network might be wild, causing the algorithm to bounce around on the error surface and be unable to settle. On the other hand, if η is too small, the calculation will be slow to converge and learning will take too long. A sensible compromise is to use a value for η which diminishes as training proceeds, so that in the early stages a coarse but far-reaching exploration of the error surface can be performed, while in the later stages more gentle adjustments to the weights allow them to be fine-tuned. To implement this, the learning rate, which can take values between 0.0 and 1.0, slowly reduces in size as training proceeds (Figure 2.26).

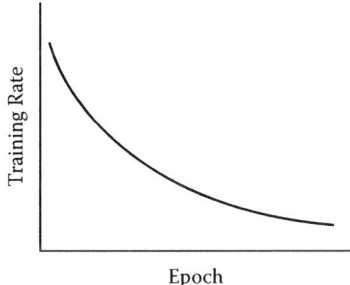

FIGURE 2.26
Typical variation of the training rate with epoch number.

2.10 Momentum

Networks with a sufficiently large number of nodes are able to model any continuous function. When the function to be modeled is very complex, we can expect that the error function, which represents the way that the discrepancy between the target output and the actual output varies with the values of the connection weights, will be highly corrugated, displaying numerous minima and maxima. We have just argued that toward the end of training connection weights should be adjusted by only small amounts, but once adjustments to the weights become small there is a real danger that the network may get trapped, so that the connection weights continue to adjust as each sample is presented, but do so in a fashion that is cyclic and so makes no progress. Each weight then oscillates about some average value; once this kind of oscillation sets in, the network ceases to learn.

This problem is more severe in multilayer networks that, though more powerful and flexible than their single-layer counterparts, are also vulnerable to trapping because the error surface, whose dimensionality equals the number of connection weights, is complex. The scale of the problem is related to the size of the network. Each connection weight is a variable whose value can be adjusted; in a large network, there will be scores or hundreds of weights that can be varied independently. As the set of connection weights defines a high dimensional space, the greater the number of weights, the more minima and maxima are likely to exist in the error surface and the more important it is that the system be able to escape from local minima during training.

To reduce the chance that the network will be trapped by a set of endlessly oscillating connection weights, a momentum term can be added to the update of the weights. This term adds a proportion of the update of the weight in the previous epoch to the weight update in the current epoch:

$$w_{ij}(n+1) = w_{ij}(n) + \eta \delta_j(n)x_{ij} + \alpha[w_{ij}(n) - w_{ij}(n-1)] \qquad (2.30)$$

α is the *momentum coefficient*, $0 \leq \alpha < 1.0$.

The effect is to give momentum to the updating of the weights as the algorithm travels across the error surface. Behaving like a ball bearing rolling over a wavy surface, the network, therefore, is more likely to travel through a local minimum on the error surface and reach the other side of it rather than becoming trapped within it.

Momentum brings an additional benefit. Not only does it reduce the chance that training will falter and the weights start to oscillate when the network encounters a local minimum, it also helps to speed up the rate at which the network weights converge. If the weight changes at one particular node have the same sign for the presentation of several successive sample patterns, momentum will amplify this movement, pulling the connection weights more rapidly across a region of the search space in which the gradient is fairly uniform. The increase in convergence speed that momentum brings about in this case is approximately $\eta/(1 - \alpha)$.

On the other hand, if successive changes to a particular connection weight tend to have opposite signs, the connection weight may be close to its optimum value; the use of momentum serves to dampen the oscillations in the value of the weights, thus slowing down movement and stabilizing the weight.

2.11 Practical Issues

2.11.1 Network Geometry and Overfitting

The network geometry, i.e., the number of nodes and the way that they are interconnected, is an important factor in network performance. Although big brains generally outperform small ones, this does not mean that it is necessarily wise to aim for the largest network that is computationally tractable. Instead, for any given dataset there is an optimum geometry. A very small network will contain too few connection weights to adequately represent all the rules in the database, but if the network is large and there are numerous weights, the network will be vulnerable to *overfitting*.

Overfitting arises when the network learns for too long. For most students, the longer they are trained the more they learn, but artificial neural networks are different. Since networks grow neither bored nor tired, it is a little odd that their performance can begin to degrade if training is excessive. To understand this apparent paradox, we need to consider how a neural network learns.

Figure 2.27 shows how the boiling points of a few compounds are related to their molecular weight. A straight line is a reasonable predictor of the relationship between molecular weight and boiling point, but the solid curve is rather better. This curve does not pass precisely through every data point, but is a good approximation to the overall trend.

We would expect that an ANN whose role was to predict the boiling point of a compound from its molecular weight should learn something close to this trend line. After a suitable period of training, the network will learn an approximation to this line, but if training is continued, it will try to fit the data points more closely because, by so doing, it can reduce the error (Figure 2.27, dashed line). The dashed line is a better fit to the data in the sense that the sum of the squared deviations has diminished, but it does not constitute an improved *model*. The dashed line in Figure 2.27 may not correctly predict the boiling points of molecules, especially above about 150 K. The problem is that the data have been *overfitted*. Rather than a smooth and slowly varying curve, which scientific intuition tells us would best represent the data, a curve has been found that, in trying to get as close as possible to all data points, has departed from the trend line.

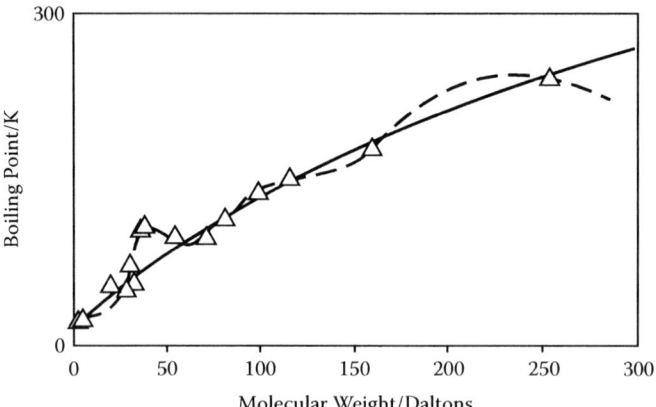

FIGURE 2.27
The relationship between the boiling point of a material in degrees Kelvin and its molecular weight.

Overfitting is a potentially serious problem in neural networks. It is tackled in two ways: (1) by continually monitoring the quality of training as it occurs using a *test set*, and (2) by ensuring that the geometry of the network (its size and the way the nodes are connected) is appropriate for the size of the dataset.

2.11.2 The Test Set and Early Stopping

Training an ANN can be a slow process, especially if the dataset is large. Tens of thousands of epochs may be required for the algorithm to converge on suitable connection weights. To achieve optimum performance and avoid overfitting, training must be continued for as long as the network continues to improve, but should be terminated as soon as no further improvement is likely.

Figure 2.28 shows how the sum over all samples of the Euclidean distance, E_d, between target output and actual output varies as a function of the number of epochs during a typical training session. E_d for the training set (solid line) falls continuously as training occurs, but this does not mean that we should train the network for as long as we can.

As the network learns, connection weights are adjusted so that the network can model general rules that underlie the data. If there are some general rules that apply to a large proportion of the patterns in the dataset, the network will repeatedly see examples of these rules and they will be the first to be learned. Subsequently, it will turn its attention to more specialized rules of which there are fewer examples in the dataset. Once it has learned these rules as well, if training is allowed to continue, the network may start to learn specific samples within the data. This is undesirable for two reasons. Firstly, since these particular patterns may never be seen when the trained network is put to use, any time spent learning them is wasted. Secondly,

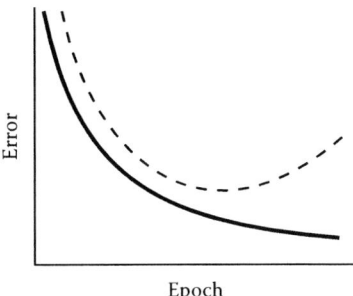

FIGURE 2.28
Variation of error in the training set (solid line) and test set (broken line) with epoch.

as the network adjusts its connection weights to learn specific examples, its knowledge of the general rules will be diluted and, therefore, the overall quality of its knowledge will diminish.

A Goldilocks step is required—training must continue until the network has learned just the right amount of knowledge, but no more. Judging when the network has a satisfactory grasp of the general rules, but has not yet turned its attention to learning specific patterns is not as difficult as it may seem. A convenient way to avoid the perils of overfitting is to divide the dataset into two parts: a *training set* and a *test set*. The training set provides the patterns that the network uses to establish the correct connection weights; the test set is used to measure how far training has progressed. Both sets should contain examples of all the rules. For large sample databases, this can be achieved by selecting the members of the two sets from the database at random.

The network is trained using the training set and, as it learns, the error on this set decreases, as shown by the solid line in Figure 2.28. Periodically, the network is shown the test set and the total error for samples in that set is determined. When the network is shown the test set, the connection weights remain untouched, whether or not performance on the set is satisfactory; the test set is used only to assess the current ability of the network. As the network learns, performance on both sets will improve, but once the network starts to learn specific examples in the training set, the network's understanding of the general rules will begin to degrade, and the prediction error for the test set will rise. The optimal stopping point is then at the minimum of the broken curve in Figure 2.28, at which point the error on the test set is at its lowest.

This method for preventing overfitting requires that there are enough samples so that both training and test sets are representative of the dataset. In fact, it is desirable to have a third set known as a *validation set*, which acts as a secondary test of the quality of the network. The reason is that, although the test set is not used to train the network, it is nevertheless used to determine at what point training is stopped, so to this extent the form of the trained network is not completely independent of the test set.

2.11.3 Leave One Out

If the database of sample patterns is small, it may not be realistic to divide it into independent training and test sets. "Leave one out" cross validation may then be required. The network is trained on the entire sample data-base, with the exception of a single sample that acts as a test for the trained network. The performance of the network predicting this single test sample gives a measure of the quality of the trained network, which is checked by continuing the training after a second sample is withdrawn (and the first one returned) so that it can be used to test it. This is carried out for all members of the database in turn. The process has a high computational cost, so would not be used if the dataset were large; however, it can be helpful when the dataset is too small to provide both a training and a test set.

2.12 Traps and Remedies

2.12.1 Network Size

There are examples in the scientific literature of networks that contain dozens of nodes and hundreds of connections being trained using a set of patterns that contains fewer samples than the number of connection weights. It is no surprise when such a network, in due course, settles down to a negligible training error because there are more than enough variable parameters, in the form of the connection weights, to allow each pattern to be recognized individually by the network.

   The degree to which it has learned general rules rather than simply learned to recognize specific sample patterns is then difficult to assess. In using a network on a new dataset, it is, therefore, important to try to estimate the complexity of the data (in essence, the number of rules that will be necessary to satisfactorily describe it) so that a network of suitable size can be used. If the network contains more hidden nodes than are needed to fit the rules that describe the data, some of the power of the network will be siphoned off into the learning of specific examples in the training set.

2.12.2 Sample Order

It is also important that the order in which the samples are presented to the network is randomized at the start of each epoch, especially if the dataset is small. The only aim in life that the network has is to learn patterns. If pattern 71 in the database is always presented to the network immediately after pattern 16, this is itself a pattern, which, though of no significance in the interpretation of data, will influence the way the network learns. These order effects are very rarely of interest in the training of a network, so must be avoided by randomizing the order in which patterns are presented.

2.12.3 Ill-Conditioned Networks and the Normalization of Data

Sometimes it seems that a network just cannot come to grips with a problem, no matter how long the training. It learns some features of the dataset, but never reaches an acceptable level of performance. If the network is sufficiently large that it should be able to describe the variability in the data satisfactorily but still does not function properly, the network may be *ill-conditioned*. Parts of such a network require a high learning rate to make progress, while other parts require a low learning rate. Because of this conflict in learning rates, it may be impossible for the network to converge and, therefore, it cannot learn effectively.

This inability of all parts of the network to learn at the same rate can have several causes. The most common is that a mismatch exists either between the absolute magnitude of the input data and the target data or between the different inputs into the network. Finding the correlation between molecular structure and boiling point provides an example. We might anticipate that the boiling point, being determined partly by intermolecular forces, would depend not only on the molecular weight, but also on the dipole moment in the molecule. The molecular weight of a compound and its boiling point (in Kelvin) are both numerical values in the range of 20 to 350 for molecules of modest size, but the difference in electronegativity between two atoms is usually less than 1 (Table 2.1).

The good correlation between molecular weight and boiling point that is apparent in Figure 2.27 suggests that molecular weight is the principal factor in determining the boiling point, but the molecular dipole could account for some of the variation about the trend line. Let us guess that the electronegativity difference accounts for perhaps 10 percent of the dependence of boiling point on molecular structure. We could build a neural network whose aim is to predict the boiling point of a liquid of diatomic molecules taking as input both the molecular weight and the electronegativity difference between the atoms.

If the molecular weight and the electronegativity difference both influence the boiling point, the signal that reaches a hidden node from the electronegativity input should be comparable to the signal that reaches the node from the molecular weight input. Because the differences in electronegativity are small, the connection weights on the dipole moment input must be larger than those on the molecular weight input, otherwise the electronegativity difference will have a negligible effect on the output of the network. However, if all the connection weights are initialized at the start of the run to values in the range −1.0 to +1.0, the input signal reaching the node from the molecular weight input will be so large that the output from the node will be very insensitive to input (Figure 2.22 and Figure 2.23). Thus, the node passes on very little information to the output about the incoming signal, so training is slow and inefficient. The normal remedy is to normalize both the input data and the targets before training commences so that inputs have a mean of zero and standard deviation of 1. This ensures that the output from the

TABLE 2.1

Boiling points, molecular weights, and electronegativity differences for some diatomic molecules.

Compound	Molecular Weight (Daltons)	Difference in Electronegativity	Boiling Point (°C)
Bromine	159.8	0	58.8
Bromine chloride	115.4	0.2	5
Bromine fluoride	98.9	1.02	−20
Carbon monoxide	28	0.89	−191.5
Chlorine	70.9	0	−101
Chlorine fluoride	54.5	0.82	−100.8
Deuterium	4	0	−249.7
Deuterium chloride	37.5	0.96	−81.6
Fluorine	38	0	−188
Hydrogen	2	0	−252.8
Hydrogen bromide	80.9	0.76	−67
Hydrogen chloride	36.5	0.96	−84.9
Hydrogen iodide	127.9	0.46	127
Iodine	253.8	0	184.4
Nitrogen	28	0	−195.8
Nitric oxide	30	0.4	−151.8
Oxygen	32	0	−183

Source: Atkins, P. and De Paula, J., *Physical Chemistry*, 8th ed., Oxford University Press, Oxford, U.K., 2006. With permission.

sigmoidal function does not become insensitive to the input, which would happen if the integrated input signal was very large or very small.

2.12.4 Random Noise

A quite different way to reduce overfitting is to use random noise. A random signal is added to each data point as it is presented to the network, so that a data pattern:

$$\{x_1, x_2, x_3, \ldots, x_n\}$$

becomes

$$\{x_1 + \text{rand}_1(), x_2 + \text{rand}_2(), x_3 + \text{rand}_3(), \ldots, x_n + \text{rand}_n()\}$$

where $\text{rand}_1()$, $\text{rand}_2()$ ... are independent random numbers chosen afresh each time a pattern is used in training. This pollution of good data with random noise looks ill-advised, but, provided that it only slightly perturbs the input pattern, the presence of this background noise will not mask the general rules in the dataset since those rules are manifest in many samples. On the other hand, the network is now less likely to learn a specific pattern because that pattern will be different on each presentation.

2.12.5 Weight Decay

An overfitted function shows a high degree of curvature because it attempts to pass through as many sample points as possible rather than just following a trend line (e.g., the dashed line in Figure 2.27). A direct, though blunt-edged, method for limiting overfitting is to reduce this curvature. High curvature is generally associated with large connection weights, so a downward scaling of the weights should reduce curvature; this can be accomplished by allowing each network weight to decay every epoch by a constant fraction:

$$w_{ij} = w_{ij}(1 - d) \qquad (2.31)$$

in which d is a decay parameter in the range $0.0 > d > 1.0$ and is usually very close to zero. Although this procedure does reduce curvature, the decay of the weights constantly pushes the network away from the weights on which it would like to converge, thus the benefits of reduced overfitting must be balanced against some loss of quality in the finished network.

2.12.6 Growing and Pruning Networks

A mixture of experience, guesswork, and experiment is required to determine the optimum network geometry. If the number of rules that describe the data is unknown, as is usually the case, a network of suitable geometry may be found by incremental training. A small network is constructed and trained. If, on completion of training, network performance is unsatisfactory as judged by performance on the test set, the network size is increased by adding one node to the hidden layer and the network is retrained. The process is repeated until performance is judged to be satisfactory.

An alternative to expanding a small network is to start with a large network and work backward, gradually pruning out nodes and the links to them, then retraining the network, continuing the process until the performance of the network starts to deteriorate.

Neither of these approaches is entirely satisfactory. Growing a network by adding nodes and retraining from scratch is conceptually simple, but tedious, if the dataset is complex so that the final network is large. Pruning nodes from a large network is tricky to implement. The difficulty is in knowing which nodes to remove when the time comes to shrink the network because different nodes within a single layer have different connection weights and, therefore, are of differing degrees of value within the network. The removal of an important node may have a very destructive effect on performance, but the importance of a node is not determined by the magnitude of the connection weights alone. In addition, judging when performance "begins to deteriorate" is a qualitative decision and it may be hard to assess at what point the diminution in performance becomes damaging. The various methods available for network pruning have intriguing names, such as "optimal brain damage," but are beyond the scope of this book.

A promising alternative to incremental networks or pruning is a growing cell structure network, in which not only the size of the network but also its geometry are evolved automatically as the calculation proceeds. Growing cell structures, which form the subject of Chapter 4, are effective as a means of creating self-organizing maps, but their use in generating feedforward networks is in its infancy.

2.12.7 Herd Effect

Certain types of datasets can give rise to a herd effect in training (Figure 2.29). Imagine a dataset of patterns that illustrate two different rules, one of which is manifest in many of the training patterns, the other in a far smaller number. Near the start of training, the network will concentrate on discovering the more widespread rule because it sees many examples of it. By adjusting the connection weights to fit this rule, it can reduce the error signal most rapidly.

FIGURE 2.29
The herd effect.

All sections of the network will simultaneously try to adjust their weights so as to find this rule. Until this rule has been satisfactorily modeled, the second rule will be largely ignored, but once the first rule has been dealt with, errors arising from the second rule predominate and all sections of the network will then switch attention to solving that rule. In so doing, the network may start to forget what has been learned about the first one. The network, thus, switches herd-like between a focus on one rule and on the other and can oscillate for some time before gradually settling into a mode in which it addresses both rules.

Various strategies can be used to distract the herd. The simplest is to run the same set of data through several different network geometries, changing the number of hidden layers as well as the number of hidden nodes in each layer and see what provides the best network measured by performance on the test set. This simple strategy can be effective if the time required in training the network is low, but for complex data many different networks would need to be tried and the time required may be excessive. Fortunately, the herd effect, if it cannot be avoided entirely, can generally be overcome by allowing training to run for an extended period.

2.12.8 Batch or Online Learning

The approach described in this chapter, in which a single pattern is extracted from the database, the gradient of the error function with respect to the weights is found and then used immediately to update the weights, is known as *online* or *stochastic* learning (or, less commonly, *sequential mode* learning or *perpattern* learning). An alternative to online learning is to update the weights only after every sample in the training set has been shown to the network once and the errors for all samples have been accumulated. This type of learning is known as *batch learning*.

Both methods are in use, but online learning is commonly faster than batch learning; in other words, the network is often found to converge to an acceptable set of weights more rapidly using online learning. The error signal obtained by feeding a single sample through the network is a (very) noisy approximation to the error signal that would be obtained by summing the error for every sample. The noise that online learning emulates is an advantage, since it helps the algorithm to avoid becoming trapped in local minima. Furthermore, each time a sample is shown to the network, the connection weights can improve slightly, while in batch learning the network has to continue to work with the "old" weights throughout each epoch before they are updated.

If the training set is not fixed because new data are being added as training is performed, or if the samples in the training set are changing with time, as is often the case when neural networks are used in control applications, online learning is essential.

Feedforward networks of the sort described in this chapter are the type most widely used in science, but other types exist. In contrast to a feedforward network, which represents only a function of its current state, in a *recurrent* or *cyclic* network, connections exist from nodes that are closer to the output layer back to nodes that are farther away, so "later" output is returned to "earlier" nodes; the network now contains a short-term memory. Recurrent networks show interesting, complicated, and rather unpredictable behavior, but their behavior is not yet well understood and there have been rather few examples of their use in science. Kohonen or self-organizing maps are discussed in the next chapter, while growing cell structures form the topic of Chapter 4.

2.13 Applications

Artificial neural networks are now widely used in science. Not only are they able to learn by inspection of data rather than having to be told what to do, but they can construct a suitable relationship between input data and the target responses without any need for a theoretical model with which to work. For example, they are able to assess absorption spectra without knowing about the underlying line shape of a spectral feature, unlike many conventional methods.

Most recent scientific applications involve the determination of direct relationships between input parameters and a known target response. For example, Santana and co-workers have used ANNs to relate the structure of a hydrocarbon to its cetane number,[4] while Berdnik's group used a theoretical model of light scattering to train a network that was then tested on flow cytometry data.[5]

QSAR studies are a fertile area for ANNs and numerous papers have been published in the field. Katritzky's group has a range of interests in this area, particularly related to compounds of biological importance; see for example Reference 6. Some QSAR studies have been on a heroic scale. Molnar's group has used training sets of around 13,000 compounds and a total database containing around 30,000 to try to develop meaningful links between cytotoxicity and molecular descriptors.[7]

ANNs are the favorite choice as tools to monitor electronic noses,[8] where the target response may be less tangible than in other studies (although, of course, it is still necessary to be able to define it). Many applications in which a bank of sensors is controlled by a neural network have been published and as sensors diminish in size and cost, but rise in utility, sensors on a chip with a built-in ANN show considerable promise. Together, QSARs and electronic noses currently represent two of the most productive areas in science for the use of these tools.

The ability of ANNs to model nonlinear data is often crucial. Antoniewicz, Stephanopoulos, and Kelleher have studied the use of ANNs in the estimation of physiological parameters relevant to endocrinology and metabolism.[9]

They used data that simulated the pathway of mammalian gluconeogenesis with [U-C-13] glucose as a tracer, calculating isotopic labelling patterns in the nine steps in the pathway. In a comparison of ANN with other approaches, including multiple linear regression and principal component regression, they found that the ANN model performed better than any other model, provided that a sufficiently large training set was provided. This outperformance was ascribed to the presence of nonlinearities in the data and the superior ability of ANNs to handle these.

2.14 Where Do I Go Now?

Numerous books have been written on the topic of artificial neural networks; most are written for, or from the point of view of, computer scientists and these are probably less suited to the needs of experimental scientists than those written with a more general audience in mind.

Two of the most accessible books in the area are *Neural Computing: An Introduction*, by Beale and Jackson[10] and *Neural Networks for Chemists*, by Zupan and Gasteiger.[11] Neither book is a recent publication, but both provide an introduction that is set at a suitable level if you have had little previous contact with ANNs.

2.15 Problems

1. Applicability of neural networks

   Consider whether a neural network would be an efficient way of tackling each of the following tasks:

   a. Predicting the direction of the stock market (1) in the next twenty-four hours, (2) over the next two years.

   b. Predicting the outcome of ice hockey matches.

   c. Predicting the frequency of the impact of meteorites on the Earth.

   d. Deriving rules that link diet to human health.

   e. Automatic extraction of molecular parameters, such as bond lengths and bond angles, from the gas-phase high resolution spectra of small molecules (which are of particular importance in interstellar space).

   f. Linking the rate of an enzyme-catalyzed reaction to the reaction conditions, such as temperature and concentration of substrate.

2. Boiling point prediction

Table 2.1 lists the molecular weight, boiling point, and difference in electronegativity for a number of small compounds. Write an ANN that can use this information to predict the boiling point of diatomic molecules, testing it using "leave one out" validation or by creating a test set by locating information on further diatomic molecules not shown in Table 2.1. (A good source of information of this type is the NIST [National Institute of Standards and Technology] Chemistry WebBook.[12]) How successful is your completed network in predicting the boiling point of molecules that contain more than two atoms? (This problem can be used to explore the ability of an ANN to extrapolate, i.e., how it deals with input data that lie beyond the range covered by the training set, which may be limited. While we would expect the model to deal well with molecules whose boiling points were below, say, 350 K, we should be less confident in its predictions for molecules with much higher boiling points, if none was in the dataset.)

3. Iris data set

The Iris data set is a well-known set of experimental data that has been used to test the performance of ANNs and certain other types of AI programs. The dataset can be found on the Internet and a copy is also provided on the CD that accompanies this text. On the CD is a simple ANN that learns this dataset, and this program may be of help in understanding how an ANN works. In this exercise, write an ANN from scratch and compare its performance against the program on the CD; you should find that your program learns the data satisfactorily within, at most, a few hundred epochs. Try incorporating a momentum term. Does this make a noticeable improvement to the performance of the network? Consider why momentum changes (or does not change) performance on this dataset (hint: how complex is the set of data? In other words, how difficult is it likely to be to learn?).

References

1. Sharda, R. and Delen, D., Predicting box-office success of motion pictures with neural networks, *Exp. Sys. Apps.*, 30, 243, 2006.
2. Rumelhart, D.E., Hinton, G.E., and Williams, R.J., Learning representations by back-propagating errors, *Nature*, 323, 533, 1986.
3. Atkins, P. and De Paula, J., *Physical Chemistry*, 8th ed., Oxford University Press, Oxford, U.K., 2006.

4. Santana, R.C. et al., Evaluation of different reaction strategies for the improvement of cetane number in diesel fuels, *Fuel*, 85, 643, 2006.
5. Berdnik, V.V. et al., Characteristics of spherical particles using high-order neural networks and scanning flow cytometry, *J. Quant. Spectrosc. Radiat. Transf.*, 102, 62, 2006.
6. Katritzky, A.R. et al., QSAR studies in 1-phenylbenzimidazoles as inhibitors of the platelet-derived growth factor, *Bioorg. Med. Chem.*, 13, 6598, 2005.
7. Molnar, L. et al., A neural network-based classification scheme for cytotoxicity predictions: Validation on 30,000 compounds, *Med. Chem. Letts.*, 16, 1037, 2006.
8. Fu, J., et al., A pattern recognition method for electronic noses based on an olfactory neural network, *Sens. Actuat. B: Chem.*, 125, 489, 2007.
9. Antoniewicz, M.R., Stephanopoulos, G., and Kelleher, J.K., Evaluation of regression models in metabolic physiology: Predicting fluxes from isotopic data without knowledge of the pathway, *Metabolomics* 2, 41, 2006.
10. Beale, R. and Jackson, T., *Neural Computing: An Introduction*, Institute of Physics, Bristol, U.K., 1991.
11. Zupan, J. and Gasteiger, J., *Neural Networks in Chemistry and Drug Design*, Wiley-VCH, Chichester, U.K., 1999.
12. National Institute of Standards and Technology (NIST). Chemistry WebBook. http://webbook.nist.gov/chemistry.

3

Self-Organizing Maps

There are many situations in which scientists need to know how alike a number of samples are. A quality control technician working on the synthesis of a biochemical will want to ensure that each batch of product is of comparable purity. An astronomer with access to a large database of radiofrequency spectra, taken from observation of different parts of the interstellar medium, might need to arrange the spectra into groups to determine whether there is any correlation between the characteristics of the spectrum and the direction of observation.

If sample patterns in a large database are each defined by just two values, a two-dimensional plot may reveal clustering that can be detected by the eye (Figure 3.1). However, in science our data often have many more than two dimensions. An analytical database might contain information on the chemical composition of samples of crude oil extracted from different oilfields. Oils are complex mixtures containing hundreds of chemicals at detectable levels; thus, the composition of an oil could not be represented by a point in a space of two dimensions. Instead, a space of several hundred dimensions would be needed. To determine how closely oils in the database resembled one another, we could plot the composition of every oil in this high-dimensional space, and then measure the distance between the points that represent two oils; the distance would be a measure of the difference in composition. Similar oils would be "close together" in space,

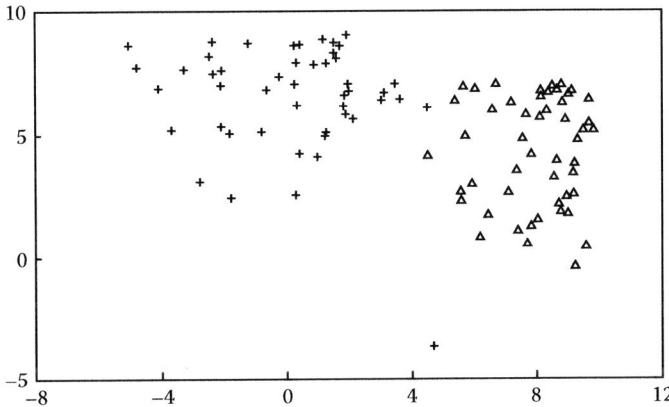

FIGURE 3.1
Clustering of two-dimensional points.

while dissimilar oils would be "far apart" and we could check for clusters by seeking out groups of points that lie close together.

This sounds rather alarming. Few of us are entirely comfortable working with four dimensions; something a hundred times larger is a good deal worse. Fortunately, a self-organizing map (SOM) can take the pain away. Its role is to identify similar samples within a large set and group them together so that the similarities are apparent. Data that are represented by points in these spaces, whether they cover a few or several hundred dimensions, are handled in a simple way and the algorithm also provides a means to visualize the data in which these clusters of points can be seen, so we need not worry about the difficulties of spotting an object in a space that spans hundreds of dimensions.

3.1 Introduction

There are good reasons why an oil company might want to know how similar two samples of crude oil are. Operating margins in the refineries that process crude oil are low and the most profitable product mix is critically dependent on how the nature of the feedstock is taken into account when setting the refinery operating conditions. A minor change in the properties of a feedstock, such as its viscosity or corrosivity, may affect the way that it should be processed and the relative quantities of the chemicals that the refinery produces. The change in product mix may, in turn, significantly increase the refinery profit or, if the wrong choices of operating conditions are made, remove it completely. Thus, operating conditions are fine-tuned to the composition of each oil and if two successive batches of oil are not sufficiently similar, adjustments will need to be made to those conditions.

It is not only in oil processing that it is valuable to identify similarities among samples. In fact, the organization of samples into groups or classes with related characteristics is a common challenge in science. By grouping samples in this way, we can accomplish a couple of objectives, as is illustrated in the first two examples below.

Example 1: Protective Gloves

One way that we can benefit from information on how samples are clustered is to use data from previously categorized samples to provide information concerning a sample never before seen.

Suppose that a database is available that contains safety information for a range of toxic chemicals. For each chemical the database specifies, among other things, the best material to use for disposable gloves that provide protection against the chemical. The relationship between the chemical and the physiological properties of a toxic substance and the glove material that is most suited to provide protection against it is not random.

> Within the database there may be many chemicals for which butyl rubber gloves provide good protection and we could reasonably anticipate that at least some structural similarities would exist among these chemicals that would help us rationalize the choice of butyl rubber.
>
> It is found that many nitro compounds fall into the group for which butyl rubber is an appropriate choice, so if a new nitro-containing compound had been synthesized and we wished to choose a glove to provide protection, inspection of the members of the class would suggest butyl rubber as a suitable candidate. In this application, we are using the observation of similarities within a class (the presence of many nitro compounds) as a predictive tool (best handled using butyl rubber gloves).

In some respects this is a trivial application. In order to select a protective glove for a new nitro compound, all we would do in practice would be to check to see what material provides good protection against known nitro compounds and assume that this material would be appropriate; we do not need a computer to tell us how to do this. But the reason that the procedure in this case is simple is that we already have a means to group compounds by noting the presence or absence of particular functional groups. If the link between structure and protective material were subtler, a more sophisticated way to determine the appropriate material would be required.

Example 2: Protective Gloves—A Rationalization

A second advantage of grouping patterns is that by considering the characteristics that members of the class display, we might gain an insight into the reasons for any clustering. This is a more sophisticated use of clustering than that outlined in Example 1 because now we are looking not just to use the clustering, but also to understand why it occurs. On discovering that butyl rubber appears to provide protection against nitro compounds, we might not only use this information in choosing a glove to protect against newly synthesized chemicals, but also investigate what characteristics of butyl rubber make it particularly effective against nitro compounds. By considering such a correlation, it may be possible to develop a deeper understanding of the mechanism by which the gloves provide protection and thereby develop superior gloves.

We have already met one tool that can be used to investigate the links that exist among data items. When the features of a pattern, such as the infrared absorption spectrum of a sample, and information about the class to which it belongs, such as the presence in the molecule of a particular functional group, are known, feedforward neural networks can create a computational model that allows the class to be predicted from the spectrum. These networks might be effective tools to predict suitable protective glove material from a knowledge of molecular structure, but they cannot be used if the classes to which samples in the database are unknown because, in that case, a conventional neural network cannot be trained.

3.2 Measuring Similarity

To determine whether sample patterns in a database are similar to one another, the values of each of the n parameters that define the sample must be compared. Because it is not possible to check whether points form clusters by actually looking into the n-dimensional space—unless n is very small—some mathematical procedure is needed to identify clustered points. Points that form clusters are, by definition, close to one another, therefore, provided that we can pin down what we mean by "close to one another," it should be possible to spot mathematically any clusters and identify the points that comprise them.

This is the role of the SOM.* The SOM has two purposes: (1) it measures the clustering among samples by calculating how far apart points are, and (2) it reveals the clustering by "projecting" the clusters onto a two-dimensional surface (Figure 3.2).** Samples that are close in high-dimensional space will also be close if they are projected onto fewer dimensions, provided that the projection process is chosen appropriately.

Example 3: Clustering Animals

We shall start by clustering animals and scientists (separately). Living creatures display a remarkable diversity; some species have few features in common, while others are very alike. An elephant is not much like an amoeba, beyond the fact that they are both alive, but leopards and cheetahs share many characteristics.

We can group animals into clusters according to how similar they are, comparing their key characteristics; the more closely those characteristics match, the more similar are the two species. The choice of characteristics is inherently arbitrary, but we might include the number of legs, covered in fur or feathers, the speed of movement, size, color, carnivore/vegetarian, and so on:

> Lion: <legs = 4, covering = fur, speed = fast, size = big,...>
> Anaconda: <legs = 0, covering = skin, speed = moderate, size = moderate,...>
> Slug: <legs = 0, covering = skin, speed = very slow, size = small, ...>
> Dolphin: <legs = 0, covering = skin, speed = fast, size = big,...>

These nonnumeric descriptions are not entirely satisfactory if we are to calculate how alike two species are. To properly compare a "very slow" animal with another whose speed is "moderate" would require fuzzy logic, which forms the subject of another chapter, but at this stage

---

* Also known as a *Self-Organizing Feature Map* or SOFM, or a *Kohonen map* after its inventor.
** The projection need not necessarily be onto two dimensions, but projections onto a different number of dimensions are less common.

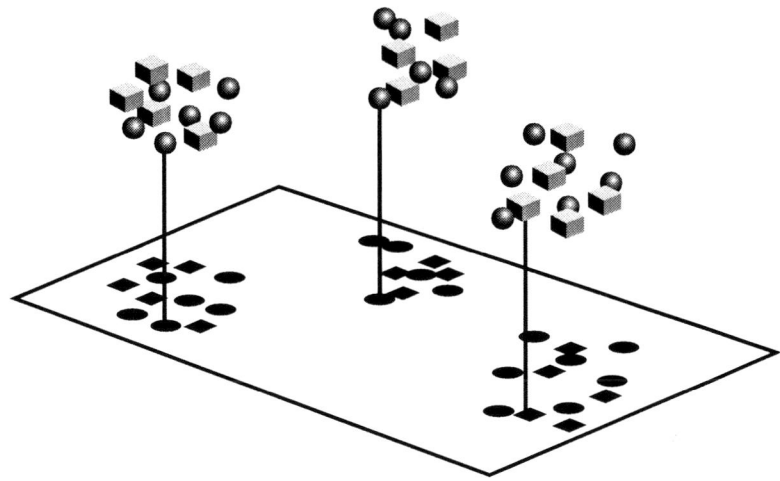

FIGURE 3.2
Projection of points that are clustered in three dimensions onto a two-dimensional plane.

it is the ideas that matter rather than rigor, so we shall replace these qualitative judgments with numerical approximations.

$$\text{Lion: } <\text{legs} = 4, \text{covering} = 1, \text{speed} = 7, \text{size} = 8, ...>$$
$$\text{Anaconda: } <\text{legs} = 0, \text{covering} = 0, \text{speed} = 3, \text{size} = 5, ...>$$
$$\text{Slug: } <\text{legs} = 0, \text{covering} = 0, \text{speed} = 1, \text{size} = 1, ...>$$

The degree to which one animal is like another can then be measured by calculating the Euclidean distance between their sets of properties.

$$d^2_{ij} = c_1 \times (\text{legs}_i - \text{legs}_j)^2 + c_2 \times (\text{covering}_i - \text{covering}_j)^2 + \qquad (3.1)$$

$$c_3 \times (\text{speed}_i - \text{speed}_j)^2 + ...$$

in which the constants $c_1, c_2, c_3, ...$ are chosen to reflect how important we think each factor ought to be in determining similarity.

Once the "distances" between the animals have been calculated using equation (3.1), we lay the animals out on a piece of paper, so that those that share similar characteristics, as measured by the distance between them, are close together on the map, while those whose characteristics are very different are far apart. A typical result is shown in Figure 3.3. What we have done in this exercise is to squash down the many-dimensional vectors that represent the different features of the animals into two dimensions.

This two-fold procedure—calculate similarity among the samples, then spread the samples across a plane in a way that reflects that similarity—is exactly the task for which the SOM is designed.

FIGURE 3.3
A map that organizes some living creatures by their degree of similarity.

3.3 Using a Self-Organizing Map

Before considering in more detail how to prepare a SOM, we shall make a short detour to learn how the information that the SOM provides can be used.

In Figure 3.3, most of the animals that live in hot climates (lions, tigers, elephants, giraffes, wallabies, kangaroos) are close to each other on the map. This clustering of animals from warm environments has occurred even though the temperature of the environment in which the animals live is not a characteristic of the animals themselves, so it did not form part of the input data.

Noting this clustering of animals that live in similar climates, we might consider whether the characteristics of these animals provide any clue as to why they share the same sort of environment. Do they all have long tails? Spots? Three legs? If the dataset were made larger so that it also included many animals from colder regions of the world, correlations would begin to emerge. We would eventually discover that land-based mammals that live in a cold climate have thick fur. Of course, this is hardly a novel discovery; we learned at an early age that polar bears have plenty of fur. But the possibility that clustering may provide a tool that allows us to infer relationships between characteristics (fur length, which was fed into the map) and some aspects of the animals' behavior (their ability to live in a cold environment, which was not fed into the map) indicates a potential way in which we might use the map.

The map can also be used as a predictive tool. If the details of another animal (a unicorn, perhaps) are fed in, this animal will find a place on the map near the animals that it most strongly resembles. By noting where the animal appears and taking into account what we know about animals in the same region of the map, we may be able to discover previously unknown information (for example, that unicorns prefer a temperate climate).

3.4 Components in a Self-Organizing Map

The SOM has a particularly simple structure; only three components are needed:

1. A number of *nodes* arranged in a regular lattice; each node stores a set of *weights*.*

2. A rule to identify the node whose weights most closely resemble a pattern drawn at random from the sample database.

3. An algorithm to adjust node weights to improve the ability of the network to classify samples.

3.5 Network Architecture

Although the SOM is a type of neural network, its structure is very different from that of the feedforward artificial neural network discussed in Chapter 2. While in a feedforward network nodes are arranged in distinct layers, a SOM is more democratic—every node occupies a site of equal importance in a regular lattice.

In the simplest type of SOM, nodes are strung like beads along a ribbon (Figure 3.4).

In two dimensions, the nodes occupy the vertices of a regular lattice, which is usually rectangular (Figure 3.5). This layer of nodes is sometimes known as a *Kohonen layer* in recognition of Teuvo Kohonen's (a Finnish academician and researcher) work in developing the SOM.

The nodes in a SOM are drawn with connections between them, but these connections serve no real purpose in the operation of the algorithm. In contrast to those in the ANN, no messages pass along these links; they are drawn only to make clear the geometry of the network.

We saw in chapter 1 that Artificial Intelligence algorithms incorporate a memory. In the ANN the memory of the system is stored in the connection weights, but in the SOM the links are inactive and the vector of weights at each node provides the memory. This vector is of the same length as the dimensionality of points in the dataset (Figure 3.6).

FIGURE 3.4
The layout of nodes in a one-dimensional SOM.

---

* The nodes in a SOM are sometimes referred to as neurons, to emphasize that the SOM is a type of neural network.

Not only are the lengths of the pattern and weights vectors identical, the individual entries in them share the same interpretation.

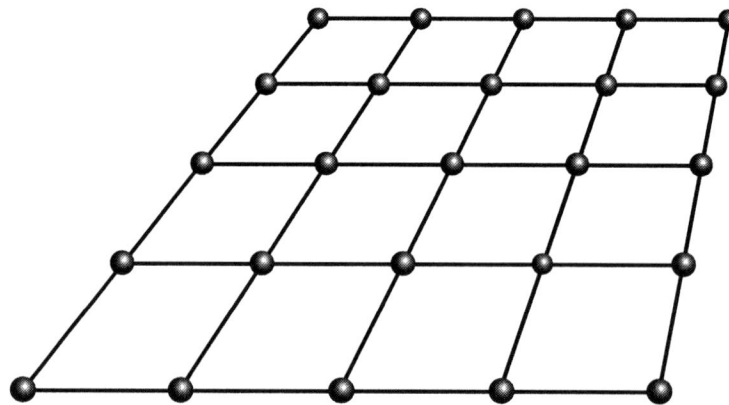

FIGURE 3.5
The layout of nodes in a two-dimensional SOM.

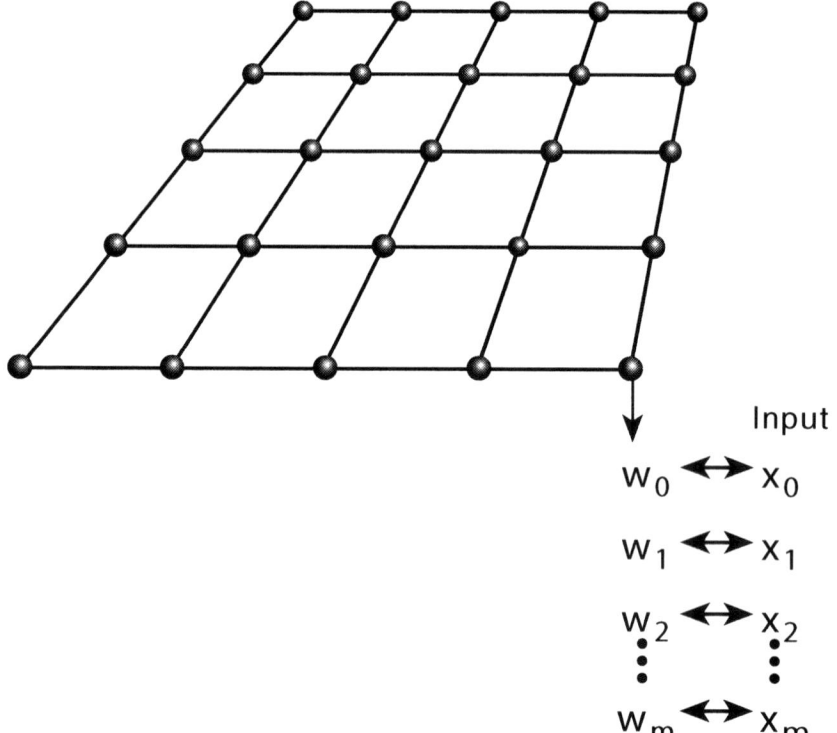

Input

$$w_0 \longleftrightarrow x_0$$

$$w_1 \longleftrightarrow x_1$$

$$w_2 \longleftrightarrow x_2$$

$$\vdots \qquad \vdots$$

$$w_m \longleftrightarrow x_m$$

FIGURE 3.6
Node weights in a two-dimensional SOM. Each node has its own independent set of weights.

FIGURE 3.7
A SOM that organizes scientists according to their physical characteristics.

Example 4: Clustering Einstein

Suppose that each sample pattern to be assessed by the SOM consisted of the height, hair length, and waistline of prominent scientists; the weights vector at every node would then be made up of three values, to be interpreted as average height, average hair length, and average waistline, respectively. The weights at the nodes are not chosen to precisely equal an entry in the database, but instead they become a blend of all of the entries. The weights at each node are independent; therefore, the waistline stored at one node will differ from the waistline at another node (Figure 3.7).

Through a process of training, the weights evolve so that each node forms a prototypical blend of groups of input patterns. Just as with the clustering of animals, scientists of similar characteristics will be positioned close together on the map.

3.6 Learning

The aim of the SOM is to categorize patterns. In order to do this successfully, it must first engage in a period of *learning*. The class to which a sample pattern within the database belongs is unknown, thus it is not possible to offer the algorithm any help in judging the significance of the pattern. Learning is therefore *unsupervised* and the SOM must find out about the characteristics of the sample patterns without guidance from the user. That the SOM can learn to organize data in a meaningful way when it has no idea what the data mean is one of the more intriguing aspects of the self-organizing map.

Training of a SOM is an iterative, straightforward process:

Unsupervised Learning in a SOM

1. Set the learning rate, η, to a small positive value $0 < \eta < 1$. Fill the weights vector at each node with random numbers.
2. Select a sample pattern at random from the database.
3. Calculate how similar the sample pattern is to the weights vector at each node in turn, by determining the Euclidean distance between the sample pattern and the weights vector.
4. Select the *winning node*, which is the node whose weights vector most strongly resembles the sample pattern.
5. Update the weights vector at the winning node to make it slightly more like the sample pattern.
6. Update the weights vectors of nodes in the *neighborhood* of the winning node.

We shall consider each of these steps below.

3.6.1 Initialize the Weights

All the knowledge of the SOM is contained in the node weights. A SOM that has never seen any sample patterns knows nothing. To reflect this, at the start of training every weight in the map is given a random value. Although these values are random, they are nevertheless chosen to lie within the range of data covered by the sample patterns because the node weights will eventually become a blend of sample data. If the initial weights were well beyond reasonable values (a waistline of 90 inches, perhaps), training would take longer because the weights would have to scale themselves up or down to get into the range of values covered by the sample patterns before they could begin to adopt values that give rise to meaningful clustering. However, the quality of the final map is not a function of the initial values chosen for the weights.

Example 5: An Untangling SOM

The ability of the SOM to learn from a random starting point is illustrated by Figure 3.8, in which a square network of one hundred nodes is trained using data points selected randomly from within a unit square.

The way in which training has been performed in this example and the interpretation of the maps will be discussed in section 3.7.3. Briefly, in the maps shown in Figure 3.8, each node is drawn at the point defined by its two weights, interpreted as x- and y-coordinates, and nodes that are neighbors on the underlying lattice are connected by a line. At the start of training, the random nature of the weights is apparent: the x and y values are unrelated to the position of a node on the lattice, thus the nodes form a tangled mess. As sample patterns, consisting of (x, y)

pairs chosen at random from within a 1×1 square, are shown to the SOM, it gradually learns about the nature of the sample data. Since the input data are spread evenly across the square, by the end of training (Figure 3.8b) the nodes have spread out and are positioned at roughly equal distances apart.

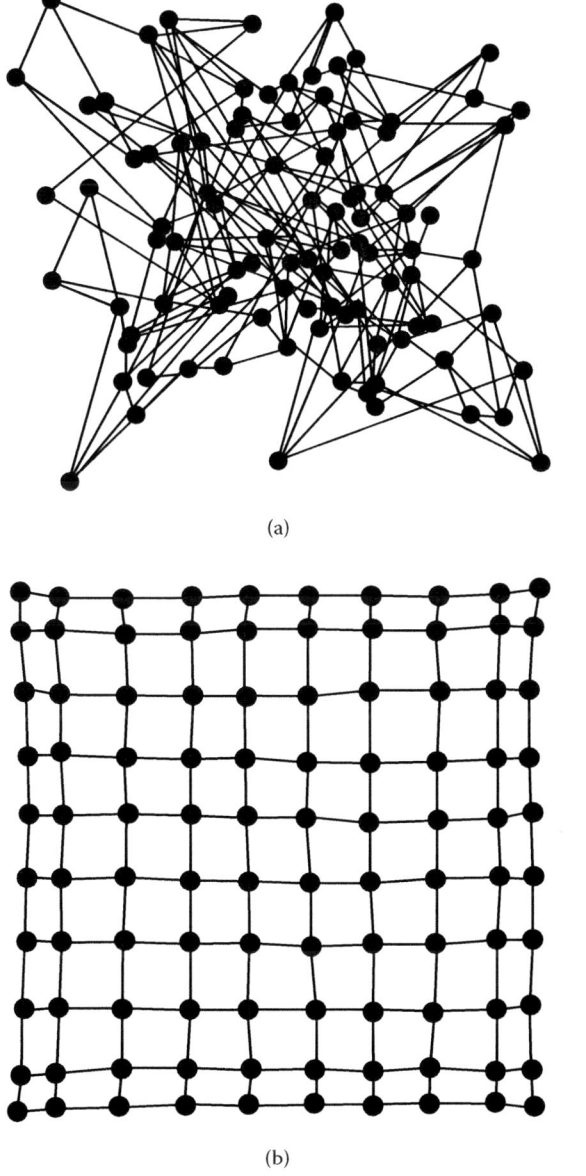

(a)

(b)

FIGURE 3.8
A 10×10 SOM trained on values taken at random from a 1×1 square; (a) at the start of training, (b) when training is complete.

3.6.2 Select the Sample

Now that the SOM has been constructed and the weights vectors have been filled with random numbers, the next step is to feed in sample patterns. The SOM is shown every sample in the database, one at a time, so that it can learn the features that characterize the data. The precise order in which samples are presented is of no consequence, but the order of presentation is randomized at the start of each cycle to avoid the possibility that the map may learn something about the order in which samples appear as well as the features within the samples themselves. A sample pattern is picked at random and fed into the network; unlike the patterns that are used to train a feedforward network, there is no target response, so the entire pattern is used as input to the SOM.

3.6.3 Determine Similarity

Each value in the chosen sample pattern is compared in turn with the corresponding weight at the first node to determine how well the pattern and weights vector match (Figure 3.9). A numerical measure of the quality of the match is essential, so the difference between the two vectors, d_{pq}, generally defined as the squared Euclidean distance between the two, is calculated.[*]

$$d_{pq} = \sum_{j=1}^{n} (w_{pj} - x_{qj})^2 \qquad (3.2)$$

Both the sample vector and the node vector contain n entries; x_{qj} is the j-th entry in the pattern vector for sample q, while w_{pj} is the j-th entry in the weights vector at node p. This comparison of pattern and node weights is made for each node in turn across the entire map.

Example 6: Calculating the Distances between Sample and Weights Vector
If the input pattern and the weights vectors for the nodes in a four-node map were

$$Input = \begin{bmatrix} 7 \\ 1 \\ 2 \\ 9 \end{bmatrix} \qquad w_1 = \begin{bmatrix} 6 \\ 4 \\ -3 \\ 9 \end{bmatrix} \quad w_2 = \begin{bmatrix} -5 \\ 0 \\ 0 \\ 7 \end{bmatrix} \quad w_3 = \begin{bmatrix} 2 \\ 7 \\ -4 \\ -1 \end{bmatrix} \quad w_4 = \begin{bmatrix} 7 \\ -9 \\ 2 \\ 9 \end{bmatrix} ,$$

the squared Euclidean distances between the input pattern and the weights vectors would be

---

[*] The distance between the node weights and the input vector calculated in this way is also known as the node's *activation level*. This terminology is widespread, but counterintuitive, since a high activation level sounds desirable and might suggest a good match, while the reverse is actually the case.

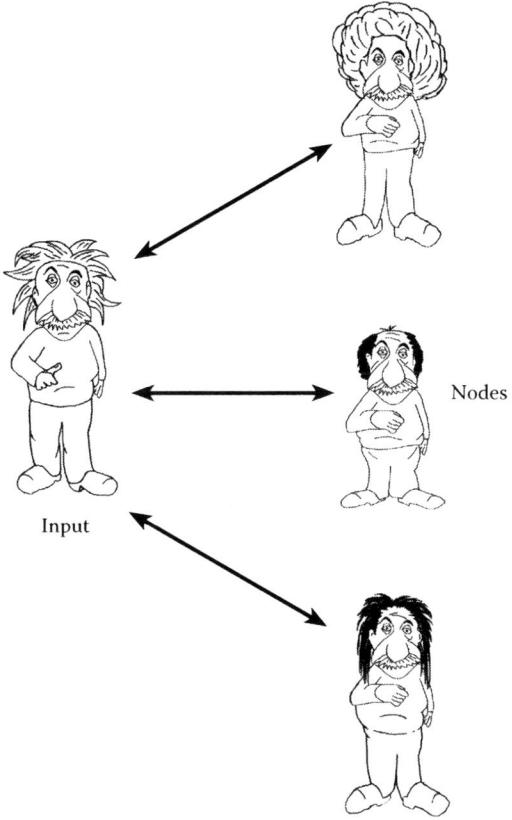

FIGURE 3.9
The input pattern is compared with the weights vector at every node to determine which set of node weights it most strongly resembles. In this example, the height, hair length, and waistline of each sample pattern will be compared with the equivalent entries in each node weight vector.

$$(7\text{-}6)^2 + (1\text{-}4)^2 + (2 - (\text{-}3))^2 + (9 - 9)^2 = 1 + 9 + 25 + 0 = 35$$
$$(7\text{-}(\text{-}5))^2 + (1\text{-}0)^2 + (2\text{-}0)^2 + (9\text{-}7)^2 = 144 + 1 + 4 + 4 = 153$$
$$(7\text{-}2)^2 + (1\text{-}7)^2 + (2\text{-}(\text{-}4))^2 + (9\text{-}(\text{-}1))^2 = 25 + 36 + 36 + 100 = 197$$
$$(7\text{-}7)^2 + (1\text{-}(\text{-}9))^2 + (2\text{-}2)^2 + (9\text{-}9)^2 = 0 + 100 + 0 + 0 = 100$$

3.6.4 Find the Winning Node

Once the squared distance d_{pq} has been calculated for every node, the *winning node* (also known as the *best matching unit*, or BMU) is identified. The winning node is the one for which d_{pq} is smallest, so, taking into account all values in the vector, the sample pattern more strongly resembles the weights vector at this node than the vector at any other; in the example above, this is node 1. If the match is equally good for two or more nodes, the winning node is chosen at random from among them.

Since the node weights are initially seeded with random values, at the start of training no node is likely to be much like the input pattern. Although the match between pattern and weights vectors will be poor at this stage, determination of the winning node is simply a competition among nodes and the absolute quality of the match is unimportant.

3.6.5 Update the Winning Node

The next step is to adjust the vector at the winning node so that its weights become more like those of the sample pattern.

$$w_{pj}(n+1) = w_{pj}(n) + \eta(n)[x_{qj} - w_{pj}(n)] \qquad (3.3)$$

In equation (3.3), $w_{pj}(n)$ is the j-th weight of node p at cycle n, $w_{pj}(n + 1)$ is the weight after updating, $\eta(n)$ is the learning rate, which lies in the range $0 < \eta(n) < 1$. The effect of this updating is to modify the weights vector at the winning node by blending into it a small amount of the sample pattern, thus making it slightly more like the sample.

Example 7: Updating the Weights Vector

In Example 4, the weights at node 1,

$$\begin{bmatrix} 6 \\ 4 \\ -3 \\ 9 \end{bmatrix}$$

provided the closest match to the input pattern of

$$\begin{bmatrix} 7 \\ 1 \\ 2 \\ 9 \end{bmatrix}$$

If the learning rate was 0.05, the node weights after updating would be

$$\begin{bmatrix} 6+0.05\times1 \\ 4+0.05\times(-3) \\ -3\times0.05\times5 \\ 9+0.05\times0 \end{bmatrix} = \begin{bmatrix} 6.05 \\ 3.85 \\ -2.75 \\ 9 \end{bmatrix}$$

As the adjustments at the winning node move each of its weights slightly toward the corresponding element of the sample vector, the node learns a little about this sample pattern and, thus, is more likely to again be the winning node when this pattern is presented to the map at some later stage during training.

Initially, the node weights are far from their optimum values. To bring them rapidly into the right general range, the weights are at first changed rapidly through the use of a large learning rate. As training progresses, the weights become a better match to sample data, so the changes can be made smaller and the algorithm settles into a fine-tuning mode. The learning rate is, therefore, a diminishing function of the number of cycles that have passed.

3.6.6 Update the Neighborhood

There is little to be gained by adjusting only the weights vector at the winning node. Were this done, the weights of all nodes in the network would evolve completely independently; each node would be a spectator to this updating except on those rare occasions when it was the winning node. Eventually most, possibly all, of the node weights would be just copies of a particular entry in the sample database, thus the SOM would have accomplished nothing of value, merely (very tediously) duplicating the sample data. If training followed this course, there would be no need to arrange the nodes in a lattice or, indeed, to specify any geometry at all because each node would be unaware of the existence of all the others.

Because we require that the map bring together similar samples in neighboring regions, as was the case in the clustering of animals, the weights at SOM nodes that are close together must be related in some way. This is achieved by adjusting the weights not only at the winning node, but also at other nodes that are nearby; these are nodes that are in the *neighborhood* of the winning node.*

The neighborhood of the winning node is a circular region centered on it (Figure 3.10). At the start of training, the neighborhood, whose size is chosen

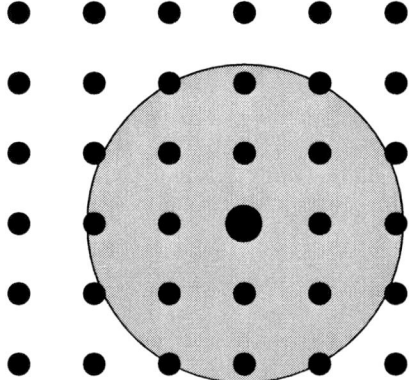

FIGURE 3.10
The neighborhood around a winning node (shown shaded).

---

* The neighborhood is usually deemed to include the winning node.

by the user, is large, possibly large enough to cover the entire lattice. As training proceeds, the neighborhood shrinks and adjustments to the node weights become more localized (Figure 3.11). This encourages different regions of the map to develop independently and become specialists in recognizing certain types of sample patterns. After many cycles, the neighborhood shrinks until it covers only the nodes immediately around the winning node, or possibly just the winning node itself.

As the weights at every node in the neighborhood are adjusted, it will be apparent that if, firstly, the neighborhood is very large and, secondly, the weights at every node are adjusted by the same amount, the weights vectors at every node across the entire lattice will eventually become very similar, if not identical. A SOM in this state would be no more informative than one in which all weights were completely uncorrelated. Therefore, to prevent this happening, the weights are changed by an amount that depends on how

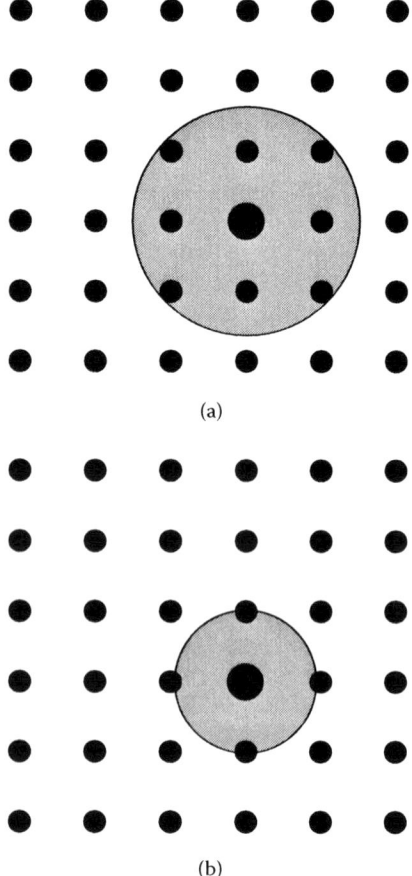

(a)

(b)

FIGURE 3.11
The size of the neighborhood around the winning node decreases as training progresses.

close a node in the neighborhood is to the winning node. The greatest change to the weights is made at the winning node; the weights at all other nodes in the neighborhood are adjusted by an amount that depends on, and generally diminishes with, their distance from that node.

$$w_{pj}(n+1) = w_{pj}(n) + f(d) \times \eta(n)[x_{qj} - w_{pj}(n)] \tag{3.4}$$

In equation (3.4), $f(d)$ is a function that describes how the size of the adjustment to the weights depends on the distance that a particular node in the neighborhood is from the winning node. This function might be the reciprocal of the distance between the winning node and a neighborhood node, measured across the lattice, or any other appropriate function that ensures that nearby nodes are treated differently from those that are far away. We shall consider several possible forms for this function in section 3.7.3.

Once the winning node has been identified, adjustments to the weights ripple out across the map from it. Because the adjustments are less pronounced far from the winning node, this process leads to the development of similar weights among nodes that are close to each other on the map, which will be a crucial step on the road to meaningful clustering. The gradual reduction of the neighborhood size as training proceeds helps to preserve this knowledge, which would otherwise continually be degraded as the weights of nodes in one part of the map were repeatedly pushed away from the values on which they would like to settle by updates forced upon them from winning nodes in other regions of the map.

3.6.7 Repeat

The winning node has been found, its weights have been updated, as have those of the nodes in its neighborhood. This completes the updating of the network that follows the presentation of one input pattern. A second pattern is now selected from the sample database and the process is repeated. Once all patterns have been shown to the SOM, a single cycle has passed; the order of samples in the database is randomized and the process begins again. Many cycles will be required before the weights at all the nodes settle down to stable values or until the network meets some performance criterion. Once this point is reached, the winning node for any particular sample pattern should be in the same region of the map in every cycle; we say that the sample "points to" a region of the map or to a particular node.

Training brings about two sorts of changes to the set of weights. At the level of the individual node, the vector of weights gradually adjusts so that each vector begins to resemble a blend of a selection of sample patterns. On a broader level, the vectors at nodes that are close together on the map start to resemble one another, and this development of correlation between nodes that are close together is illustrated in the one-dimensional SOM whose training is depicted in Figure 3.12.

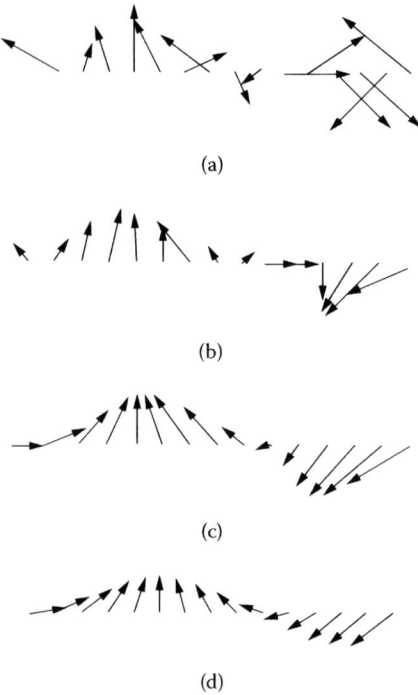

FIGURE 3.12
Training a one-dimensional SOM using a set of random angles. The starting configuration is shown at the top.

Example 8: Training a One-Dimensional Map

The database used for training this SOM is very simple. It consists of integer values distributed randomly within the range 1 to 360, so each sample pattern is just a single number and, therefore, the data are one-dimensional. This range of values is chosen because it makes display of the network weights particularly straightforward. Any data pattern, and any network weight, can be interpreted as a pointer whose angle with the vertical axis is given by the value of the data point. At the start of training (Figure 3.12a), each of the fifteen nodes is seeded with a random number in the range 1 to 360. After a few dozen cycles, Figure 3.12b, the weights in neighboring nodes are beginning to align and, by the end of training, the weights have settled down to stable values; the smooth progression of weights from one end of the SOM to the other is clearly evident.

Although it is clear that from a random starting point the final weights have become highly ordered, you may wonder why it is that, if the input data contain values that cover the range from 1 to 360, this range is not fully reflected in the final map, in which no arrows are pointing directly down. We might have expected that at the end of training the weights at neighboring nodes would differ by approximately 24°, so that the whole range of possible values of angles was covered, since the weights at fifteen nodes can be chosen ($15 \times 24 = 360°$). The reason that has not happened is that in this run the neighborhood around each node extends across the complete set of nodes, so even at the end of the run the weight of one node is influenced by that of others around it. The final arrow at the right-hand end of the map, for example, might "want" to become vertical so as to be able to properly represent an input pattern of 360°, but is constrained to be a bit like its left-hand neighbor because each time that neighbor is the winning node, some adjustment is made to the weight at the rightmost node also. The weight at the neighbor is in turn affected to some extent by the weight at its left-hand neighbor and so on. If we reduce the size of the neighborhood with time so that the neighborhood is eventually diminished to a size of two (which includes the winning node and a single neighbor), we see a greater range of weights, as expected, though the full range of angles is still not covered (Figure 3.13).

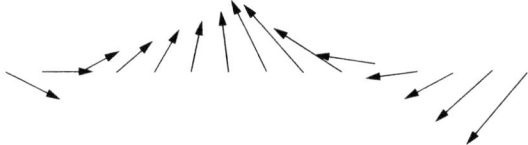

FIGURE 3.13
The result of training a one-dimensional SOM using a set of random angles. The final neighborhood includes only the winning node and one neighbor.

Example 9: Training a Two-Dimensional Map

Two-dimensional SOMs are more widely used than those of one dimension because the extra flexibility that is provided by a second dimension allows the map to classify a greater number of classes. Figure 3.14 shows the result of applying a 2D SOM to the same dataset used to create Figure 3.12 and Figure 3.13; the clustering of similar node weights is very clear.

Figure 3.15 shows the result of running the same dataset through a SOM of the same geometry, but starting from a different set of random node weights. The two maps are strikingly different, even though they have been created from the same set of data. The difference is so marked that it might seem to throw into doubt the value of the SOM. How can we trust the algorithm if successive runs on identical data give such different output?

This question of reproducibility is an important one; we can understand why the lack of reproducibility is not the problem it might seem to be by considering how the SOM is used as a diagnostic tool. A sample pattern that the map has not seen before is fed in and the winning node (the node to which that sample "points") is determined. By comparing the properties of the unknown sample with patterns in the database that point to the same winning node or to one of the nodes nearby, we can learn the type of samples in the database that the unknown sample most strongly resembles.

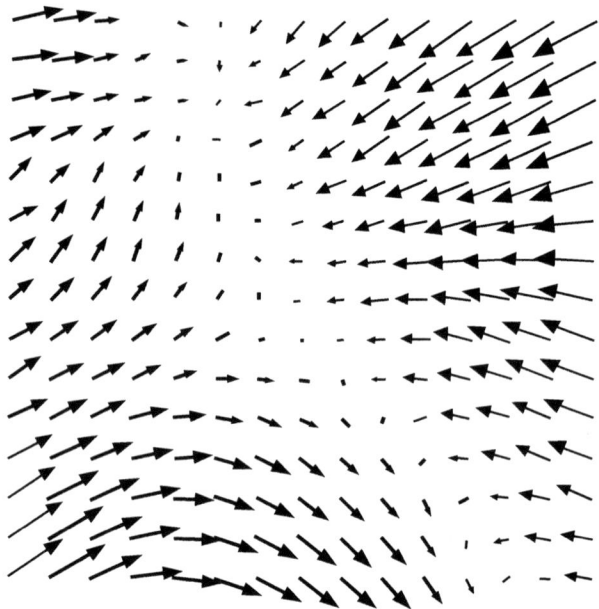

FIGURE 3.14
The result of training a two-dimensional SOM with a set of angles in the range 1 to 360°.

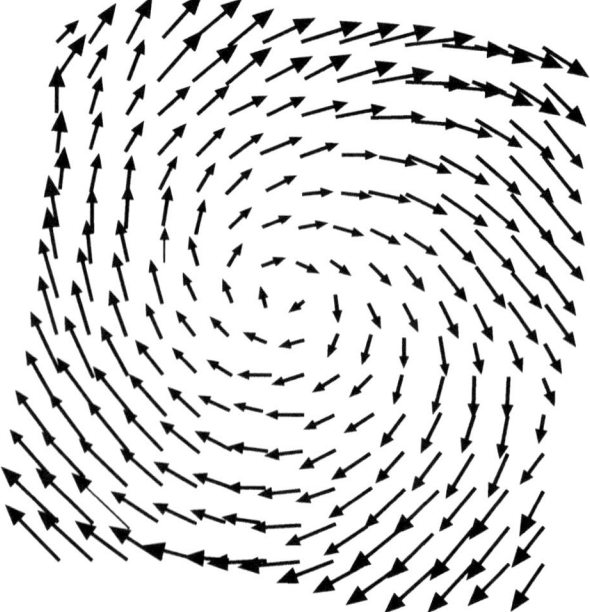

FIGURE 3.15
The result of training a two-dimensional SOM with a set of angles in the range 1 to 360°. The data used for Figure 3.14 and Figure 3.15 are identical. The same geometry was used on each occasion, but with two different sets of random initial weights.

Let us do this using the map shown in Figure 3.14. We feed in some value that was not contained in the original sample dataset, say 79.1°, and find the winning node, in other words, the node whose weight is closest to the value 79.1°.

The weight at the winning node for the sample pattern 79.1° will have a value that is close to 79°. The figure reveals that this node is surrounded by other nodes whose weights correspond to similar angles, so the pattern "79.1" will point to a region on the map that we might characterize as being "angles around 80°." If 79.1° were fed into the second map, Figure 3.15, the position of the node on the map to which it would point, as defined by the node's Cartesian coordinates, would be different from the position of the winning node in Figure 3.14, but the area in which that winning node is situated could still be described by the phrase "angles around 80°," thus the pattern is still correctly classified.

We conclude that the absolute position of the node on the map to which the sample pattern points is not important; neither of the maps in Figure 3.14 and Figure 3.15 is better than the other. It is the way that samples are clustered on the map that is significant. It is, in fact, common to discover when using a SOM that there are several essentially equivalent, but visually very different, clusterings that can be generated.

A comparable situation may arise in the training of an ANN. If an ANN is trained twice using the same dataset, starting from two different sets of random connection weights, the sets of weights on which the algorithm eventually converges will not necessarily be identical in the two networks, although if training has been successful, the output from both networks in response to a given input pattern should be very similar. Provided that the resulting networks function reliably and accurately, this unpredictability in the way that the ANN and the SOM train does not cast into doubt their value.

3.7 Adjustable Parameters in the SOM

3.7.1 Geometry

The geometry of the SOM, which includes both the size of the network and the topological relationship of one node to another, must be chosen at the start of a run. So far, we have assumed a square lattice on which to place the nodes, but other space-filling geometries can be used, such as a hexagonal or a triangular lattice in two dimensions or a tetrahedral lattice in three (Figure 3.16). The choice of whether to use a triangular, hexagonal, or rectangular arrangement of nodes is mainly a matter of personal taste; in most applications this choice does not have a large influence on the potential performance of the SOM.

By contrast, the number of nodes that the map contains is a key factor in determining the quality of the trained map and, thus, the performance of the

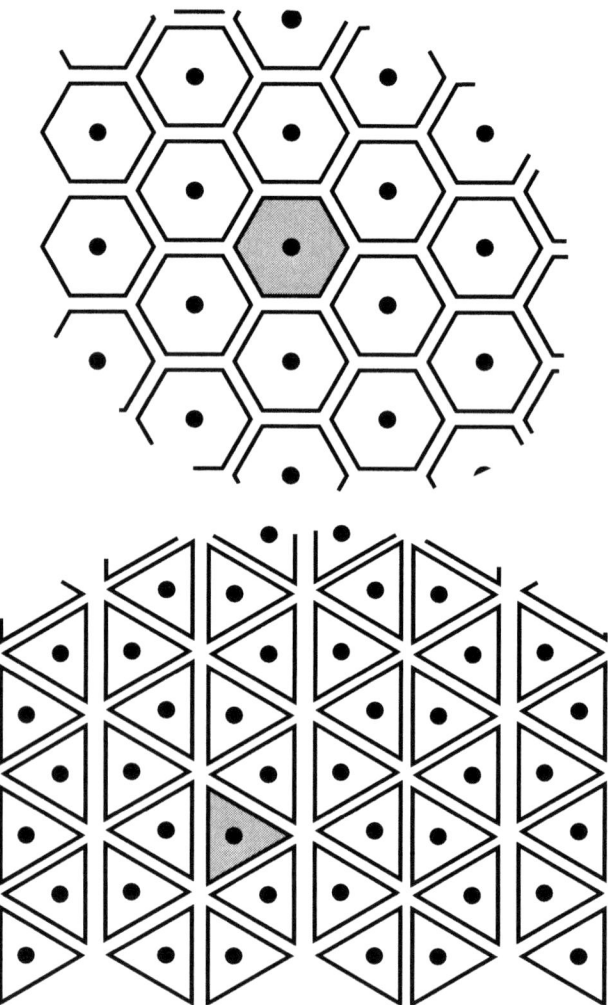

FIGURE 3.16
Hexagonal and triangular alternatives to a square lattice for the layout of nodes.

SOM. If the dataset contained examples of a dozen different classes, a 3×3 map would contain insufficient nodes to allow each class to be allocated to a separate node, thus separation of the classes on the trained map would be incomplete. Most or all of the node vectors would be a composite of patterns that included at least two classes and it would not be possible for the map to cleanly separate members of all twelve classes. A small map will be able to cluster only the simplest of data, so it is inappropriate for a dataset that shows a high degree of variety.

At the other extreme, although very large maps offer the maximum opportunity for samples to spread out, biggest is not always best, as we noted with artificial neural networks. The aim of the SOM is both to cluster samples and to separate classes. If the number of nodes in the map exceeds the number of sample patterns, all sample patterns will have their own node by the end of training. Because the neighborhood shrinks as the cycles pass, in the later stages of training there will be little interaction between the winning node and any neighbors other than those that are very close. The node weights will then adjust independently and the forces that lead to the partial homogenization of weights in neighboring regions of the map will die away.

This does not mean that no clustering occurs if the number of nodes is large compared to the number of classes. Later in this chapter we shall see an example in which the number of nodes has been set to a large value as an aid in visualization. Some broad-brush clustering is formed when the neighborhood is large, but if the map contains too many nodes, clustering may be less well defined than would be the case in a map of smaller dimension. There is, therefore, for any given dataset some optimum size of map (or more realistically, a range of sizes) in which enough room is available on the map that very different patterns can move well away from each other, but there is sufficient pressure to place some patterns close together so that clustering will still be clearly defined. This optimum size depends on the number of classes that exist within the data.

3.7.2 Neighborhood

The adjustment of weights is spread across a neighborhood. The idea of a neighborhood may seem simple enough, but it is only fully defined once we have specified two features: its extent and how the updating of the node weights should depend on the location of the node within it. The size of the neighborhood is determined by choosing some cut-off distance beyond which nodes lie outside the neighborhood, whatever the geometry of the lattice. This distance will diminish as training progresses.

3.7.3 Neighborhood Functions

The size of the adjustment made to the node weights is determined by a neighborhood function. Using the simplest plausible function, the amount of adjustment could be chosen to fall off linearly with distance from the node (Figure 3.17). Beyond some cut-off distance from the winning node, no changes are made to the weights if this function is used.

A Gaussian neighborhood function is frequently used.

$$w(t) = \frac{1}{\sigma\sqrt{2\pi}} \times e^{-\frac{t^2}{2\sigma^2}} \tag{3.5}$$

where $t = x - \mu$.

The Gaussian function does not show the abrupt change in value at the edge of the neighborhood that the linear function does because there is no longer any edge other than the boundary of the map; instead the adjustment

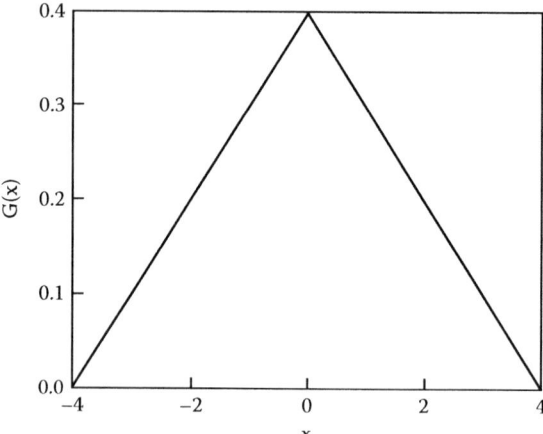

FIGURE 3.17
A linear neighborhood function; x denotes the number of nodes to the right or left of the winning node.

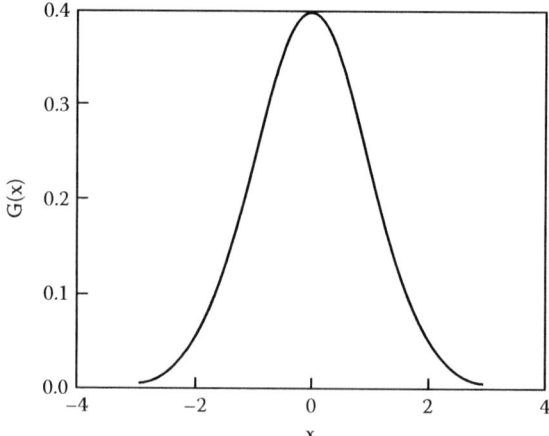

FIGURE 3.18
A Gaussian neighborhood function.

can extend arbitrarily far from the winning node (Figure 3.18). The width of the function can be adjusted to make the weight changes more or less strongly focused around the winning node. To increase computational speed, this function is applied to only those nodes that fall within a neighborhood of limited size, even though the function itself extends to infinity.

Functions that determine how large the updates to the weights should be, such as the Gaussian, fall off with distance from the node. An almost equivalent procedure to lessening the size of the neighborhood as the cycles pass is to use a neighborhood function that becomes more localized as the algorithm runs so that the weights at distant nodes are only slightly perturbed later in training.

In a small map, the Gaussian function may lead to weight changes in the neighborhood being larger than needed. To narrow down the region in which large weight changes are made, back-to-back exponentials may be used (Figure 3.19).

Each of the functions shown in Figure 3.17 to Figure 3.19 adjusts the weights at every node within the neighborhood in a way that increases the similarity between the weights vector and the sample pattern. However, in the completed SOM the weights at nodes in regions of the map that are far apart should be very different and this suggests that the weights of nodes that are distant from the winning node, yet still in the neighborhood, should perhaps be adjusted in the *opposite* direction to the adjustment that is applied to the winning node and its close neighbors.

In the Mexican Hat function (Figure 3.20), the weights of the winning node and its close neighbors are adjusted to increase their resemblance to the sample pattern (an excitatory effect), but the weights of nodes that are

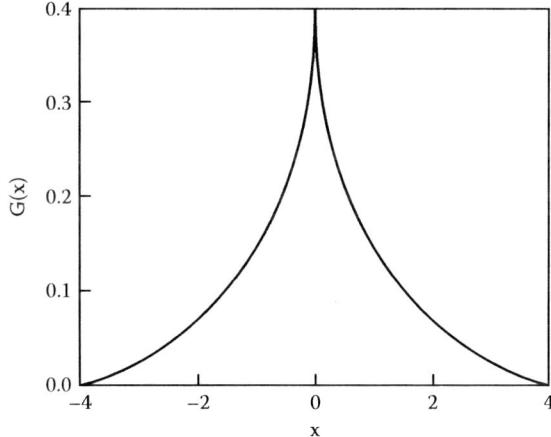

FIGURE 3.19
Back-to-back exponentials used as a neighborhood function.

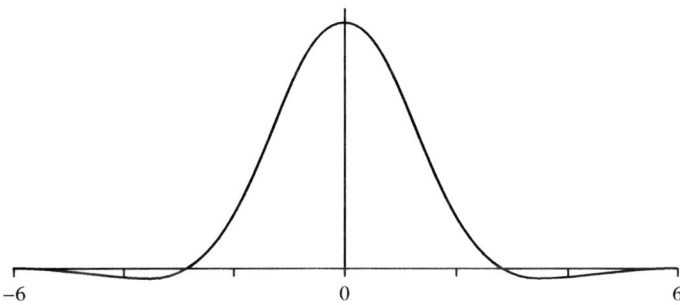

FIGURE 3.20
The Mexican hat neighborhood function.

farther away are adjusted in the opposite direction, making them less like the sample pattern (an inhibitory effect). This long-distance inhibition tends to emphasize differences in weights across the network.

The Mexican Hat is the second derivative of a Gaussian function:

$$\Psi(t) = \frac{1}{\sigma^3 \sqrt{2\pi}} (1 - \frac{t^2}{\sigma^2}) \times e^{-\frac{t^2}{2\sigma^2}} \text{ (Mexican hat)} \tag{3.6}$$

Example 10: Two-Dimensional Input Data

The SOM displays intriguing behavior if the input data are drawn from a two-dimensional distribution and the SOM weights are interpreted as Cartesian coordinates so that the position of each node can be plotted in two dimensions. In Example 5, the sample pattern consisted of data points taken at random from within the range [$x = 0$ to 1, $y = 0$ to 1]. In Figure 3.21, we show the development of that pattern in more detail from a different random starting point.

Each node is drawn at a position defined by its two weights, interpreted as an x- and a y-coordinate, respectively. Connecting lines are then drawn to join nodes that are next to each other in the SOM lattice. Thus, if the first and second SOM nodes, with lattice positions [0, 0] and [0, 1], have initial weights (0.71, 0.06) and (0.98, 0.88), points are drawn at ($x = 0.71$, $y = 0.06$) and ($x = 0.98$, $y = 0.88$) and connected with a line. The points occupy the available space defined by the range of x and y coordinates. Because the data points are positioned at random within a 1×1 square, the network nodes are initially spread randomly across that same space.

Since at the start of a run, the weights are random, nodes that are side-by-side on the SOM lattice may have very different weights, so most will be far apart in the representation given above—the initial layout appears (and is) random. However, as the cycles pass, the algorithm forces the weights at neighboring nodes to become more alike, so nodes that are close on the lattice, but which started with very different weights, are sucked together and the initial random tangle unravels to form a regular pattern. In the final map, the nodes are equally spaced, apart from a little noise, except for the outermost set of nodes, which appear to be too close to their neighbors. As was the case with the arrows in Figure 3.12, the final neighborhood includes more than just the winning node, so nodes at the edge of the map, which have no "outside" neighbor, but do have "inside" neighbors, are pulled in toward the interior of the map, thus reducing the gap between the outermost nodes and the first layer of interior nodes.

A one-dimensional SOM is less effective at filling the space defined by input data that cover a two-dimensional space (Figure 3.22) and is rather vulnerable to entanglement, where the ribbon of nodes crosses itself. It does, however, make a reasonable attempt to cover the sample dataset.

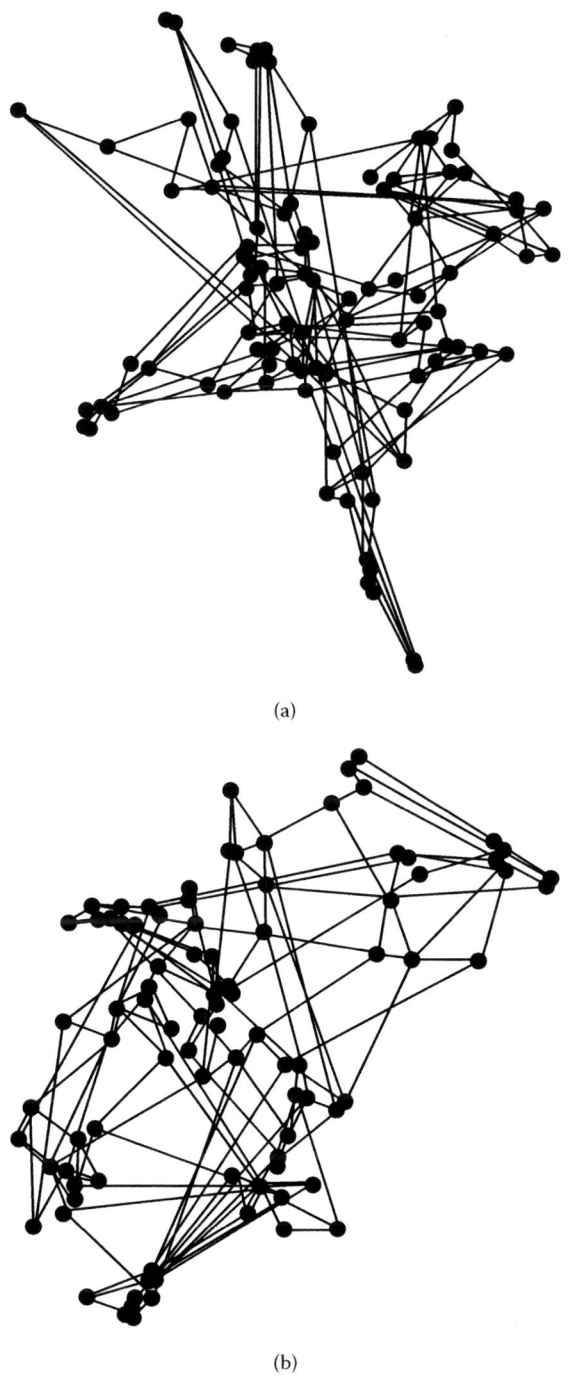

(a)

(b)

FIGURE 3.21
The node weights in a square SOM after (a) generation 27, (b) generation 267, (c) generation 414, (d) generation 2,027, and (e) generation 100,000.

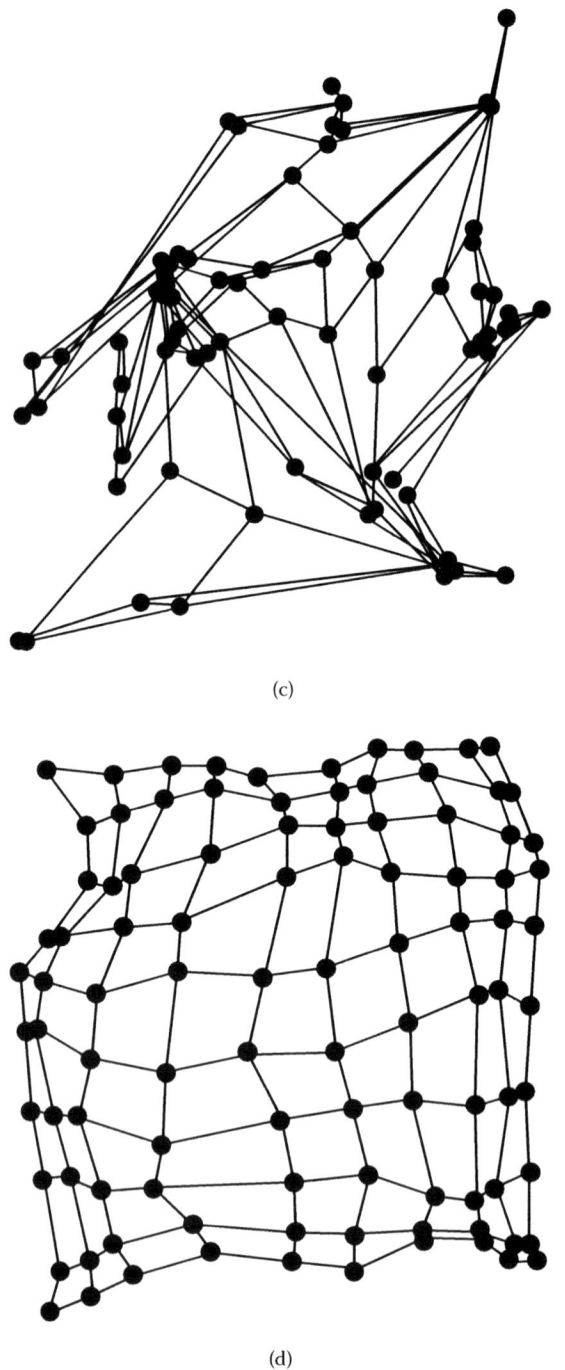

(c)

(d)

FIGURE 3.21 (CONTINUED)
The node weights in a square SOM after (a) generation 27, (b) generation 267, (c) generation 414, (d) generation 2,027, and (e) generation 100,000.

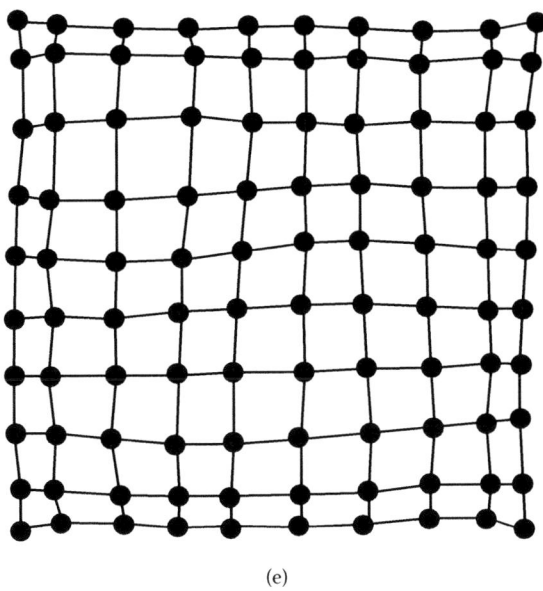

(e)

FIGURE 3.21 (CONTINUED)
The node weights in a square SOM after (a) generation 27, (b) generation 267, (c) generation 414, (d) generation 2,027, and (e) generation 100,000.

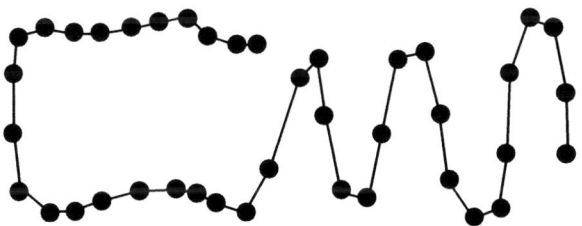

FIGURE 3.22
Points in a one-dimensional SOM attempting to cover the space defined by a 2 × 1 rectangle.

If the input data are not spread evenly across the x/y plane, but are concentrated in particular regions, the SOM will try to reproduce the shape that is mapped by the input data (Figure 3.23), though the requirement that a rectangular lattice of nodes be used to mimic a possibly nonrectangular shape may leave some nodes stranded in the "interior" of the object.

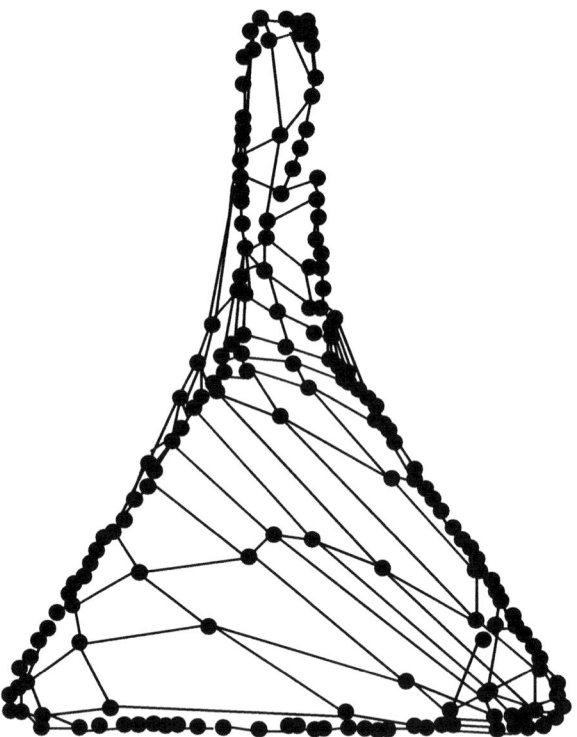

FIGURE 3.23
A familiar shape from the laboratory, learned by a rectangular SOM. Sample patterns are drawn from the outline of an Erlenmeyer flask.

3.8 Practical Issues

3.8.1 Choice of Parameters

It is a common feature of most AI methods that flexibility exists in the way that we can run the algorithm. In the SOM, we can choose the shape and dimensionality of the lattice, the number of nodes, the initial learning rate and how quickly the rate diminishes with cycle number, the size of the initial neighborhood and how it too varies with the number of cycles, the type of function to be used to determine how the updating of weights varies with distance from the winning node, and the stopping criterion.

With this number of factors influencing the development of the map, it is not possible to specify precisely what their values should be in all cases. The most suitable values will depend on how many features are present in the sample patterns and how diverse they are, but some general guidelines can be given. In the large majority of applications, a two-dimensional map is used; these are more flexible than one-dimensional maps, yet are simple to visualize and program. Rectangular networks are marginally simpler computationally than hexagonal or triangular lattices, but the choice between them is usually not important.

The number of nodes determines the degree of separation of classes that can be achieved. As mentioned above, if there are many nodes, clustering may be weak, although this can be partly offset by extending the training period during which a large neighborhood is used. If the number of classes cannot be estimated in advance, it may be necessary to prepare several maps of different sizes to test the degree to which they are able to separate samples.

The initial learning rate is usually chosen to be high, typically around 0.9, because the network weights then scale quickly into the correct range, but the rate is subsequently lowered to a small value within the range 0.05 to 0.005 by the start of a fine-tuning phase in which the network weights are allowed to converge on stable values.

If the dataset is large and contains many examples of each class, only a few hundred cycles may be needed for the SOM to develop a rough model of the data so that fine-tuning can begin. On the other hand, if the dataset is of modest size, if the data are complex, and if there are few examples of each class, tens of thousand or hundreds of thousands of cycles may be required with a high training rate before both training rate and neighborhood can be lowered to allow the map to start to refine the weights.

The initial neighborhood radius should be large and is often chosen to be sufficiently large so that it includes the entire map. The neighborhood may be reduced in parallel with the learning rate, but it is more common to reduce it more rapidly to a final value of 1 (which includes the winning node and its immediate neighbors) or, less usefully, 0 (the winning node alone).

3.8.2 Visualization

At the start of this chapter, we noted that the SOM has two roles: to cluster data and to display the result of that clustering. The finished map should divide into a series of regions, each corresponding to a different class of sample pattern.

Even when the clustering brought about by the SOM has been technically successful, so that samples in different classes are allocated to different areas of the map, there is no guarantee that an informative, readily interpretable display will be produced; there may still be work to be done. The reason is that, if each node stores many weights, not all of those weights can be plotted simultaneously on a two-dimensional map, thus decisions must be made about how the weights can best be used to visualize the data. If the underlying data are of high dimensionality, some experimentation may be required to get the most helpful map.

When the weights are one-dimensional, as in the angles data (see Figure 3.12), a display that shows the node weight as an arrow is effective. If the data are two-dimensional, interpreting the two values at each node as either a vector with magnitude and direction and so providing a display comprised of arrows of different lengths, or interpreting the two values as Cartesian coordinates and drawing each node at the position determined by those coordinates can be helpful.

It is unlikely that both the Cartesian coordinate display and a vector display will be equally informative. Figure 3.23 and Figure 3.24 show two different ways of displaying maps trained on the same data. Both figures show a degree of organization, but there is no doubt which one gives a better picture of the underlying data.

When the vectors at each node contain many weights, a different approach is required to display the final vectors in a readily interpretable way. The simplest procedure is to plot a separate map for each weight, showing each node in a color that is determined by that particular weight. This method is commonly used in the scientific literature and, if the number of weights at each node is small, can be very informative. However, if there are numerous weights, it is unlikely that the plotting of a separate map for every weight will convey much useful information. The SOM clusters the data according to the total similarity of the weights and sample patterns, thus it is optimistic

FIGURE 3.24
The nodes weights for the Erlenmeyer interpreted as a force field.

to expect that clustering will be evident when a display is chosen in which only one weight out of many is used. If we have clustered animals on the basis of thirty different characteristics, we should not expect that all the animals with long stripy tails will be gathered together in the same cluster.

This problem can be partly overcome by using several weights simultaneously to create the display. A visually appealing way to do this is to choose the same group of three weights at each node and interpret them as RGB (red, green, blue) values, then to plot on the map a blob of the appropriate color at the position of the node.

In a black and white representation, plots of this sort are less informative (color versions of Figure 3.25, Figure 3.26, and Figure 3.27 are available on the CD that accompanies this book), but even in grey scale the separation of the nodes into classes can be seen in these figures. In the following figures, all of which relate to a single dataset created using trigonometric functions, two types of display are used. In Figure 3.25a and Figure 3.26a, the map is generated by interpreting the weights directly as RGB values; a reasonable degree of clustering is apparent.

Clustering may become more obvious if, rather than using just three values to determine the color to be plotted, every entry in the weights vector is used, no matter how long the vector may be. This can conveniently be done by relating the color in which a node is drawn to the *difference* between the weights of a node and the weights of nodes in its immediate neighborhood, as determined by the Euclidean distance. This method of display highlights the boundaries between different regions on the map because it is in these areas that the change in node weights is most marked. In the examples shown, the neighborhood over which this distance is determined, without the central node, includes a total of twenty-four units.

Figure 3.25a/b shows the SOM at an early stage of development and Figure 26a/b the completed map. Figure 27a/b shows another run on the same set of data. The two completed maps are quite different in appearance, an effect that we noted before with the two-dimensional arrow set, but tests show that they have similar ability in the classification of samples.

The dataset used to create these maps is of quite low complexity and we would normally choose a far smaller map to analyze it because the smaller the map the shorter the processing time and the more reliable the clustering. However, in black and white a map made up of many nodes is more revealing than a map that contains few nodes because of the better visual definition that a large number of nodes provides. Some of the structure that is apparent in Figure 3.25 to Figure 3.27, especially in Figure 3.25b and Figure 3.26b, is partly a consequence of this choice of size. In the later stages of training, there are regions of the map that contain no winning nodes and that are not near to winning nodes. As the neighborhood shrinks, nodes in these regions eventually cease to be updated because they are never near to a winning node; their weights, therefore, become frozen. The final weights of these

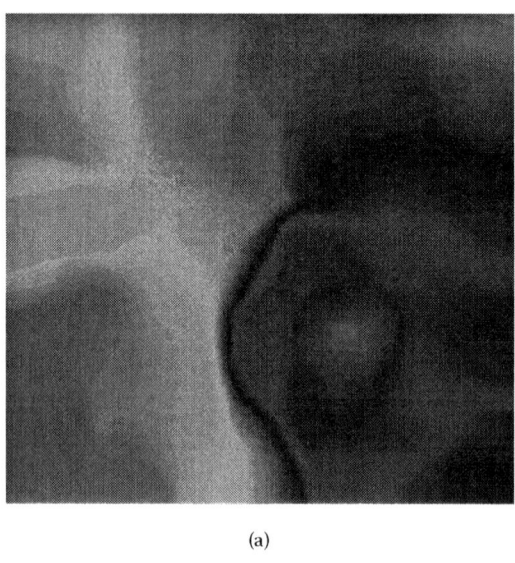

(a)

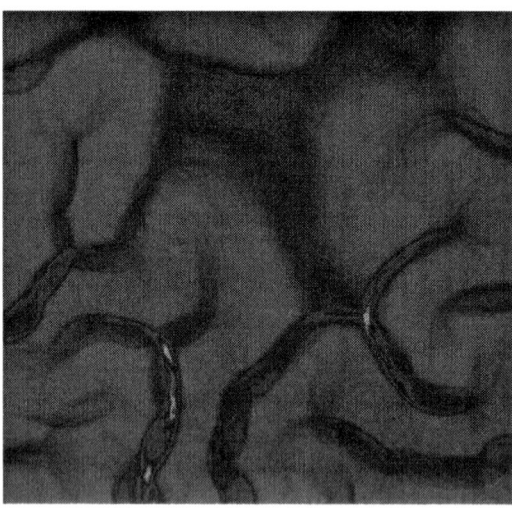

(b)

FIGURE 3.25
A SOM trained on some trigonometric data. In Figure 3.25a, the SOM is shown with the first three weights interpreted as RGB values. Figure 3.25b shows the same map with the color of each point determined by the difference in weights between one node and those in its immediate neighborhood.

nodes represent the structure of the map at an *early stage of training*, not when the map has converged, so while some parts of the map are fully representative of the dataset, others are not. This does not mean that the map cannot be used, but a run with a much smaller map size will reveal the extent to which

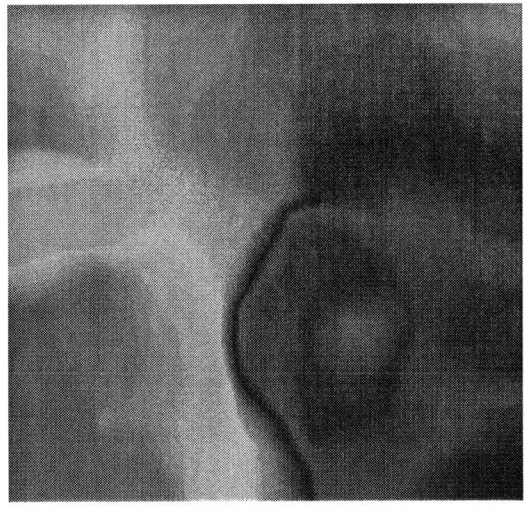

(a)

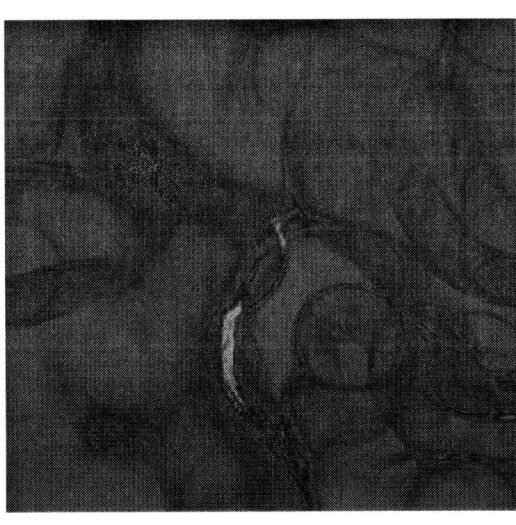

(b)

FIGURE 3.26
The same SOM as in Figure 3.25, but at a later stage of development.

the structure seen in these figures reflects real differences in the data. It will be discovered that a map of far smaller size is, in fact, optimum.

3.8.3 Wraparound Maps

Although all the nodes in a SOM should be equivalent, nodes at the edge of the map are in a different environment than those near the middle. The

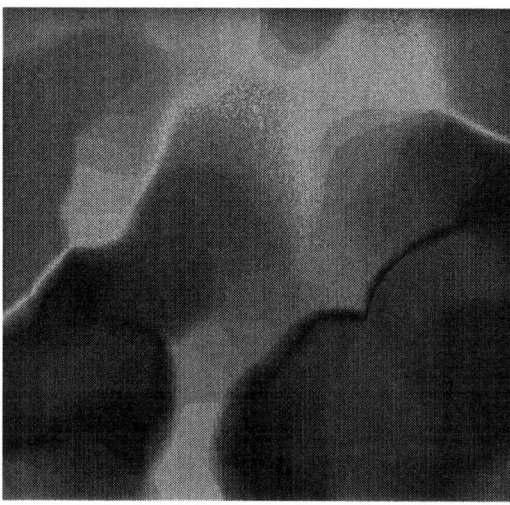

FIGURE 3.27
A SOM trained on the same dataset as the SOM shown in Figure 3.25 and Figure 3.26, but starting from a different random set of initial weights.

two nodes at the extremities of a one-dimensional SOM have an "inside" neighbor, but no "outside" neighbor, while all other nodes have neighbors on both sides. In a two-dimensional SOM, nodes in the interior of the map have neighbors in every direction, while those on an edge lack neighbors in one direction and those at a corner lack neighbors in two directions.

This asymmetry may have an effect on the development of the map. If there are few examples of a particular class in the dataset or if the characteristics of some sample patterns are markedly different from the characteristics of most other samples, development of the map may be eased if these unusual samples find their way to the edge of the map where they have fewer neighbors. The remaining samples, which share a wider range of characteristics, then have the whole of the rest of the map to themselves and they can spread out widely to reveal the differences between them to the maximum degree permitted by the size of the map.

There is no guarantee, though, that odd samples will gravitate to the edges of the map. If instead they settle somewhere in the middle, they may become trapped by other very different samples and, as the size of the neighborhood diminishes, can form a barrier that divides the map into portions, therefore preventing samples that may be quite alike (but initially find themselves in different parts of the map) from ever getting together. This is most readily envisaged in a one-dimensional map. If at an early stage in training, two similar samples point to nodes that are near the opposite ends of the map and are separated by a small group of samples that are very different, the similar samples may never be able to find a region on the map in which they can meet.

To reduce this trapping of samples with similar characteristics in different parts of the map, it is common practice to wrap the map around on itself by linking the periphery. In one dimension the map then becomes a ring (which is, nevertheless, still one-dimensional) (Figure 3.28).

Wrapping of a two-dimensional map at both sides and top-to-bottom creates a torus (Figure 3.29). The nodes on the left-hand edge of the map then find that they are neighbors to those on the right hand edge; top and bottom rows are similarly neighbors. This again places all nodes in an equivalent position and helps to prevent node vectors on the map from becoming trapped at a local minimum during training.

3.8.4 Maps of Other Shapes

Two-dimensional maps are easy to visualize and are readily programmed. If the dataset is complicated, a larger two-dimensional map might be replaced

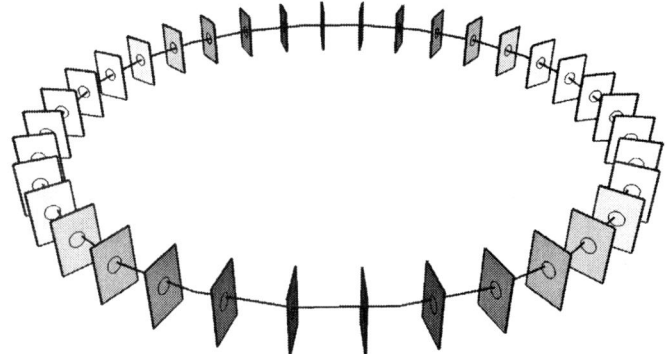

FIGURE 3.28
A (one-dimensional) ring of nodes.

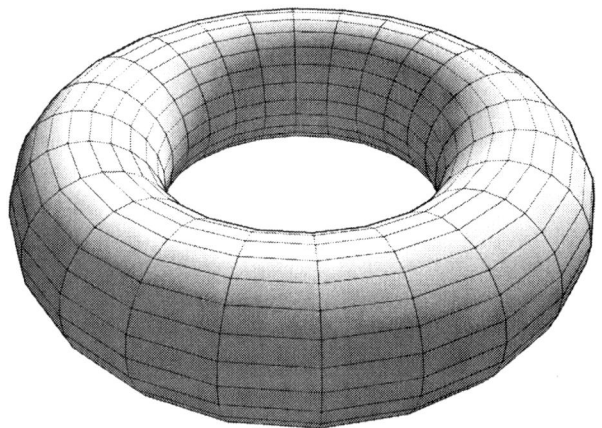

FIGURE 3.29
A torus, which is the result of linking opposite sides of a two-dimensional rectangular SOM.

by a smaller three-dimensional map, which we could investigate using a three-dimensional tool. (EJS [Easy Java Simulations], on the accompanying CD, allows users who have an elementary knowledge of Java to create interactive color three-dimensional displays in a simple fashion.)

Wraparound in three dimensions is more complicated to program and very much more complicated to visualize. In one dimension, we accomplished wraparound by making neighbors of the two most extreme points in the map. In two dimensions, we needed to join outer edges of the map, but, in three dimensions, exterior faces at the extremes of the grid must be connected. The additional computational bookkeeping required to work in three dimensions may cancel out any extra flexibility that it provides in the evolution of a cubic or tetrahedral SOM.

There has been some recent work on the use of spherical SOMs. Just as a SOM shaped as a ring is one-dimensional (each node only has neighbors to the left and right), so a spherical SOM resembles a torus and is two-dimensional (neighbors to the left and right, and also to the top and bottom, but not above and below), so a spherical SOM should be faster in execution than a genuinely three-dimensional SOM.

3.9 Drawbacks of the Self-Organizing Map

The primary difficulty in using the SOM, which we will return to in the next chapter, is the computational demand made by training. The time required for the network to learn suitable weights increases linearly with both the size of the dataset and the length of the weights vector, and quadratically with the dimension of a square map. Every part of every sample in the database must be compared with the corresponding weight at every network node, and this process must be repeated many times, usually for at least several thousand cycles. This is an incentive to minimize the number of nodes, but as the number of nodes needed to properly represent a dataset is usually unknown, trials may be needed to determine it, which requires multiple maps to be prepared with a consequent increase in computer time.

A second difficulty is that regions that correspond to the same class may be split between different parts of the map. As the map settles and the neighborhood around the winning node shrinks, the separation may become embedded. Splitting can lead to the number of classes in the dataset being overestimated or to the development of regions of the map that are not fully representative of elements of the dataset. Again repeated runs may help to identify and solve this problem. We see an example in the two-dimensional SOM in Figure 3.30, which, like the SOM in Figure 3.21, has been fed with data from a unit square. The SOM has (literally) tied itself into a knot and the size of the neighborhood by this stage of training is so small that it will not be able to untangle itself. While the SOM represents points on the right-

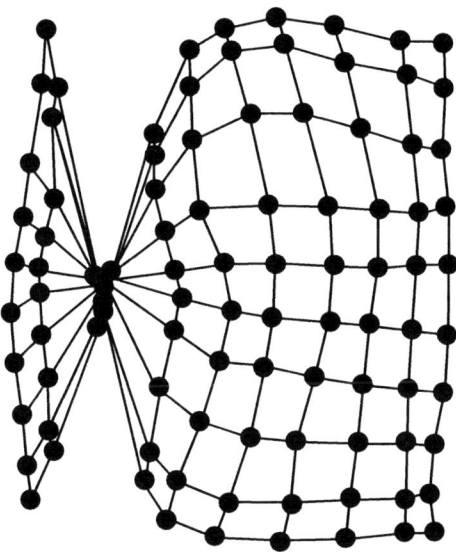

FIGURE 3.30
A tangled map.

hand side moderately well, it is clear that it is unable to do so toward the left. Similar problems can arise with three- and higher-dimensional data and are more difficult to detect.

3.10 Applications

How might we use a SOM in a scientific context? Suppose that we prepared a map in which the molecular descriptors of many compounds that have been used as herbal remedies provided the input patterns. The map would group these materials according to the values of the molecular descriptors. If some correlation existed between molecular structure and therapeutic activity, we might hope to find that remedies that were active against migraine, for example, were clustered together in some region of the map. The clustering might then provide information about what molecular features we might expect in a new drug that could be effective in treatment of the same condition.

Scientists, of course, do not create a map out of curiosity; the SOM must have a purpose. The map created by a SOM may be used in several ways, which usually lie within one of four broad categories:

Classification: Assigning a pattern to one of several classes.

Association: Retrieving a complete copy of a data item if only incomplete data can be fed in.

Projection: Crushing of high-dimensional data into two dimensions.

Modeling: Developing a simplified relationship between a high-dimensional vector and a simpler representation.

A typical use of the SOM is in the classification of crude oil samples, mentioned at the start of this chapter. If crude oils were grouped by composition, we might find that oils cluster together because they have a similar composition having been derived from wells in the same field. More interestingly, we might find that clustered oils in some cases come from quite different fields, but from similar geological formations. The most powerful way to characterize the very complex mixtures that are crude oils is to use mass spectrometry. This provides a detailed fingerprint for each sample, but the data are complex. Fonseca et al. have used GC-MS (gas chromatography-mass spectrometry) data as input into a SOM to characterize the geographic origin of oils, with a success rate of about two-thirds of their samples on the basis of a single SOM.[1] Recognizing that an individual SOM may not be fully optimized, they also investigated the use of ensembles of maps and were able to lift the identification rate to about 90 percent using this technique. In an additional step, they investigated whether a SOM would still be able to identify the geographic origin of the samples after a period of simulated weathering, obtaining encouraging results.

There have also been studies of the use of SOMs in tasks that are well beyond the capabilities of most other methods. Kohonen and his group have used SOMs to investigate the self-organization of very large document collections on the basis of the presence of words or short phrases.[2] The significance of work such as this is more in the general principles that underlie it rather than in the specific application. As the volume of scientific information available through the Internet grows, sophisticated tools are needed that can unearth publications that may be relevant to a particular, specialized area of research. Because of the volume of scientific research, few scientists now have the time to finger through *Chemical Abstracts* or similar publications ("finger through" is, in fact, often no longer an option because of the move to electronic publishing). Thus, tools that are able to combine a broad survey of many millions of documents with a focused output of results are increasingly valuable.

It has been noted elsewhere in this book that increasing computing power has led to the linking of AI methods with other computational tools, either a second AI method to form a hyphenated tool or a simulation to generate data, which can then be analyzed via an AI tool. Recent work on the conformation analysis of lipids by Murtola, Kupiainen, Falck, and Vuttalainen[3] falls in the latter category. Lipid bilayers are of central interest in biochemistry and other areas of science because of the broad range of processes in which they are involved. The structural properties of lipids are crucial in determining the behavior of biological membranes and are an important consideration in the rapidly growing field of biosensors.

Molecular dynamics (MD) simulations of the behavior of lipid bilayers using software such as GROMACS (GROningen MAchine for Chemical Simulations) are now feasible because of growth in computing power. These simulations generate "snapshots" of the state of the system at different times as the lipid molecules move within their environment, but an individual snapshot may be a poor reflection of the properties of the lipid over an extended time period, so some way of extracting time-averaged values is required.[3] Determining just what it is that the MD calculation is telling us may then reduce to a choice between calculating the average value of some parameter, such as the number of hydrogen bonds in which the lipid layer is involved, over the period of the simulation, or of selecting discrete times or periods of time at which the lipid conformation is inspected in order to infer its properties.

A promising alternative is to feed the conformations delivered by the MD simulation into a SOM so that the SOM can break the many conformations into a number of groups, thus allowing a more detailed analysis of the simulation data.

The SOM is mainly used as a qualitative tool, to identify groups rather than to provide some numerical estimate of similarity. Nevertheless, these methods also show promise as quantitative tools. Recent unpublished work by Leung and Cartwright[4] on the fluorescence of dyes has shown that the SOM can be used as a sensitive way to determine quantitatively the composition of mixtures containing several fluorescent species. When a sample of unknown composition is fed into the trained map, the node to which the sample points indicates the concentration of each species by the value of each of the different weights (Figure 3.31).

An area of increasing interest is the use of SOMs to extract information from large volumes of data. This is of particular importance in the life sciences, as automated techniques in areas such as gene sequencing, nuclear magnetic resonance, or mass spectrometry can generate enormous datasets whose size renders them difficult to analyze. Wang, Azuaje, and Black[5] have considered the use of SOMs in the context of several biological problems, most notably the assessment of electrocardiogram beat data and DNA splice-junction sequences, with encouraging results.

In an application area so far barely touched upon in science, the SOM can be used to regenerate damaged data. While missing data presents a problem in training as it does for all neural network models, it also presents an opportunity. Incomplete data cannot be used in the training of the SOM as their use would bias the development of the map. However, if some data points are not available for an unknown sample, this fragmentary data may still be fed into a trained map, with the winning node selected by comparison only of the data points for which data are available. The SOM can then be used to predict the missing data.

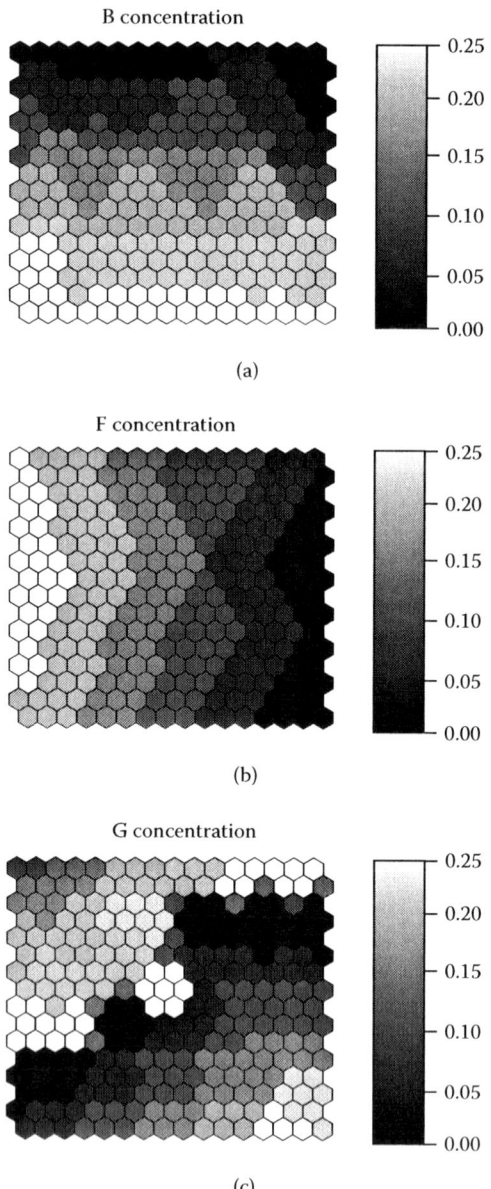

(a)

(b)

(c)

FIGURE 3.31

Three layers of a SOM trained on ternary mixtures of fluorescent dyes. Each layer corresponds to the concentration of a different dye. (Leung, A. and Cartwright, H. M., unpublished work.)

3.11 Where Do I Go Now?

That the SOM is often called a Kohonen map indicates the degree to which Kohonen and his co-workers have helped to define the field. Papers by Kohonen provide a rapid route into work with SOMs, but Zupan and Gasteiger's book *Neural Networks for Chemists: An Introduction*[6] offers a broader look at the techniques and should be helpful for anyone starting work in this area.

3.12 Problems

1. The Periodic Table

   The Periodic Table forms one of the most remarkable, concise, and valuable tabulations of data in science. Its power lies in the regularities that it reveals, thus, in some respects, it has the same role as the SOM. Construct a SOM in which the input consists of a few properties of some elements, such as electronegativity, atomic mass, atomic radius, and electron affinity. Does the completed map show the kind of clustering of elements that you would expect? What is the effect of varying the weight given to the different molecular properties that you are using?

2. Colorful arrows

   Not every value in a sample pattern needs to be given equal importance. For example, we might feel that the waistline of scientists was a more important factor in sorting them than their hairstyle. Write a SOM that sorts out angles, as in Figure 3.12, but this time each sample point should have both a value for the angle and a randomly chosen RGB color. Investigate the effect on the degree of ordering on a large map (say, 30 × 30) of varying the weight attached to each of these two characteristics from 0 to 100 percent. What happens in a large map if the weights of red, green, and blue are allowed to vary independently?

3. Sorting whole numbers

   Interesting effects may be obtained by using a SOM to sort integers. Create a SOM in which each input pattern equals the factors of an integer between 1 and 100. A simple way to do this would be to create a list of all the prime factors below 100 and prepare a binary vector that indicates whether the number is a factor. Alternatively, a binary input could be chosen that indicates which integers are divisors of the chosen number. For example, the divisors of 24 are 1, 2, 3, 4, 6, 8, and 12, so we could specify its divisors as the vector:

$$\{1, 1, 1, 1, 0, 1, 0, 1, 0, 0, 0, 1, 0, \ldots\}$$

You will need to consider how best to display the SOM weights.

References

1. Fonseca, A.M., et al, Geographical classification of crude oil by Kohonen self-organizing maps, *Analytica Chimica Acta*, 556, 374, 2006.
2. Kohonen, T., Spotting relevant information in extremely large document collections, *Computational Intelligence: Theory and Applications,* Lecture Notes in Computer Science, (LNCS), Springer, Berlin, 1625, 59, 1999.
3. Murtola, T., et al., Conformational analysis of lipid molecules by self-organizing maps, *J. Chem. Phys.*, 125, 054707, 2007.
4. Leung, A. and Cartwright, H.M., unpublished work.
5. Wang, H.Y., Azuaje, F., and Black, N., Interactive GSOM-based approaches for improving biomedical pattern discovery and visualization, *Proceedings Computer and Information Science,* Lecture Notes in Computer Science (LNCS), Springer, Berlin, 3314, 556, 2004.
6. Zupan, J. and Gasteiger, J., *Neural Networks for Chemists: An Introduction*, Weinheim, Cambridge, U.K., 1993.

4

Growing Cell Structures

As we saw in the previous chapter, self-organizing maps (SOMs) are a powerful way to reveal the clustering of multidimensional samples. The two-dimensional SOM is often able to provide an informative separation of samples into classes and the learning in which it engages requires no input from the user, beyond the initial selection of parameters that define the scale of the mapping and the way that the algorithm operates.

However, the method suffers from two notable disadvantages. First, a self-organizing map is slow to train. During each epoch, every data point in every sample pattern in the database must be compared in turn with the corresponding weight in the vector at every node. Table 4.1 shows how quickly the total number of comparisons required in the training of a SOM grows as the scale of a problem increases.

If, as an approximate figure, we assume that the comparison between a data point from the sample pattern and the corresponding node weight takes 10^{-7} sec, (which includes a contribution from the bookkeeping required for executing the rest of the algorithm), several days will be required to train SOM B. This length of time may already be unacceptably great, but the situation could

TABLE 4.1

Number of Pattern–Weight Comparisons Required in Training a SOM, as a Function of the Size of the Problem

SOM A (a small-scale problem):

No. of samples:	100
No. of points per sample:	10
Map size:	5×5
Epochs:	10000
No. of comparisons required:	2.5×10^8

SOM B (a large-scale problem):

No. of samples:	400
No. of points per sample:	100
Map size	40×40
Epochs	5×10^4
No. of comparisons required:	3.2×10^{12}

get even worse because the more samples there are in the database and the greater their dimensionality, the more features probably exist that distinguish one sample from another. As the number of features increases, the number of cycles needed for the SOM to converge will increase in proportion, so it might be impossible to run the algorithm to completion, except on a parallel processor machine.

A second difficulty that arises in the use of a SOM relates to the number of classes into which it can reliably divide a dataset. This is closely linked with the size of the map. A small map will be able to separate data into only a limited number of classes; on the other hand, a larger map, though able to separate more classes, is more expensive to train. It is important to choose a suitable number of nodes to form the map if the correct degree of clustering is to be revealed, but the number of classes in the database is almost certainly unknown at the start of a run; thus it is possible only to guess at what might be an appropriate map size. A computationally expensive period of experimentation may be required, during which maps of different sizes are created and tested in order to find one that shows the desired degree of clustering. Although this process is a reasonable way to determine the size of a map that best reveals clustering, it is time-consuming and is still dependent on a measure of guesswork in determining what level of clustering is appropriate.

A method exists that largely overcomes the problems of computational expense and uncertainty in the size of the map. This is the growing cell structure algorithm, which we explore in this chapter.

4.1 Introduction

Constructive clustering is an ingenious alternative to the creation of many SOMs of different size to find out which is the most effective in separating classes. Several constructive clustering algorithms exist. They are all related to the SOM, but they have the crucial difference from it that the size of the map is not specified in advance; instead, the geometry is adjustable and evolves as the algorithm learns about the data, eventually settling down to a dimension that, one hopes, best suits the complexity of the data that are being analyzed.

Initially a map of minimal size is prepared that consists of as few as three or four nodes. Since the map at this stage is so small, it is very quick to train. As training continues and examples of different classes are discovered in the database, the map spreads itself out by inserting new nodes to provide the extra flexibility that will be needed to accommodate these classes. The map continues to expand until it reaches a size that offers an acceptable degree of separation of samples among the different classes. As in a SOM, on the finished map, input patterns that are similar to one another should be mapped onto topologically close nodes, and nodes that are close in the map will have evolved similar weights vectors.

Methods in which the geometry of the map adjusts as the algorithm runs are known as *growing cell algorithms*. Several growing cell methods exist; they differ in the constraints imposed on the geometry of the map and the mechanism that is used to evolve it.

In the *growing grid* method, the starting point is a square 2 × 2 grid of units. During a preliminary *growth phase*, the map expands by the periodic addition of complete rows or columns of units, thus retaining rectangular geometry (Figure 4.1).

When a new row or column is inserted into the grid, all statistical information previously gathered about winning units is thrown away before learning resumes. This simplifies the algorithm, but increases computation time. As the grid expands, the neighborhood around each cell also grows; therefore, unlike a SOM, the number of units whose weights are adapted on presentation of each sample pattern increases as the cycles pass. Once a previously selected grid size has been reached, or some performance criterion has been met, growth of the network is halted and the algorithm enters a fine-tuning mode, during which further adjustments are made to the node weights to optimize the network's interpretation of the dataset.

A *growing neural gas* has an irregular structure. A running total is maintained of the *local error* at each unit, which is calculated as the absolute difference between the sample pattern and the unit weights when the unit wins the competition to match a sample pattern. Periodically, a new unit is added close to the one that has accumulated the greatest error, and the error at the neighbors to this node share their error with it. The aim is to generate a network in which the errors at all units are approximately equal.

Similar aims underlie the growing cell structure (GCS) approach, which relies on the use of lines, triangles, or, in more general terms, "dimensional hypertetrahedrons," (which, as we shall see, are far easier to use than the name suggests).

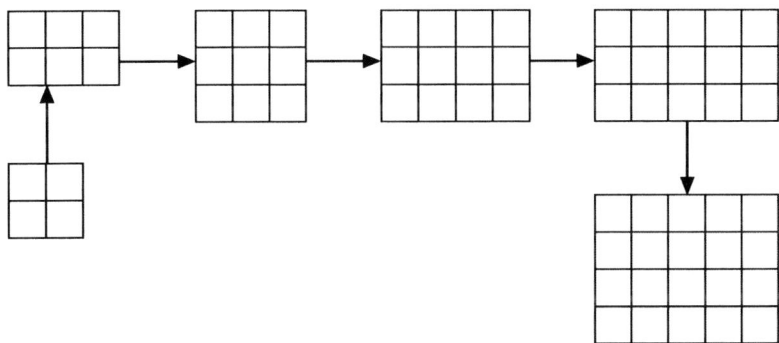

FIGURE 4.1
Expansion of a growing grid.

At present, it appears that the most productive types of constructive clustering in the physical and life sciences will be the growing neural gas and the GCS methods; in this chapter we focus on the latter. Although this method has notable advantages over the SOM, scientific applications of the GCS have only recently started to appear. There is a little more to the method than a SOM because of the need to grow the network as well as train it, but lack of familiarity with the technique rather than a lack of power explains the present paucity of applications in science because GCSs have nearly all the advantages of the SOM with few of the drawbacks.

4.2 Growing Cell Structures

The growing cell structure algorithm is a variant of a Kohonen network, so the GCS displays several similarities with the SOM. The most distinctive feature of the GCS is that the topology is self-adaptive, adjusting as the algorithm learns about classes in the data. So, unlike the SOM, in which the layout of nodes is regular and predefined, the GCS is not constrained in advance to a particular size of network or a certain lattice geometry.

A GCS can be constructed in any number of dimensions from one upwards. The fundamental building block is a *k*-dimensional simplex; this is a line for *k* = 1, a triangle for *k* = 2, and a tetrahedron for *k* = 3 (Figure 4.2). In most applications, we would choose to work in two dimensions because this dimensionality combines computational and visual simplicity with flexibility. Whatever the number of dimensions, though, there is no requirement that the nodes should occupy the vertices of a regular lattice.

Whereas in the SOM the weights are held at each *node*, a node in the GCS is often referred to as a *unit*, terminology that we shall follow in this chapter. However, the terms are used interchangeably.

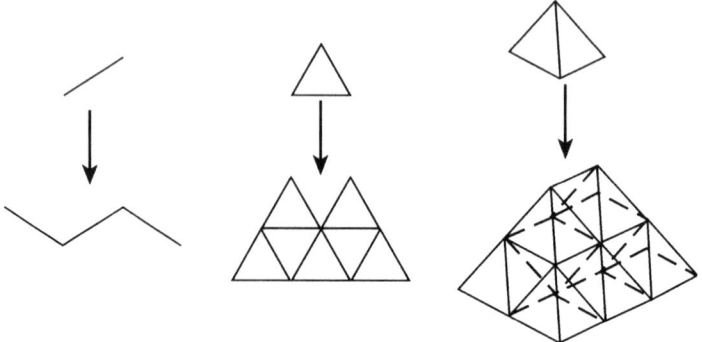

FIGURE 4.2
The building blocks of one-, two-, and three-dimensional growing cell structures.

The extension of the GCS to four or more dimensions is feasible computationally, but is challenging to visualize and is rarely attempted. In a fundamental difference with the SOM, the number of nodes that the network should contain when it is complete is not specified in advance; instead, training continues until the network has grown to a size at which it has reached some target of performance.

4.3 Training and Evolving a Growing Cell Structure

The training of a GCS takes place concurrently with the evolution of its geometry. The adjustment of the node weights resembles the equivalent process in the SOM, but some extra bookkeeping is needed so that, when the time comes to expand the network by adding a unit, the new unit can be inserted into the network at the position in which it can be of the greatest benefit. In summary, the GCS algorithm is as follows:

4.3.1 Preliminary Steps

Generally, a two-dimensional GCS offers the best combination of computational simplicity and flexibility in use, so we shall assume that dimensionality in what follows; the extension to three dimensions, or contraction to just one, is computationally straightforward. The building block from which the network is constructed in two dimensions is a triangle, thus the starting network consists of three interconnected units. Each unit stores a vector whose dimensionality equals that of the samples, just as was the case in the SOM. Again, as in the SOM, each network weight has the same interpretation as the corresponding element in the sample pattern, so if the first entry in a sample pattern is the height of a distinguished scientist, the corresponding weight at each node in the network is also interpreted as a height.

At the start of the run, each vector is filled with random numbers. These are chosen to lie in the range covered by the corresponding pattern, to minimize training time (Figure 4.3).

4.3.2 Local Success Measures

The most significant difference between the GCS and the SOM is the way in which a GCS is able to grow. In this section, we consider the mechanism by which this happens.

Expansion of the network is guided by the principle that, when a new unit is added, it should be inserted at the position in the network where it can be of the greatest value; this is determined by the *local measure of success*. There are several ways to define this location. We will consider the two most popular: signal counters and local error. The difference between signal counters

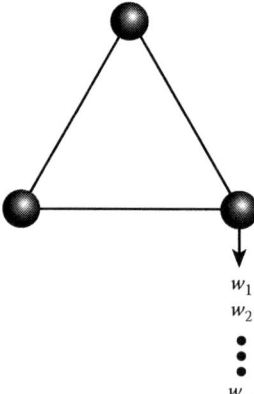

FIGURE 4.3
The starting configuration for a two-dimensional GCS. Each unit stores a vector of weights.

and local error is that the first measures how frequently a unit is chosen from among all the units in the map to represent a sample pattern, while the second measures the quality of the match between unit weights and sample pattern when a unit has won the competition to represent a sample. In both cases, as the network is trained, a record is kept of how successful each part of it has been. When the time comes to expand the network, a new unit is inserted into the network at the location where it can be most useful.

4.3.2.1 Signal Counter

When we use the signal counter as a local measure of success, as sample patterns are fed into the network, a count is kept of the number of times that each unit is selected as the best batching unit (BMU). The signal counter, thus, is incremented each time that the weights vector at a particular unit provides a better match to the sample pattern than can be managed by the vector at any other unit. Periodically, a new unit is added next to whichever unit at that stage in the training has the largest signal counter. The rationale is that, in order to achieve the most effective separation of sample patterns into classes, the patterns should be spread evenly across the map. As a consequence, when the network is unable to distinguish between many samples because they all point to the same BMU, an additional unit should be inserted beside that unit in order to produce a more fine-grained separation of samples.

4.3.2.2 Local Error

The second method that may be used to measure local success is to monitor the accumulated error at each unit. This measures not how frequently the unit wins, but how closely the weights at the unit match the sample pattern when it does win. Every time a unit wins the competition to represent a sample pattern, the squared distance between the unit weights and the input pattern is added to a local error variable E_a at that unit.

$$E_a = E_a + \sum_j (x_{jk} - w_{ja})^2 \tag{4.1}$$

In equation (4.1), x_{jk} is the j-th data point in sample pattern k and w_{ja} is the j-th weight at network unit a. After many sample patterns have been fed through a GCS that is being trained, some units will have accumulated a much larger local error than others; this might be for either of the following reasons:

1. The unit wins the competition to be the BMU only rarely and, when it does win, the match between its weights and the sample pattern is poor. Although the unit rarely is the BMU, the poor match when it is chosen means that its local error grows to a large value. If the match at this unit is the best for a particular pattern, even though the match is poor, it is evident that no unit in the network is capable of representing the pattern adequately.

   The database cannot contain many samples that are similar to this particular pattern, for if it did the unit would match many patterns and, in due course, the weights at the BMU would match them well; therefore, the poor match indicates that those samples for which this is the winning unit must be very different from each other and also few in number. As no unit in the network represents the samples well, the error in this region of the map is high and at least one new unit is required to represent a part of the variability in the samples.

Alternatively:

2. The unit is the BMU for a large number of samples and is a good match to nearly all of them. It has accumulated a large error because it is so often the winning unit; the small errors gradually add up. Since this unit wins frequently, the network is failing to differentiate between many of the patterns in the database, so a new unit should be added near the unit that has the greatest error to allow the network to better distinguish among these samples.

If alternative 2 applies, the same unit might be selected by the local error measure for insertion of a new unit as would be picked by the signal counter because, in both cases, the unit is frequently chosen as BMU. Alternative 1, however, picks out units that have a low signal counter rather than a high one. It follows that the course of evolution of a GCS will depend on the type of local measure of success that is used.

4.3.3 Best Matching Unit

Once a dimensionality for the map and the type of local measure to be used have been chosen, training can start. A sample pattern is drawn at random from the database and the sample pattern and the weights vector at each unit are compared. As in a conventional SOM, the winning node or BMU is the unit whose weights vector is most similar to the sample pattern, as measured by the squared Euclidean distance between the two.

After the winning unit has been identified, say u_b, the local measure at that unit is updated. If the local measure is a signal counter, the signal counter at the BMU is incremented by 1:

$$\tau_b = \tau_b + 1 \tag{4.2}$$

If the local measure is the error, this is increased by adding to it the Euclidean distance between the sample pattern and the weights at the unit.

4.3.4 Weights

The next step is to update the network weights. The weights at the winning unit are updated by an amount Δw_{ja}, as in a standard SOM:

$$\Delta w_{ja} = \varepsilon_b (x_{jk} - w_{ja}) \tag{4.3}$$

ε_b is a learning rate, which determines how rapidly the weights at the BMU move toward the sample pattern.

The weights of neighboring units must also be updated, but the way that this is done differs from the procedure used in a SOM. In that algorithm, all nodes within some defined distance of the winning node are considered to be neighbors, thus the updating may cover many nodes or even the whole lattice. In addition, the size of the adjustment made to the weights depends on the distance between the winning node and the neighbor. By contrast, in the GCS, the lattice is irregular, so distance is not an appropriate way to identify neighbors and does not enter into the updating algorithm. The neighborhood is instead defined to include only those units to which the winning unit is directly connected. The weights of the neighbors are adjusted by a smaller amount than are the weights of the winning unit, determined by the parameter ε_n.

$$\Delta w_{jn} = \varepsilon_n (x_{jk} - w_{jn}) \tag{4.4}$$

This simpler definition of a neighborhood means that there is no need to define a neighborhood function of the sort used in the SOM because all of the small number of neighbors in a GCS are treated identically. Because the neighborhood is of only limited size, updating of weights in that neighborhood is rapid.

4.3.5 Local Measures Decay

When signal counters are used as the local measure of success, they are not allowed to grow without bound. Either after each pattern is shown to the network or at the end of each cycle, all signal counters are decreased by a small fraction:

$$\tau_n = \tau_n \times \alpha \qquad (4.5)$$

α is a real-valued number in the range $0 < \alpha < 1.0$ and is generally very close to 1, so that the signal counter declines only slowly.

This slight, but regular, chipping away at the value of the counter serves two purposes. Suppose that a unit won the competition to represent sample patterns many times in the early period of training, but now wins the competition only rarely. It is clear that its value to the network, though it might once have been considerable, has fallen to a low level. Because this region of the network is of limited value, it should not be chosen as an area for the insertion of a new unit and the reduction in the signal counter helps to ensure this.

The slow decay of the signal counter, unless it is boosted by fresh wins, serves a second purpose. While a large counter indicates a suitable region of the map for the insertion of a new unit, a very small value indicates the opposite. The unit may be of so little value to the network that it is a candidate for deletion (section 4.5). Unlike the SOM, not only can units be added in the GCS, they can also be removed, so the signal counter can be used to identify redundant areas of the network where pruning of a unit may enhance efficiency.

If the local error is used to measure success, it too is decreased by a small amount after each sample pattern has been processed or each cycle is complete.

$$E_n = E_n(1 - \beta) \qquad (4.6)$$

β is a decay parameter whose value is close to zero. The local error is allowed to decay for reasons that are similar to those that applied to the signal counter. If a unit gained a large local error early in the training, but now represents sample patterns well, a large error would no longer correctly reflect the current value of the unit in the network. Therefore, the local error decays with time so that it reflects the recent value of the unit rather than its historical value.

4.3.6 Repeat

Once the weights at the winning node and its neighbors have been updated, another pattern is selected at random from the database and the BMU once again determined. The process continues for a large, predetermined number of cycles, λ, at which point, unless the performance of the network is judged to be satisfactory, a unit is added.

4.4 Growing the Network

After a specified number of presentations of all sample patterns in the dataset, the unit with the greatest signal count or greatest local error gets the right to insert a new unit. Suppose that this unit with the greatest local success measure is unit u_m. If a new unit is created next to it, with similar but not identical weights, some of the sample patterns that pointed to u_m will be enticed away, thereby lowering the number of wins that u_m manages in the future and spreading the signal counters more evenly across the map. If the measure of success is the local error, by adding a new unit next to u_m, the number of wins that u_m will make in the future will fall, thus its error will gradually diminish.

The new unit must be inserted between u_m and one of the units to which it is connected because the structure of the network as a series of connected simplexes must be maintained. To determine which of the neighbors this is, the weights at each of the direct neighbors are checked and the one whose weights are most different from the weights at u_m, as measured by the Euclidean distance between them, is selected. The new node is inserted into the network and joined:

1. To the unit u_m
2. To the neighbor
3. To any other nodes that already have connections to both u_m and to the neighbor

Finally, the direct connection between u_m and its neighbor is removed.

Two further steps are required. First, the new unit must be given an initial set of weights. While these could be set to random values, we recall that, on a completed SOM, units that are close together on the lattice have similar weights vectors. The same observation applies to the GCS, and this suggests that the weights at the new unit should be set to the average of the corresponding weights in unit u_m and the neighbor that was found to be least like it:

$$w_{i,new} = (w_{i,win} + w_{i,neigh}) / 2 \qquad (4.7)$$

The reason for selecting the unit that is most unlike u_m is that we require a wide range of weights vectors across the map if it is to adequately represent the full variety of sample patterns. By choosing to average the weights at the two units that are least alike, we introduce the greatest possible degree of variety into the map.

Finally, the local measures at u_m and all its neighbors are reduced, with the new unit taking a share from every node to which it is attached. Several recipes exist for doing this. Typically, we determine the similarity between the weights at the new node and the weights of each unit to which it is joined and then assign to it a share of the neighbors' signal counters in proportion to the degree of similarity. The share is inversely proportional to the number of neighbors that are sharing, so that the total local success measure over all nodes remains unchanged. Thus, were the weights at the new node to be identical to each of its five new neighbors (an unlikely event), each neighbor would give up one-sixth of its signal counter so that all six units subsequently had the same signal counter.

If the local error is used instead of the signal counter, it is common to set the initial local error at the new unit to the average of the errors at all units to which it is connected:

$$E_{new} = \frac{1}{n} \sum_{i=1,n} E_i \qquad (4.8)$$

The local error at each neighbor is then reduced by the amount of error that it has relinquished.

The effect of this expansion process is illustrated in Figure 4.4, in which all points in the sample database are drawn from a donut-shaped object. The outline of that object is quickly reproduced, with increasing fidelity as the number of units increases. It is notable that fewer units lie in the interior of the object once the map is complete than was the case when the SOM was trained on a set of points that defined a conical flask. This is a consequence of the flexibility of the geometry of the lattice in which the GCS grows.

As with any computational method, it is important that the parameters that govern operation of the method are chosen appropriately. In the GCS, one of the key parameters is the frequency with which new units are added. Each time a unit is added, the map should be given sufficient time to adjust all the unit weights so that the resulting map is a fair representation of the data. If this is not done, new units may be added in inappropriate positions, which will compromise the quality of the map.

Figure 4.5 illustrates this. It shows several stages in the development of a GCS that has been trained with the same dataset as was used to prepare Figure 4.4. In this case, though, a new unit has been added every twenty cycles, which is much too frequently, and it can be seen that, although the resulting map bears some resemblance to the underlying data, the fit to that data is poor.

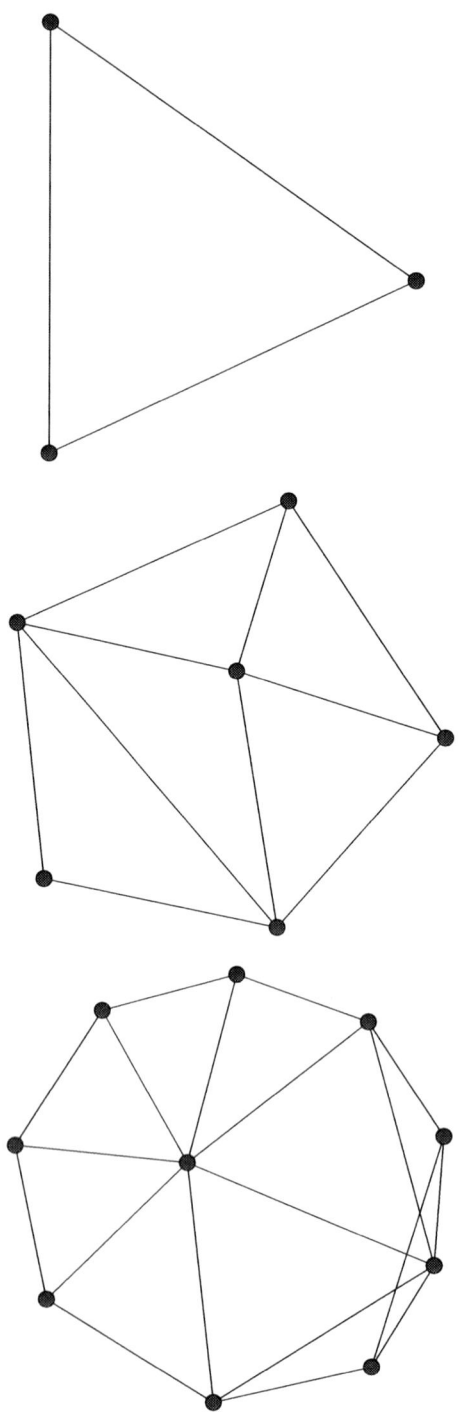

FIGURE 4.4
The evolution of a GCS fitting data to a donut-shaped object.

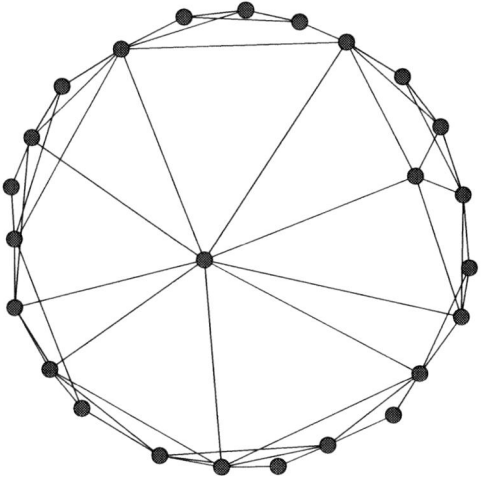

FIGURE 4.4 (CONTINUED)
The evolution of a GCS fitting data to a donut-shaped object.

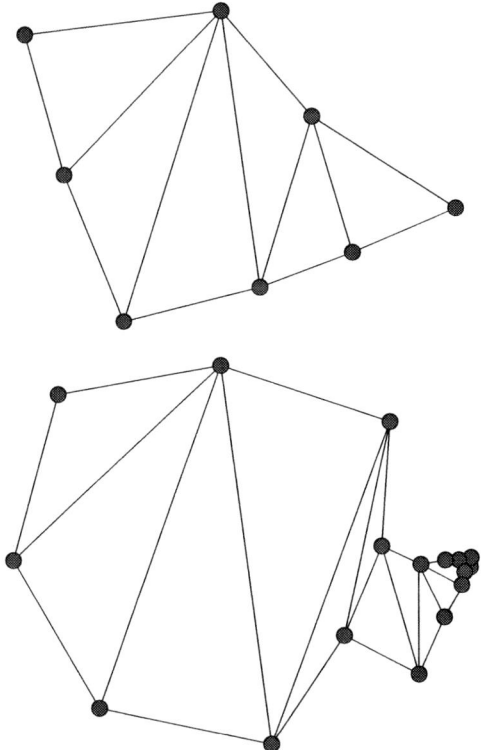

FIGURE 4.5
The evolution of a GCS fitting data to a donut-shaped object. A new node has been added at the rate of one unit per twenty cycles.

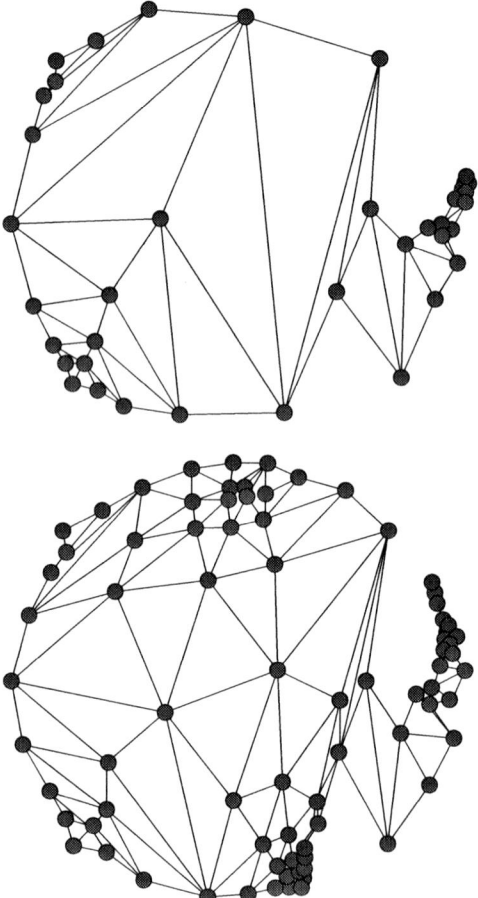

FIGURE 4.5 (CONTINUED)
The evolution of a GCS fitting data to a donut-shaped object. A new node has been added at the rate of one unit per twenty cycles.

4.5 Removing Superfluous Cells

In a completed map, every unit should have a similar probability of being the winning unit for a sample picked at random from the dataset. However, as the map evolves and the weights vectors adjust, the utility of an individual unit may change. Because the signal counter or the local errors are reduced every epoch by a small fraction, the value of this measure for units that are very rarely selected as BMUs will diminish to a value close to zero, indicating that these units contribute little to the network and, therefore, can be pruned out.

Deletion of cells is simple to perform. All connections to the cell(s) of least value in a network, as determined by the local success measures, are severed.

The local success measures of the excised cells are then shared out among their neighbors, either equally, or in inverse proportion to the difference in weights between the cell that is removed and each neighbor to which it was joined.

Pruning units out of a network that has only recently been grown sounds like a curious tactic. Why expand the network in the first place if later on bits of it will be removed? However, the ability to delete units as well as add them maximizes the utility of the network by spreading knowledge as evenly as possible across the units.

In addition, deletion of units allows a network to divide into multiple independent portions, without interconnections, and thus describe more accurately datasets in which one part consists of sample patterns that are very unlike the remainder. Once the performance of the network is not noticeably improved by the addition of a unit, the expansion and contraction may be brought to a halt. The network can then be used as an interpretive tool in exactly the same fashion as a SOM.

4.6 Advantages of the Growing Cell Structure

The previous section covered the mechanics of the GCS. Since the GCS is used in applications for which we might otherwise use the SOM, we shall briefly consider the advantages of the GCS and the conditions under which the GCS can compete effectively with, or surpass the performance of, SOMs. These include:

1. The finished network automatically reflects the characteristics of the data domain: Not only do the network weights evolve so that they describe the data as fully as possible, but so also does the network geometry. The size of the network is not chosen in advance and as topology is determined by the algorithm and the dataset in combination, it is more likely to be appropriate than the geometry used for a SOM, especially in the hands of an inexperienced user, who might find it difficult to choose an appropriate size of network or suitable values for the adjustable parameters in the SOM.

2. Constructive clusters are faster to train than conventional SOMs: In the early stages of training, the GCS is very small. Because there are few units, each pass through the database is fast. The training time per cycle increases as the map grows, but it is only when the map has grown to a size comparable to the equivalent SOM that training times per cycle approach those for the SOM. Even then, because the neighborhood only includes units that are directly connected to the BMU, the GCS may be faster to train than the SOM. In addition, as a map of optimum geometry should be generated at the first attempt,

it is not necessary to prepare multiple maps to determine which one is best.

3. Fewer parameters are needed to define evolution of the map: In particular, decay schedule parameters are not needed because the size and the shape of the neighborhood vary as the algorithm runs. Nor is it necessary to decide what form of neighborhood function to use, as all neighbors of the BMU are in an identical position.

4. It is possible to interrupt a learning process part way through if more data become available or if some data are revised: The GCS is, therefore, well suited to the development of incremental or dynamic learning systems.

5. The GCS should generate a map of minimum size: While the number of nodes on an n × n SOM is unavoidably the square of an integer value and thus may contain more nodes that are actually required for the development of a fully optimized map, the number of nodes in a GCS is not restricted in this way, thus the smallest and, therefore, most economical, map that describes the data can be constructed.

4.7 Applications

Wherever a SOM has been used to analyze scientific data, a GCS could be used instead. However, the GCS has a further advantage only alluded to above. Not only does the GCS include only as many units as are needed to properly describe the dataset, but, in addition, when units are added, they are positioned in the regions of the map in which they can be of greatest value, i.e., in the region where the dataset is most detailed. As an example, the GCS has been used to generate meshes to describe complex three-dimensional objects from a list of scanned data points.[1] The parts of an object where there is little detail and which can therefore be described by a small number of data points will be mapped by only a few units, while those regions in which the surface is very varied will be represented by a high density of units, and, therefore, with high fidelity. Averaging is built in, so if the signals are noisy, automatic smoothing takes place.

The GCS shows a great deal of potential as a tool for visualization. Wong and Cartwright have investigated the use of GCS to help in the visualization of large high-dimensionality datasets,[2] and Walker and co-workers have used the method to analyze biomedical data.[3] Applications in the field are starting to increase in number, but at present the potential of the method far exceeds its use.

4.8 Where Do I Go Now?

Little work has been published so far in this field, but Bernd Fritzke has led the development of the GCS. His published papers and citations of them provide an excellent starting point for those who wish to explore further.

4.9 Problems

1. In the last chapter, a SOM was trained with data points that lay on the outline of an Erlenmeyer flask. Construct a GCS that is trained using points drawn from the outline of a figure 8.

2. Repeat exercise 1, but use two identical figures 8s with a gap between them. Check that, as training proceeds, the removal of units which are rarely the BMU leads eventually to two separate regions on the map that are not connected, each of which defines the points in one object only.

3. The CD that accompanies this book contains a well-studied set of data known as the Iris dataset. A brief description of the data is included on the CD. Write a GCS to analyze the Iris data. If you have SOM software available, compare the performance and execution time of the SOM and the GCS on this dataset.

References

1. Ivrissimtzis, I.P., et al., unpublished work. http://www.mpi-sb.mpg.de/ivirs-sim/neural.pdf.
2. Wong, J.W.H. and Cartwright, H.M., Deterministic projection by growing cell structure networks for visualization of high-dimensionality datasets. *J. Biomed. Inform.*, 38, 322, 2005.
3. Walker, A.J., Cross, S.S., and Harrison, R.F., Visualization of biomedical datasets by use of growing cell structure networks: A novel diagnostic classification technique, *Lancet*, 354, 1518, 1999.

5

Evolutionary Algorithms

The processes of evolution have been remarkably successful in creating animals that are well adapted to their environment. By continually adjusting the population through genetic mutation, evolutionary change works toward the perfect individual to fill a particular environmental niche. Not every new individual is well suited to its environment, of course, and the fitness of new individuals is repeatedly tested as the animals try to survive, prosper, and reproduce.

In an evolutionary algorithm, we use software to perform the same trick. These algorithms mimic in a computer the natural processes of selection and genetic change, but instead of trying to simulate the behavior of populations in nature, the role of evolutionary algorithms is to provide a general computational method for solving optimization problems. The most promising evolutionary algorithms form the subject of this chapter.

5.1 Introduction

It is a common experience in science when we try to solve a new problem to find that an exact solution proves elusive. If no closed-form solution to the problem seems to be available, some process of iterative improvement may be needed to move from the best approximate solution to one of acceptable quality. With persistence and perhaps a little luck, a suitable solution will emerge through this process, although many cycles of refinement may be required (Figure 5.1).

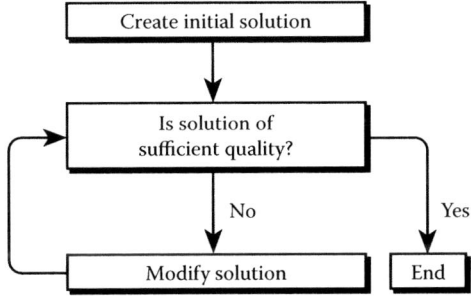

FIGURE 5.1
The process of iterative improvement of an initially modest solution to a problem.

113

When we are engaged in this type of iterative process, it is almost always just a single solution to the problem that is sought. If this is the case, it seems self-evident that, to minimize calculation time, our efforts should be limited to the identification and improvement of this one solution, not several. What benefit could there be in refining dozens of potential solutions when only one is required? Yet, if the search space is large and the problem is complex, we may discover solutions of high quality more rapidly if we process many in parallel and then choose the best one from among them, than if we focused our efforts on a single candidate. The solution eventually reached may even be superior to that arrived when we refine only one, so this approach offers potential advantages both in terms of speed of search and quality of the result.

Parallel refinement of multiple potential solutions is the method adopted by evolutionary algorithms. In this chapter, we discover why, when tackling optimization problems, it can be more productive to spread our efforts across a group of solutions rather than to concentrate on one.

5.2 The Evolution of Solutions

Evolutionary algorithms draw their inspiration from the evolutionary processes that operate in nature. The individuals in a large animal population display a range of characteristics. Animals are of different sizes and colors; there may be variations in the lengths of their legs or the thickness of their fur. On a timescale that is long compared to the animals' lifespan, a substantial turnover of genetic material occurs; as animals die and others are born, the average characteristics of the population drift as a result of genetic mutations. These mutations are random and, if the population is under only minor environmental pressure so an animal in which a random mutation has occurred is neither significantly more fit nor less fit than its brethren, many of those mutations will be of little consequence. The global population characteristics will then wander aimlessly through gene space and, in a population lucky enough to find itself in a completely benign, food-filled environment, the number of individuals will grow exponentially (Figure 5.2). The behavior of such a population is largely uninteresting and is of no value to us in solving scientific problems.

When the resources in the environment are finite and individuals must compete for them, the situation is very different. Not only do the dynamics of population growth change completely because the population cannot expand indefinitely without bumping up against the limitations of the resources, but more interestingly for our purposes, the behavior of the population can then form the basis of a method for solving scientific problems. It is possible to use evolution as a problem-solving tool precisely because the characteristics of individual members of the population adjust in response

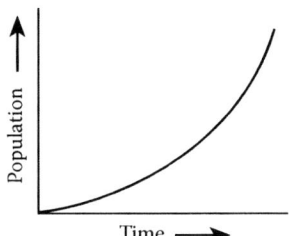

FIGURE 5.2
The growth in numbers of a population of animals in the absence of any environmental pressure.

to pressure from the environment. If the environment can in some way be made to reflect the scientific problem that we wish to solve, individuals may evolve that are suitable solutions to that problem.

In an animal population, whether a new characteristic generated by a random genetic mutation takes hold and starts to spread through the population is partly a matter of chance. A rabbit, whose genes contain a mutation that makes it more nimble than average, is a potentially useful addition to the world of rabbits, but the young rabbit might survive no longer than its first trip beyond the burrow if it attracts the attention of a hawk. In using evolution as a tool to solve scientific problems, we, therefore, can anticipate that chance will have an important part to play, just as it does in the development of animal populations.

Despite the unpredictability of the way in which a population might evolve, mutations that confer some fitness advantage are more likely to survive and proliferate (especially among rabbits) than those that are without value. In a cooling environment, animals with thicker fur are better protected than those with a lighter covering, so will be more tolerant of harsh weather conditions. If food must be hunted, animals with longer legs will be more successful in capturing their prey. Useful physical attributes, such as thicker fur or longer legs, which confer an advantage if the environment is cold or food is elusive, help the individuals that possess them to compete successfully with their naked and short-legged companions. If the advantage is marked, fast, furry animals will displace over many generations their slow, naked colleagues from the population.

At an individual level, the genetic mutations that lead to thicker fur or longer legs are entirely random, and yet the response of the population as a whole to changes in the environment is not. The pressure that the environment imposes on all members of the population gradually weeds out those individuals whose characteristics are less well suited to life within it than those of the average animal. The complete population resembles a living computer that constantly adapts its operation in an attempt to produce an individual perfectly adapted to the current environment.

Evolutionary algorithms apply these principles in a problem-solving context. A computer-based population of individuals, each of which, in a silicon world, is a potential solution to a problem, finds itself under "environmental pressure" provided by the problem to be solved. The algorithm grades these virtual individuals according to their ability to survive under this pressure and preferentially rejects the poorer ones in favor of newly created individuals of potentially higher quality. This repeated destruction of poor individuals and creation of new ones forces the population in the direction of better solutions; the answer to the scientific problem thus evolves.

Several flavors of evolutionary algorithm exist, among them evolutionary strategies, genetic programming, particle swarm optimization, and genetic algorithms; of these, the approach most widely used in science is the genetic algorithm. Apart from particle swarm optimization, most other types of evolutionary algorithm can be viewed as variants of genetic algorithms, so an understanding of this area provides a sound background from which to build an appreciation of the other varieties. Genetic algorithms form the primary focus of this chapter.

5.3 Components in a Genetic Algorithm

The steps that a genetic algorithm (GA) takes in evolving the solution to a problem are shown in schematic form in Figure 5.3.

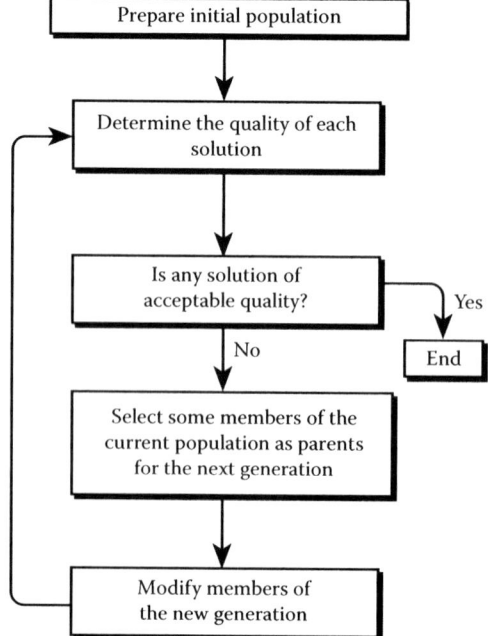

FIGURE 5.3
The genetic algorithm (GA).

A *population of solutions* is created and each member of it is tested to determine how good it is. If there is no solution in the population whose quality is acceptable, the solutions are manipulated by picking the better ones and modifying them to form a fresh population whose members are tested in turn. This cyclic process continues over many *generations,* during which the population improves until, one hopes, a solution of suitable quality emerges. The entire process, from the preparation of the first population to the final identification of a solution of near-optimum quality, makes up a GA *run.*

The GA contains the following components:

1. A method to represent potential solutions to a problem as *individuals* that can be manipulated by *evolutionary operators.*

2. A recipe that can be used to assess the quality of individuals.

3. A method to select the better individuals to act as *parents* in the creation of a new generation.

4. A mechanism for *genetic change,* so that a newly created member of the population need not be an exact copy of a previous member.

5.4 Representation of a Solution in the Genetic Algorithm

Before we can start to use the genetic algorithm, we must answer the question: What exactly is a "population of solutions?" It is easy to envisage a population of crocodiles or ants or lamas, but what does a population of solutions look like?

To define in detail a population of any sort, we should specify the significant characteristics of each individual. A particular animal might be described by a vector of properties (Figure 5.4):

Daisy = < height at shoulder, weight, breed, predominant color, spotted(?), sex, average weekly milk yield, disposition ... >

FIGURE 5.4
Daisy.

The vector of properties is a kind of chromosome, a code that defines the genotype of the individual. This sequence of values specifies the "genetic makeup" that (possibly uniquely) defines the animal. In Daisy's case, not all of the animal's characteristics are defined by her biological genetic code; we have included in the vector other characteristics that would help to pick her out from the kine (cattle) crowd. In a similar way, a single GA solution can be constructed as a "chromosome" or genome in vector form, which contains all the information needed to define the potential solution to a scientific problem.

Example 1: Optimizing Molecular Geometry

If the goal of a genetic algorithm application was to find the lowest energy arrangement of the atoms in bromochloromethane (Figure 5.5), the chromosome that defines a possible solution to the problem could be formed as an ordered list of the Cartesian coordinates of each atom:

$$S = \{x_C, y_C, z_C, x_{Br}, y_{Br}, z_{Br}, x_{Cl}, y_{Cl}, z_{Cl}, x_{H1}, y_{H1}, z_{H1}, x_{H2}, y_{H2}, z_{H2}\}$$

The vector that the genetic algorithm manipulates is known conventionally as a *chromosome* or a *string*; we shall use the latter terminology in this chapter. The individual units from which each string is constructed, a single *x-*, *y-*, or *z*-coordinate for an atom in this example, are referred to as *genes*.

Once the calculation is under way, it is useful to be able to monitor progress of the algorithm as it searches for a suitable solution. This monitoring is easier if the string that codes for each solution is readily interpretable, like the string above that defines where each atom is in space. However, it is not always possible for the physical meaning of a string to be so clear, so it is fortunate that the effectiveness of the algorithm itself is not directly affected by whether or not the solution is simple to interpret.

While floating-point values are used to construct the strings in most scientific applications, in some types of problem the format of the strings is more opaque. In the early development of the genetic algorithm, strings were formed almost exclusively out of binary digits, which for most types of problem are more difficult to interpret; letters, symbols, or even virtual objects

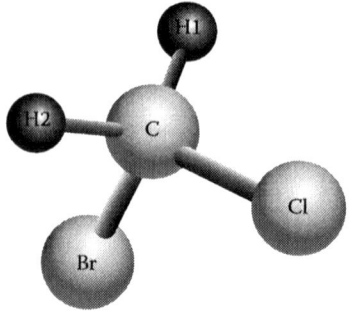

FIGURE 5.5
Bromochloromethane.

{ 0001010100011011010100 }
{ 1, 5, 66, −2, 0, 0, 3, 4 , −2, −110, −2, 1 }
{ 71.7, −11.4, 3.14159, 0.0, 78.22, −25.0, −21.4, 2.3, 119.2 }

FIGURE 5.6
Some typical genetic algorithm (GA) strings.

can also be used. In this chapter, we shall be working with strings that consist of integers (Figure 5.6).

The scientific problem that we are trying to solve enters into the algorithm in two ways. First, we must *encode* the problem, that is, choose a way in which potential solutions to it can be expressed in the form of strings that the algorithm can manipulate. Second, the algorithm must be given a way to evaluate the quality of any possible string so that it can determine how good each string is.

To understand how the GA manages to evolve a solution, it is helpful to focus on a particular example; therefore, we shall use the algorithm to solve a straightforward problem.

FIGURE 5.7
Ten randomly oriented dipoles.

> **Example 2: Energy Minimization Using the Genetic Algorithm**
> We will illustrate the operation of the GA through a simple application—determination of the lowest energy arrangement of a small group of dipoles. Suppose ten identical dipoles are spaced at equal intervals along a straight line (Figure 5.7).
>
> Each dipole is pinned down at its midpoint, but can rotate freely in the plane of the paper around that point. The GA will be used to determine the set of angles made by the dipoles with the vertical axis that yields the lowest total energy. It is not hard to think of a low-energy arrangement, but this problem forms a convenient platform through which to follow the workings of the GA.

5.5 Operation of the Genetic Algorithm

The steps in the algorithm are shown below.

1. Select genetic algorithm parameters
2. Prepare initial population
3. Determine quality of solution

4. Selection

5. Crossover

6. Mutation

We will look at each of these steps in turn.

5.5.1 Genetic Algorithm Parameters

In common with most other AI algorithms, the GA contains several variables whose values are chosen at the start of a run. Decisions must also be made about how to implement the evolutionary operators within the algorithm because there may be more than one way in which the operators can be used. We shall deal with the permissible values of these parameters and the factors that help us to choose among the evolutionary operators as they are introduced.

We first choose the *population size*, which is the number of individuals that comprise the population, n_{pop}. In contrast to an animal population, the size of a GA population is fixed once we have chosen it; consequently, when the algorithm creates a new individual, an old one must give way to make room for the newcomer. The genetic algorithm works best when the population consists of a significant number of individuals, typically in the range forty to one hundred, but in the present problem we shall use a population of ten individuals to make it easier to follow the progress of the algorithm.

5.5.2 Initial Population

Once the population size has been chosen, the process of generating and manipulating solutions can begin. In preparation for evolution, an initial population of potential solutions is created. In some problems, domain-specific knowledge may be available that can be used to restrict the range of these solutions. It might be known in advance that the solution must have a particular form; for example, we might loosely tie together opposite ends of neighboring dipoles in our example, so that, instead of being able to adopt any arbitrary relative orientation, the difference in angle between two neighboring dipoles could not be more than 60°. If we have advance knowledge of the format of a good solution, this can be taken into account when the individuals in the first population are created. Within the limits of any such constraints, each solution is created at random.

In the present problem, we shall assume that nothing is known in advance of the form of a good solution. A typical string, created at random, is given in Figure 5.8; shown below the string are the orientations of the dipoles that it codes for.

The initial population of ten random strings is shown in Table 5.1.

{103 176 266 239 180 215 69 217 85 296}

\ ← ↑ / ← ↗ / ↗ ↑ \

FIGURE 5.8
A random string from generation 0.

TABLE 5.1

Generation 0: The Initial Set of Strings, Their Energies E, and Their Fitness f

String	θ_1	θ_2	θ_3	θ_4	θ_5	θ_6	θ_7	θ_8	θ_9	θ_{10}	E	f
0	103	176	266	239	180	215	69	217	85	296	1.052	0.1169
1	357	295	206	55	266	180	166	95	131	293	0.178	0.1302
2	141	335	116	132	346	119	275	278	255	26	−1.016	0.1542
3	182	267	168	353	90	36	82	251	160	178	−0.047	0.1341
4	292	102	41	179	191	286	73	336	273	257	0.288	0.1283
5	109	5	180	287	23	301	30	185	117	101	0.059	0.1322
6	217	308	234	87	13	239	25	127	304	3	−0.374	0.1403
7	294	13	170	142	45	313	91	278	235	228	0.454	0.1257
8	102	105	198	148	61	316	15	117	323	143	−0.839	0.1501
9	69	95	278	50	216	18	0	328	357	355	0.554	0.1241

Note: Average fitness 0.1336, best fitness 0.1542, best string 2.

5.5.3 Quality of Solution

The next step is to measure the quality of each candidate solution in the initial population. "Quality" is a slippery beast; several factors may contribute to it and these could easily conflict. Since real-life scientific problems are multifaceted and full of trade-offs, distilling the quality of a string down to a single number is not always easy. A pharmaceutical company might be developing a drug for the treatment of some illness. The drug, of course, must be effective, but the more potent a drug, the greater the chance that it may have unpleasant or even potentially life-threatening side effects. The drug may be more cost-effective than competing drugs if it can be self-administered, but any cost savings made possible by relying on the patient to administer the drug to himself are irrelevant if the patient is too ill or forgetful to take the drug when it is needed. Measuring the "quality" of the drug, which the company might reasonably choose to equate with its commercial potential, is then a matter of trying to balance the conflicts between potency and toxicity, cost-effectiveness and lack of doctor control. This assessment may not be simple.

The quality of a GA string is calculated using a *fitness function* (also known as an *objective function*), which yields a quantitative measure, known as the *fitness* of the string. The fitness function must reward the better strings with higher fitness than their poorer competitors, but the exact function that is

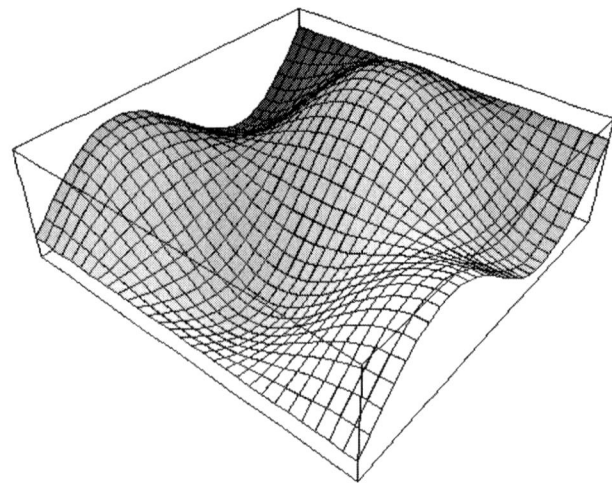

FIGURE 5.9
The surface defined by an arbitrary fitness function, across which the genetic algorithm (GA) searches for a maximum or minimum. The surface may be complex and contain numerous minima and maxima.

used in an application to link fitness to quality is not prescribed by the algorithm; instead, it is in the gift of the user to decide upon a suitable relationship for each problem.

The algorithm searches during a GA run for the fittest possible string. The fitness function defines a surface across which this search takes place (Figure 5.9); the surface is usually of considerable complexity because it is a function of all the variables that comprise the GA string.

Even for a simple problem like the alignment of dipoles, the surface is difficult to visualize. In the present problem, both the energy, which we are seeking to minimize, and the fitness of each solution, which we need to maximize, are functions of ten different angles, thus both the fitness and the energy define surfaces in ten-dimensional space (And this is a simple problem!). The task of the algorithm is then to find an arrangement of minimum interaction energy (maximum fitness) in this space.

The algorithm works to find the solution of maximum fitness, so the fitness of each solution must be related in some way to the interaction energy; this latter term is readily calculated. The energy between point charges q_1 and q_2, a distance r apart, is

$$E = \frac{q_1 q_2}{4\pi\varepsilon_0 r} \tag{5.1}$$

in which ε_0 is the vacuum permittivity. The interaction energy between two dipoles can be calculated as the sum of four pair-wise interactions (Figure 5.10.)

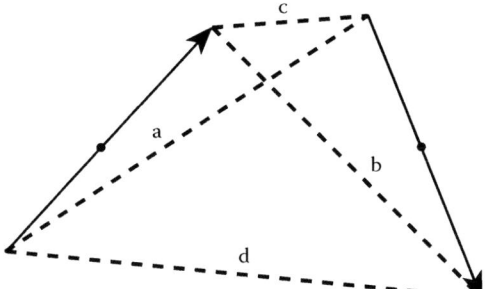

FIGURE 5.10
Distances used in the calculation of the interaction energy of two dipoles.

Interactions between two unlike ends of the dipoles are negative and, therefore, attractive, while those between two like ends are positive, and thus are repulsive. The total interaction energy is a summation over all ten dipoles, and if we assume that the calculation can be simplified by including only interactions between neighboring dipoles, the total energy can be calculated from equation (5.2).

$$E = \sum_{i=1}^{i=9} const \times \left\{ \frac{q_i q_{i+1}}{a} + \frac{q_i q_{i+1}}{b} - \frac{q_i q_{i+1}}{c} - \frac{q_i q_{i+1}}{d} \right\} \tag{5.2}$$

in which a, b, c, and d are as shown in Figure 5.10, so they vary from one pair of dipoles to the next, and the constant is $1/4\pi\varepsilon_0$. The lowest energy arrangement of the dipoles does not depend on the units in which we chose to measure the energy, so the value of the constant, in fact, is not important in this problem.

The energy and the fitness can be related through any function of our choice that assigns higher fitness to strings of lower energy. As high energy is characteristic of a poor solution, our first stab at a fitness function might be one in which an inverse relationship exists between energy and fitness:

$$f_i = \frac{c}{E_i} \tag{5.3}$$

In equation (5.3), c is a constant, E_i is the energy of string i, and f_i is its fitness. However, the simple inverse relationship expressed by equation (5.3) is actually *too* simple. The energies of attraction and repulsion have opposite signs; therefore, when attraction between some of the dipoles is perfectly balanced by repulsion among the rest, the total energy is zero and the fitness according to equation (5.3) becomes infinite, as Figure 5.11 shows. The appearance of an infinite fitness is not just computationally inconvenient, the function in equation (5.3) fails to reward the strings of lowest energy with the highest fitness as we require, allocating instead the highest fitness to those strings in

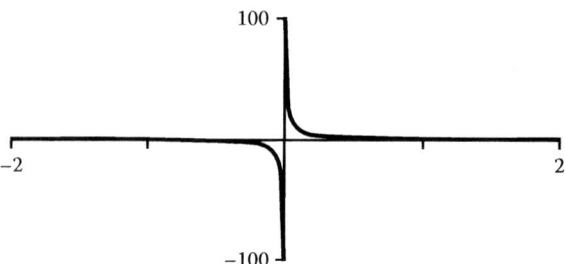

FIGURE 5.11
A possible, but unsatisfactory, relationship between string energy and string fitness, calculated using equation (5.3).

which attraction and repulsion cancel. This defect in the fitness function is easily remedied by making a minor modification to equation (5.3):

$$f_i = \frac{1}{E_i + 6.0} \tag{5.4}$$

where the value 6.0 has been chosen because the minimum energy using the distance between dipoles in this problem and the selected constant is −5.5, so the fitness now cannot become infinite. Figure 5.12 shows that the fitness calculated with equation (5.4) is better behaved; it is nowhere infinite, and the lower the total energy of the set of dipoles defined by a string, the higher the corresponding fitness.

You will notice how remarkably arbitrary this process is. The constant value 6.0 was not chosen because it has some magical properties in the GA, but because it seemed to be a value that would be "pretty good" in this particular application. The lack of sensitivity in the operation of the GA to the precise way in which the calculation is set up and run can seem unsettling at first, but this tolerance of the way that the user decides to use the algorithm is one of its important attractions, giving even novice users every chance of achieving good results.

Because the orientations of the dipoles defined by the strings in generation 0 have been chosen at random, none of these strings is likely to have a high fitness. The initial points in any GA population will be scattered randomly across the fitness surface (Figure 5.13).

The energy of each of the ten strings in the initial population is shown in Table 5.1, together with the fitness calculated from equation (5.4).

5.5.4 Selection

The fitness of a string is an essential piece of information on the road to the creation of solutions of higher quality. The first step in the process of improving the strings is to use the fitness to select the better strings from the

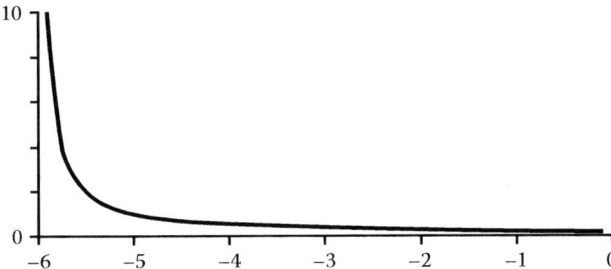

FIGURE 5.12
A possible relationship between string energy and fitness, calculated from equation (5.4).

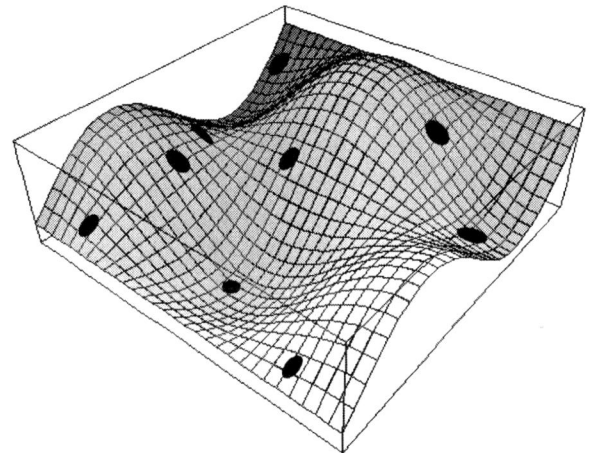

FIGURE 5.13
The strings in an initial population, having been created randomly, will have fitnesses spread widely across the fitness surface.

population so that they can act as parents for the next generation; selection employs a survival-of-the-fittest competition.

The selection process, which is run as soon as the fitnesses have been calculated, must be biased toward the fitter strings if the algorithm is to make progress. If selection did not have at least some preference for the fitter strings, the search in which the algorithm engages would be largely random. On the other hand, if the selection of strings was completely deterministic, so that the best strings were picked to be further upgraded, but the poorer strings were always ignored, the population would soon be dominated by the fittest strings and would quickly become homogeneous.

The selection procedure, therefore, is only biased toward the fitter strings; it does not choose them to the inevitable exclusion of poorer strings. To accomplish this, the selection procedure is partly *stochastic* (random).

Several alternative methods of selection exist; in this example, we shall use *Binary Tournament Selection*. Later we shall encounter other methods. Two strings are chosen at random from the current population, say, strings 3 and 6 from Table 5.1. Their fitness is compared (f_3 = 0.1341, f_6 = 0.1403) and the string with the higher fitness, in this case string 6, is selected as a parent and becomes the first entry in the *parent pool*. Another pair of strings is drawn at random from the population and the fitter of the two becomes the next entry in the parent pool. The process is repeated n_{pop} times to yield a group of parent strings equal in number to the set of strings in the starting population.

Binary tournament selection

1. Choose two strings at random from the current population.
2. Select the string with the greater fitness and copy the string into the parent pool.
3. If n_{pop} strings have been chosen, selection is complete; otherwise return to step 1.

Some strings will participate more than once in the binary tournament because the fitness of twenty randomly selected strings must be compared to ten select parents. Uniquely, the weakest string cannot win against any string, even if it is picked to participate, so it is sure to disappear from the population, leaving a place open for a second copy of one of the other strings; consequently, at least one string will appear in the parent pool more than once.

The selection of strings for the tournament is random, thus the results may vary each time the process is performed on a given starting population. This unpredictable selection of strings is the second time that random numbers have been used in the algorithm (the first was in the construction of the strings in the initial population) and random numbers will return shortly. It may seem surprising that random numbers are needed when the problem to be solved has a definite (albeit, still unknown) solution, but their use is a common feature in AI methods and, in fact, is fundamental to the operation of several of them.

The strings chosen to be parents by the binary tournament are shown in Table 5.2.*

In an animal population, individuals grow old. Aging, though it changes the outward appearance of an animal, does not alter its genetic makeup, and since the GA only manipulates the equivalent of the genes, "age" has no meaning in a standard GA. Although, in common with the evolution of a population of animals, the algorithm operates over a succession of generations, during which some individuals will be lost from the population and others created, as soon as a GA population has been constructed, the algorithm immediately sets about destroying it and giving birth to its successor.

---

* The "string numbers" shown in Table 5.1 and Table 5.2 are of no significance within the algorithm; they are shown to make it simpler for the reader to keep track of the operation of the GA.

TABLE 5.2

Strings Chosen by Binary Tournament Selection to Participate in the First Parent Pool

Parent String #	θ_1	θ_2	θ_3	θ_4	θ_5	θ_6	θ_7	θ_8	θ_9	θ_{10}	Old String #
0	217	308	234	87	13	239	25	127	304	3	6
1	69	95	278	50	216	18	0	328	357	355	9
2	292	102	41	179	191	286	73	336	273	257	4
3	141	335	116	132	346	119	275	278	255	26	2
4	141	335	116	132	346	119	275	278	255	26	2
5	109	5	180	287	23	301	30	185	117	101	5
6	357	295	206	55	266	180	166	95	131	293	1
7	217	308	234	87	13	239	25	127	304	3	6
8	102	105	198	148	61	316	15	117	323	143	8
9	294	13	170	142	45	313	91	278	235	228	7

Note: The "old string" number refers to the labeling in Table 5.1. The "parent string" number will be used in Table 5.3.

5.5.5 Crossover

The average fitness of the parent strings in Table 5.2 is higher than the average fitness of the strings in the starting population (the parent pool has an average fitness of 0.1380, compared with an average fitness of 0.1336 for the population in generation 0). Although the rise in fitness is neither guaranteed nor substantial, it is to be expected because the selection process was biased in favor of the fitter strings. We cannot be too pleased with ourselves though; this increase in fitness is not yet evidence of real progress. No new strings are created by selection and at least one string has been removed by it. If the selection process was repeated a few times, the average fitness would probably continue to improve, but the algorithm would quickly find its way down a dead-end, converging on a uniform set of strings in which each was a copy of one from the first population.

It is clear that selection on its own cannot generate a high quality solution; to make further progress, a mechanism is needed to modify strings. Two tools exist for this purpose: mating and mutation.

In most living things, reproduction is a team sport for two participants. By combining genes from a couple of parents, the genetic variety that can be created in the offspring is far greater than that which would be possible if all genes were derived from a single parent. The creation of individuals whose genes contain information from both parents induces greater variability in the population and increases the chance that individuals which are well adapted to the environment will arise. The same reproductive principle, that two parents trumps one, is used in the GA, in which mating is used to generate new strings from old strings.

The mechanism for accomplishing this is *crossover*. In *one-point crossover* (Figure 5.14), two strings chosen at random from the freshly created parent pool are cut at the same randomly chosen position into a head section and a tail section. The heads are then swapped, so that two offspring are created, each having genetic material from both parents.

141	**335**	**116**	**132**	**346**	**119**	**275**	**278**	**255**	**26**
						cut $\updownarrow$			
357	295	206	55	266	180	166	95	131	293

$$\Downarrow$$

141	**335**	**116**	**132**	**346**	**119**	**275**	95	131	293
				+					
357	295	206	55	266	180	166	**278**	**255**	**26**

FIGURE 5.14
One-point crossover of parent strings 3 and 6, cut to the right of the arrows, between genes 7 and 8.

The probability that a string is selected for crossover is determined by the *crossover probability*, p_c. The crossover probability, whose value is chosen by the user at the start of a run, is usually in the range 0.6 to 1.0 per string per generation; hence, most, possibly all, of the strings in the parent pool will be affected by crossover. In our example, strings picked once for crossover may be picked a second time, thus some strings may be crossed twice, or several times, while others may escape going under the knife.

After crossover and mutation, discussed below in section 5.5.6, the strings are as shown in Table 5.3.

5.5.6 Mutation

The crossover operator swaps genetic material between two parents, so unless the segment to be swapped is the same in both strings (as has happened with strings 0 and 7 in Table 5.3 because the two parents that were crossed were identical*), child strings differ from their parents. Now we are starting to make progress. Crossover is creating new strings, which introduces information that was not in the original population. However, crossover is

---

* Although strings 0 and 7 were identical after crossover, the first angle in string 0 was then mutated, changing from 217 to 359, so in Table 5.3 the two child strings differ at the first angle.

TABLE 5.3

The Strings from Generation 0, after Selection, Crossover, and Mutation

String#	θ_1	θ_2	θ_3	θ_4	θ_5	θ_6	θ_7	θ_8	θ_9	θ_{10}	Created from Parents	Mutated at Position
0	359	308	234	87	13	239	25	127	304	3	6/6	1
1	69	95	278	50	216	18	91	278	235	228	9/7	
2	292	102	41	179	191	131	15	117	323	143	4/8	6
3	141	335	116	132	23	301	30	185	117	101	2/5	
4	141	335	116	132	346	119	275	95	131	293	2/1	
5	109	5	180	287	346	119	275	278	255	26	5/2	
6	357	295	206	55	266	180	84	278	255	26	1/2	
7	217	308	234	87	13	239	25	127	304	3	6/6	
8	102	105	198	148	61	316	73	336	273	257	8/4	
9	294	13	170	142	45	313	0	328	357	355	7/9	

Note: The parent numbers are given in Table 5.2.

not powerful enough on its own to enable the algorithm to locate optimum solutions. This is because crossover can only shuffle between strings whatever values for angles are already present; it cannot generate new angles. We might guess that the optimum value for the angle of the first dipole is 90°. Table 5.1 shows that in no string in the initial population does the first dipole have an angle of 90°, so no matter how enthusiastically we apply crossover, the operator is incapable of inserting the value of 90° into a string at the correct position or, indeed, at any position. To make further progress, an operator is needed that can generate new information.

Mutation is this operator. At a random position in a randomly chosen string, it inserts a new, randomly selected value (more random numbers!). One of the two strings produced by crossing parents 4 and 8 is

$$292 \quad 102 \quad 41 \quad 179 \quad 191 \quad 316 \quad 15 \quad 117 \quad 323 \quad 143$$

The string is picked out by the mutation operator and mutated at position 6 (chosen at random) to give the new string

$$292 \quad 102 \quad 41 \quad 179 \quad 191 \quad \textbf{131} \quad 15 \quad 117 \quad 323 \quad 143$$

Mutation refreshes the population through the insertion of new data into a small fraction of the strings. This introduction of fresh data is essential if the algorithm is to find an optimum solution, since, in the absence of mutation, all variety in the population would eventually disappear and all strings would become identical. However, the news is not all good. Although mutation is

TABLE 5.4

The Genetic Algorithm Parameters Used in the Dipoles Problem

n_{pop}	10
Crossover type	1-point
p_c	1.0
p_m	0.2

necessary, the mutation operator also destroys information and is as likely to insert an unwanted angle into the string as a good one because the operator has no way of knowing what are "good" angles and what are "bad" ones.

If a string of high quality is selected for mutation, the random alteration made to the string may damage it, creating a string of lower quality. As a result, mutation is applied cautiously, rather than freely, at a rate specified by the *mutation rate, p_m*. In our example, $p_m = 0.2$ per string per cycle, hence, on average two strings out of ten are mutated in each generation. In many applications, the mutation rate is a good deal lower, but with the small population used in the current example, the strings would quickly become homogeneous without the use of a high mutation rate (Table 5.4).

5.6 Evolution

Updating of the population is now complete—strings were assessed for quality and the better strings chosen semistochastically through n_{pop} binary tournaments. The strings lucky enough to have been selected as parents were paired off and segments swapped between them using a one-point crossover. Finally, a couple of strings, on average, were mutated. The new generation of strings is now complete and the cycle repeats.

We can monitor progress of the algorithm by observing how the fitness of the best string in the population, or alternatively the average fitness of the whole population, changes as the cycles pass. Table 5.5 lists the fitness of the best string in the population at a few points during the first fifty generations.

The table reveals an encouraging rise in fitness, which is confirmed in Figure 5.15. This shows how the fitness of the best string in each generation varies over a longer period. It is clear that the algorithm is making steady progress, although the occasional spikes in the plot show that on a few occasions a good string found in one generation fails to survive for long, thus the fitness of the best string, rather than marching steadily upward, temporarily declines.

TABLE 5.5

The Fitness of the Best String at Several Points in the First Fifty Cycles

Generation	0	5	10	15	20	30	40	50
f_{best}	0.1542	0.2031	0.2404	0.2633	0.2633	0.2985	0.3059	0.3478

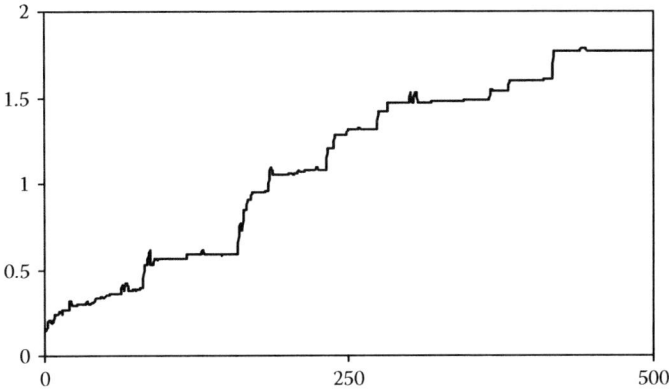

FIGURE 5.15
Fitness of the best string in the population during the first five hundred generations.

By generation 100 (Table 5.6) the dipoles are starting to align and there are hints that a good solution is starting to emerge.*

It is clear that the population is homogenizing as a result of the removal of the poorer solutions. This homogeneity brings both benefits and disadvantages. A degree of homogeneity is beneficial because it provides protection against loss of information. If there are several copies of a good string in the population, the danger of that string being lost because it was not chosen by the selection operator or has been destroyed by crossover or mutation, is reduced. As Figure 5.15 shows, the fitness of the best strings takes several short-lived steps backwards in the early part of the run, but these setbacks are rare once a couple of hundred generations have passed and several copies of the best string are present.

On the other hand, the algorithm requires a degree of diversity in the population to make continued progress. If every string in the population is identical, one-point crossover has no effect and any evolutionary progress

---

* The best solution consists of a head-to-tail arrangement of all ten dipoles. There are two such arrangements, one in which all the dipoles are pointing to the left and the other in which they are all pointing to the right. Both solutions are equally likely to be found by the algorithm and both have the same energy, but any single run will eventually converge on only one or the other. Owing to the stochastic nature of the algorithm and the fact that the solutions are equivalent, if the GA was run many times, it would converge on each solution in 50 percent of the runs.

TABLE 5.6

The Population, Generation 100

String	θ_1	θ_2	θ_3	θ_4	θ_5	θ_6	θ_7	θ_8	θ_9	θ_{10}
0	102	92	73	130	271	115	75	107	94	209
1	102	92	73	130	77	115	75	107	94	209
2	102	92	73	130	77	115	75	107	95	209
3	102	92	73	130	77	115	75	107	95	209
4	92	92	135	130	77	115	75	107	95	209
5	102	102	73	130	77	115	75	107	95	209
6	102	92	73	130	77	115	75	107	95	209
7	102	100	73	130	77	115	75	107	95	209
8	102	92	135	130	77	115	75	107	95	209
9	102	92	159	130	77	115	75	107	95	209

TABLE 5.7

The Population, Generation 1000

String	θ_1	θ_2	θ_3	θ_4	θ_5	θ_6	θ_7	θ_8	θ_9	θ_{10}
0	96	92	90	93	93	97	95	88	88	88
1	96	92	90	93	93	97	95	88	88	88
2	96	92	90	93	93	97	95	88	88	88
3	96	92	90	93	44	97	95	88	88	88
4	96	92	90	93	93	97	95	88	88	88
5	96	92	90	93	93	97	95	88	88	25
6	96	92	90	93	93	97	95	88	88	88
7	96	92	90	93	93	97	95	88	88	88
8	96	92	90	93	93	97	95	88	88	88
9	96	92	90	93	93	97	95	88	88	88

will depend purely on mutation. This balance between a high-quality, low-diversity population and a more varied population of lower-average quality is an important factor when assessing performance of a GA and we shall return briefly to it in section 5.8.4.

Table 5.7 shows that, after a thousand generations, the strings are now all of good quality, though the perfect string has yet to be discovered.

5.7 When Do We Stop?

As the GA is iterative, it has no natural stopping point. The algorithm will cycle forever if we leave for coffee and forget about it, although it must eventually run into the sand and be unable to make further progress. It would be helpful if there was something that we could use to judge when the run should be stopped, so as to avoid wasting computer time or alternatively stopping the calculation before the best solution has been found.

One way to do this is to estimate in advance (which is a scientific-sounding way of saying that we guess) how many generations should be run. Once this number of generations has passed, the run is temporarily halted and the population is inspected to see whether an individual of sufficient quality is present. If a high-quality solution has been found, the calculation is complete; if no strings are yet good enough, it is resumed. This is a reasonable approach if the fitness function is easy to evaluate so that the generations pass quickly and if we have some idea of the fitness of the optimum solution, but when the evaluation is computationally expensive, each run may take hours, or even days. If the number of generations likely to be required before a good solution emerges is misjudged, the calculation may run for much longer than necessary with a consequent waste of processing time or we may unnecessarily interrupt it many times to evaluate progress.

Some semiautomatic way of deciding when to end the calculation is needed. In favorable cases, we may be able to estimate the fitness of an optimum solution before the run is started. If so, the fitness of the best string in each generation can be compared against what we have determined to be the fitness of a perfect solution and the run can be brought to a halt once a solution of suitable fitness is generated. If we cannot estimate how fit the optimum solution would be, another approach is required. It would be unwise to wait until every string codes for the same solution, as this point never comes. The population is never entirely static because the mutation operator is always nudging the current solutions into a new state. On the other hand, if the run lasts long enough, a time will arrive when no further improvement in the fitness of the best string is likely. This suggests that we should monitor the fitness of the population as a whole and of the best string; when neither improves over many generations, the calculation can be halted.

The variation with generation number of the average fitness of the population and the fitness of the best string within that population are shown in Figure 5.16. The fairly steady rise in the fitness of the best string and the more erratic improvement of the average fitness are typical of the progress of

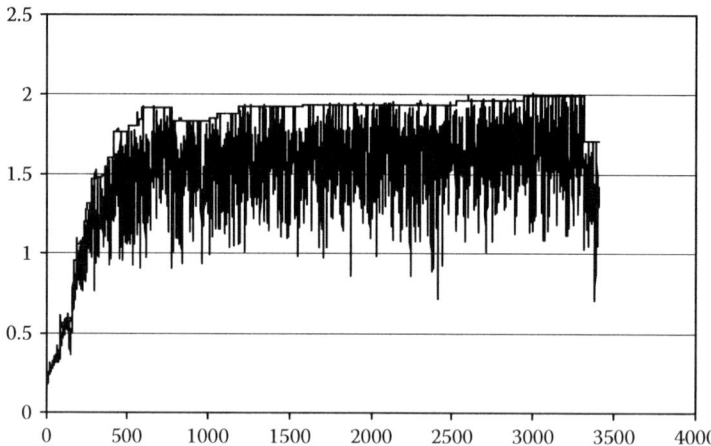

FIGURE 5.16
Variation of the fitness of the best string (upper line) and the average population fitness (lower line) with generation number.

the algorithm in a problem in which a single mutation in a string can cause a large change in string fitness. The fitness of the best string is more stable than the average fitness because after many cycles there are usually several copies of the best string in the population and, if one is damaged by crossover or mutation, another should escape unscathed.

Despite the protection that multiple copies of a good string provides, there are abrupt drops in the fitness of the best string at generations 784, 2382, and 3313. (A color version of Figure 5.16 together with the raw data are available on the CD that accompanies this text.) The drop at 784 was partly reversed within five generations, although a further 397 generations were required before the fitness of the best string fully recovered. The sudden drop at generation 2382 was reversed immediately, while that at generation 3313 was reversed and a near optimum string (the maximum fitness in this calculation was 2.0) was recreated after a further 200 generations. These sudden drops illustrate how useful it can be to continuously monitor the fitness of the population rather than guillotining the calculation when a fixed number of generations has passed. If we had arbitrarily decided that 3500 generations would give the algorithm sufficient time to find a near-optimum solution and picked the best string in generation 3500, we would have missed a better solution that appeared a few hundred generations earlier.

To prevent a good string from sneaking under the wire without us noticing it, we can use an elitism strategy (section 5.11.1) or alternatively make a note of the form of the best string every time one is created whose fitness exceeds that of any that have gone before.

5.8 Further Selection and Crossover Strategies

The progress of the GA depends on the values of several parameters that must be set by the user; these include the population size, the mutation rate, and the crossover rate. Choosing the values of these parameters is not the only decision to be made at the start of a run, however. There are tactical decisions to be made about the type of selection method, the type of crossover operator, and the possible use of other techniques to make the algorithm as effective as possible. The choice of values for these parameters and type of crossover or selection can make the difference between a calculation that is no better (or worse) than a conventional calculation and one that is successful. In this section, we consider how to choose parameters to run a successful GA and start with a look at tactics.

5.8.1 Selection Strategies

There are several methods that can be used to choose which strings will act as parents for the next generation. These differ in the proportion of the strongest and weakest strings in the current population that are selected.

5.8.1.1 Many-String Tournaments

The simplest tournament is a battle between two strings in which the stronger always wins, but this is not the only sort of competition that can be run. In a probabilistic tournament, two strings are chosen to participate and the fitter string wins, not on every occasion, but with probability p, where $1.0 > p > 0.5$. A probabilistic tournament decreases the selection pressure on the weaker strings as, even if a less-fit string finds itself competing against a stronger string, there is a still a chance that the less-fit string will win. $p = 0.5$ corresponds to a completely random choice between the two strings, which breaks the link between fitness and selection.

Tournaments can be opened up to more than two participants. In an n-string tournament, n strings are chosen at random and from this pool the best string is chosen as a parent. This step is repeated n_{pop} times to create the new population.

In contrast to a probabilistic tournament, increasing the number of strings in the tournament raises the selection pressure on the poorest strings. In a binary tournament, only the weakest string in the population has no chance of making it through to the parent pool; even if it is selected to participate in the tournament, there is no string in the population against which it can win. The modest selection pressure that a binary tournament provides helps to retain a reasonable degree of diversity within the population from one generation to the next. However, in an n-string tournament, the worst $(n - 1)$ strings are certain to be removed. In the extreme (and pointless) case when $n = n_{pop}$, only the best string would ever be chosen in the tournament; thus, in a single generation all diversity would be lost. Because of the desirability of having a diverse population, tournaments involving more than three strings are uncommon, although an n/m-tournament is occasionally used, when the best n strings participating in an m-string tournament are selected.

5.8.1.2 Roulette Wheel Selection

In *roulette wheel* (or *dartboard*) selection, each string in the population is allocated a segment on a virtual roulette wheel (dartboard) of a size proportional to the string's fitness (Figure 5.17).

The wheel is spun and the string into whose slot the virtual ball falls is copied once into the parent pool. This procedure is repeated n_{pop} times to pick the full complement of parent strings. In roulette wheel selection, every string has a nonzero chance of being picked, proportional to its fitness, so even the poorest string may be chosen as a parent. The method also preferentially selects the fitter strings and, because of these features and its simplicity, roulette wheel selection is widely used.

5.8.1.3 Stochastic Remainder Selection

A weakness of both roulette wheel selection and tournament selection is that although both are biased in favor of fitter strings as the algorithm requires, neither guarantees that the best string in one generation will be chosen as a parent for the next.

The ball might never fall into the widest slot in the roulette wheel. In fact, it is possible, though unlikely, that the roulette wheel will select only the poorest strings from a population to act as parents. Tournament selection cannot select the worst string, but it is only able to choose the best string from among

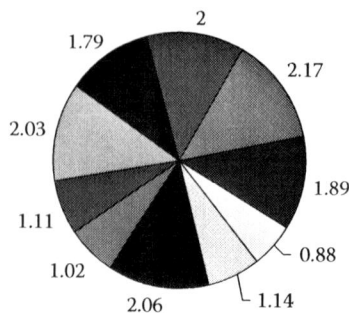

FIGURE 5.17
Roulette wheel selection. The area of the slot on the virtual roulette wheel for each string is proportional to the string's fitness.

those picked to participate in the tournament. The fittest string might never be picked and, without being invited to the dance, it cannot become a parent. If the population is comprised of a large number of average strings with just one or two strings of superior quality, the loss of the best strings may hinder progress toward an optimum solution. We can see evidence of stalled progress in Figure 5.16 at generation 784. The fitness of the best string collapses from 1.9 to less than 1.5 before making a partial recovery a few generations later. *Stochastic remainder* is a hybrid method for the selection of parents that combines a stochastic element with a deterministic step to ensure that the best string is never overlooked.

There are three steps in stochastic remainder selection. First, the fitnesses are scaled so that the average string fitness is 1.0.

$$f_{i,scaled} = \frac{f_{i,raw} \times n_{pop}}{\sum\limits_{j=1}^{n_{pop}} f_{j,raw}} \tag{5.5}$$

As a result of this scaling, all strings that previously had a raw fitness above the mean will have a scaled fitness greater than 1, while the fitness of every below-average string will be less than 1. Each string is then copied into the parent pool a number of times equal to the integer part of its scaled fitness (Table 5.8).

Thus, two copies of string 2 in Table 5.8, whose scaled fitness is 2.037, are certain to be made in the first round of selection using stochastic remainder, while at this stage no copies of string 7, whose fitness is 0.850, are made. This deterministic step ensures that every above-average string will appear

TABLE 5.8

Stochastic Remainder Selection

String	1	2	3	4	5	6	7	8	9	10
$f_{i,raw}$	0.471	1.447	0.511	0.208	1.002	0.791	0.604	0.604	0.985	0.444
$f_{i,scaled}$	0.663	2.037	0.719	0.293	1.411	1.114	0.850	0.850	1.387	0.625
$C_{i,certain}$	0	2	0	0	1	1	0	0	1	0
$f_{i,residual}$	0.663	0.037	0.719	0.293	0.411	0.114	0.850	0.850	0.387	0.625
$C_{i,roulette}$	1	0	1	0	0	0	1	1	0	1
$C_{i,total}$	1	2	1	0	1	1	1	1	1	1

Note: $f_{i,raw}$ is the raw fitness of a string, $f_{i,scaled}$ is its scaled fitness, $C_{i,certain}$ is the number of parent strings guaranteed to be made by stochastic remainder, $f_{i,residual}$ is the residual fitness after the number of parents made has been subtracted from the scaled fitness, $C_{i,roulette}$ is the number of copies made by a typical roulette wheel selector, and $C_{i,total}$ is the total number of parents prepared from the string.

at least once in the parent pool. Any string whose fitness is at least twice the mean fitness will be guaranteed more than one place in the pool.

Finally, the number of copies made of each string is subtracted from its scaled fitness to leave a *residual fitness,* which for every string must be less than 1.0. A modified roulette wheel or tournament selection is then run using these residual fitnesses to fill any remaining places in the population. In this modification, once a copy has been made of a string as a result of its selection by the roulette wheel operator, or in the tournament selection, its residual fitness is reduced to zero to prevent it being chosen again. Stochastic remainder neatly combines a deterministic step that ensures that every good string is granted the opportunity to act as a parent for the new population, with a stochastic step that offers all strings a chance, no matter how poor they may be, to also pass on their genes.

Table 5.9 shows how the number of copies of a set of strings may vary depending on the selection method chosen. Because each method contains a stochastic element, the values in the table would change if the algorithm was run a second time.

Stochastic remainder selection

1. Scale all string fitnesses so that the average fitness is 1.0.
2. For each string whose scaled fitness is at least 1.0, copy the string into the parent pool a number of times equal to the integer part of its fitness.
3. For each string, subtract the number of parents made from the scaled fitness to leave a residual fitness.
4. Run roulette wheel selection on the residual fitnesses to fill any remaining places in the parents pool, reducing the residual fitness to zero for any string as soon as it is selected to be a parent in this process.

5.8.1.4 Fitness Scaling

The efficient identification and selection of high-quality strings is fundamental to the working of the GA. Table 5.6 shows the position after the dipoles problem has run for one hundred generations. Even at this quite early stage, the strings are very alike, so they have similar fitnesses, a feature that becomes more prevalent in the middle and later stages of a GA calculation. When the fitnesses of most strings in the population are nearly identical, the capacity of the selection operator to distinguish between them diminishes and selection loses focus within this high-quality pool. Fitness scaling provides one means of helping the selection operator to overcome the difficulties that this lack of variability causes.

Fitness scaling stretches out the fitness of the set of strings so that they span a wider range (Figure 5.18). Because the fitnesses now show a greater spread, it is easier for the selection operator to notice and select the better strings.

Just as there is no prescription in the algorithm that defines the fitness function itself, we are free to choose a method for scaling that seems to be

TABLE 5.9

The Number of Copies of a Set of Strings Made by Running Different Selection Strategies

String	Fitness	Scaled Fitness	Binary Tournament	3-String Tournament	Roulette Wheel	Stochastic Remainder
1	0.471	0.663	1	0	1	1
2	1.447	2.037	2	3	2	2
3	0.511	0.719	1	0	0	1
4	0.208	0.293	0	0	1	0
5	1.002	1.411	2	3	2	1
6	0.791	1.114	1	1	1	1
7	0.604	0.850	1	1	1	1
8	0.604	0.850	0	1	0	1
9	0.985	1.387	1	1	2	1
10	0.444	0.625	1	0	0	1

Note: Since each of the selection methods contains a stochastic element, different results would be obtained if the selection were run a second time.

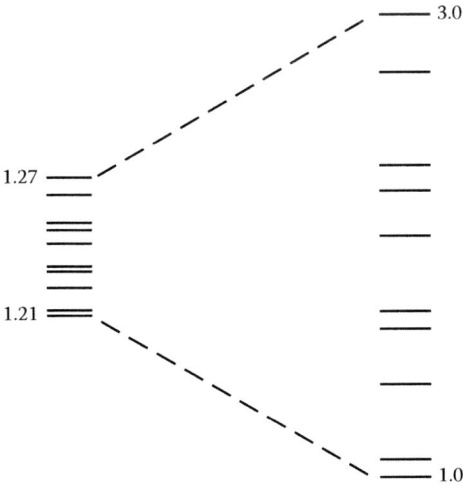

FIGURE 5.18
Fitness scaling. The original range of fitnesses from 1.21 to 1.27 is stretched to cover a wider range.

the most effective for a given problem. A typical procedure would be to scale all fitnesses so that they cover the range of, say, 1.0 to 3.0. This can be accomplished by applying the linear transformation:

$$f_{i,scaled} = 1.0 + (f_i - f_{min}) \times \frac{2.0}{(f_{max} - f_{min})} \tag{5.6}$$

where f_i is the fitness of string i, and f_{max} and f_{min} are the fitnesses of the best and worst strings, respectively, in the population. Scaling is unlikely to be necessary in the early part of a run, but may be turned on once the population begins to homogenize.

5.8.2 How Does the Genetic Algorithm Find Good Solutions?

At the start of a run, the random strings in the first population will correspond to points widely scattered across the search space, as illustrated in Figure 5.13.

As the search progresses, the points move out of low-lying areas of the surface and also become more clustered (Figure 5.19). By the end of the run, most points are at or near one of the fitness maxima.

To make informed choices concerning the values of adjustable parameters so that we can encourage the algorithm to seek out, and converge on, high-quality solutions, it is helpful to understand how evolution can locate a good solution; therefore, we shall consider in this section the mechanism of the GA.

A straightforward, though simplified, explanation of why the GA functions as an effective optimizer is as follows:

The algorithm is constantly assessing strings on the basis of their fitness. The fitness of one string may surpass that of its fellow strings because of the presence of a few high-quality genes. For example, in generation 1, string 5 is

109   5   180   287   346   119   275   278   255   26

The relative orientations of most of the dipoles in this string are unpromising, but dipoles 7 and 8 (angles shown in bold) are pointing almost in the

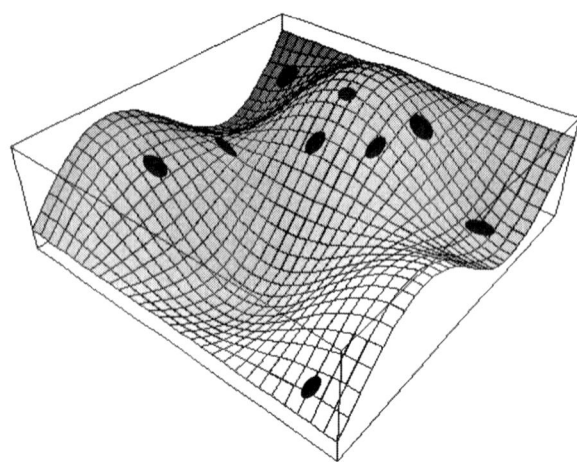

FIGURE 5.19
The growing clustering of strings in regions of higher fitness.

same direction and are roughly aligned along the axis. This head-to-tail arrangement brings opposite ends of these two neighboring dipoles into close proximity and, thus, is favorable energetically, and although the remaining dipoles in the string are arranged almost randomly, this small segment of the string serves to give the string a fitness that is slightly superior to that of a string in which every dipole adopts a random orientation. Another string in the same population, string 8, has two short segments in which dipoles are in a favorable alignment:

102 **105** 198 148 61 316 73 336 **273** **257**

Dipoles 1 and 2 are roughly aligned along the axis, as are those in positions 9 and 10, though the first pair point in the opposite direction to the second pair.

Suppose that these two strings are selected as parents and then paired together for crossover between positions 8 and 9 (Figure 5.20).

109	5	180	287	346	119	**275**	**278**	255	26
							cut ↕		
102	**105**	198	148	61	316	73	336	**273**	**257**

⇓

109	5	180	287	346	119	**275**	**278**	**273**	**257**
				+					
102	**105**	198	148	61	316	73	336	255	26

FIGURE 5.20
Action by the crossover operator that brings together two promising segments.

In the first child string, all of dipoles 7 to 10 now have a similar orientation and, consequently, we would expect that the new string will have a fitness that exceeds that of either parent.* Because of this improved fitness, the string will be more likely to be chosen as a parent when the next generation is processed by the selection operator, so both the string and the good segments that it contains will begin to spread through the population.

By contrast, if the crossover or mutation operators produce a string that contains few useful segments or none at all, either because there were no useful segments in the parent strings or because the evolutionary operators disrupted these segments, the string will have relatively poor fitness and, in due course, will be culled by the selection operator. Thus, good strings that contain chunks of useful information in the form of a group of high-quality

---

* Although the string has only four dipoles with a favorable orientation, just like one of its parents, all four are contiguous, so there are three favorable dipole–dipole interactions: 7 to 8, 8 to 9 and 9 to 10. Parent string 5 had only one such interaction and parent 8 had two.

genes tend to proliferate in the population, carrying their good genes with them, while poorer strings are weeded out and discarded. Because of the presumed importance of segments that confer above-average fitness on a string, they are given a special name: *building blocks* or *schemata* (sing. *schema*).*

This discovery of schemata sounds promising; it seems that building a high-quality string should be simple. All we need to do is to create a large number of random strings, figure out where these valuable building blocks are located in each string, snip them out, and bolt them together to form a super string. Though an appealing idea, this is not feasible. The algorithm does not set out to identify good schemata — indeed it cannot do so because the high fitness calculated by the algorithm is associated with the string in its entirety, not with identifiable segments within it. During the evolutionary process, the GA has no means by which it can pick out only good schemata and leave behind those of little value, so explicitly cutting and pasting the best parts of different strings is not possible. However, although GA selection chooses strings, not schemata, it is automatically biased toward good schemata as they are more likely to be contained within superior strings. Therefore, even though the algorithm is schemata-blind, it still, in effect, identifies the schemata needed to build good solutions.**

Following this brief discussion of how the GA finds good solutions, we consider now how to oil the gears of the algorithm to make it function as effectively as possible.

5.8.3 Crossover Revisited

The preceding outline of how the GA is able to bolt together promising segments to build up a good solution brick by brick helps to explain why the crossover and mutation operators may be effective at creating good solutions. Through crossover, promising building blocks from two different strings may be brought together in a single superior individual, while

---

* The terms schema/schemata actually apply to any building block, good, bad or indifferent, but we are generally interested only in those that raise the quality of the string.

** You may suspect that this argument is a simplification—and it is. It assumes that schemata that are of high fitness and that will, therefore, proliferate in the population are also contained in the global optimum string. During evolution, schemata will proliferate if their average fitness exceeds the average fitness of all the schemata in the population, but problems exist in which the optimum solution is built from schemata that, when found within a *typical* string, are of below average fitness; only when all the necessary schemata required to form the perfect string come together do they cast off their frog's clothing and reappear as a string of high quality.

The proliferation in the population of highly fit schemata that do not appear in the optimum string, at the expense of less fit schemata that do, is known as deception and renders the problem difficult to solve. Although there are many deceptive problems in the literature, the number of different examples is itself deceptive, as a large proportion were specifically created to be deceptive, so that ways of dealing with deception could be investigated. Nevertheless, deception is a serious challenge for the genetic programming algorithm (see later).

mutation plays its part by creating new valuable building blocks that confer high fitness on the string that contains them. However, both operators are also disruptive: If a string contains a single short building block of value, it is unlikely to be disrupted by one-point crossover, but if two building blocks are far apart in a string, crossover will probably separate these segments and, thereby, reduce the quality of the string, not enhance it. Several alternative crossover operators are available to help us circumvent this difficulty.

5.8.3.1 Uniform Crossover

In uniform crossover, many small segments rather than a single large block are swapped between strings. The genes to be moved from one string to the other are defined by a mask that consists of a list of random binary values, created afresh every time the crossover operator is applied. This mask specifies which genes in the first string should be swapped with the corresponding genes in the second. Where a 1 appears in the mask, the corresponding genes are swapped; where there is a zero, the genes remain in place. Applying uniform crossover to the two strings from Figure 5.20 using a random mask, the following child strings are formed (Figure 5.21):

	109	5	180	287	346	119	275	278	255	26
mask	0	0	0	1	0	1	1	0	1	0
	102	105	198	148	61	316	73	336	273	257

$$\Downarrow$$

	109	5	180	148	346	316	73	278	273	26
					+					
	102	105	198	287	61	119	275	336	255	257

FIGURE 5.21
Uniform crossover.

Uniform crossover is simple, but severely disruptive. The number of crossing points depends on the mask, but is on average one half of the string length. The two parents are blown apart and information from them is divided randomly between the two child strings. Uniform crossover is likely to break up any building blocks that are more than a few genes in length and, in many types of problems, this crossover operator is too destructive. Uniform crossover, therefore, is at its most effective when applied to problems where there is only limited correlation between genes that are more than one or two places apart in the string.*

5.8.3.2 Two-Point Crossover

Two-point crossover is designed to overcome some of the disadvantages of both uniform crossover and one-point crossover. A segment of length *l* is swapped between a pair of strings, as in one-point crossover, but this segment is defined by *two* cut points in each string, with the cut points chosen independently in each string. Although the length of the segment to be swapped is the same in both strings, the segment can be located anywhere in either string, so a much greater variety of child strings becomes possible. The chunk of genes to be moved may be snipped out of the middle of a string leaving both ends intact in the same child string, thus crossover no longer necessarily separates genes that are positioned far apart within the string. The effect of two-point crossover is shown below (Figure 5.22) where a block of three genes is swapped between strings.

109	5	180	287	346	119	275	278	255	26
	cut ↑	– – –	– – –	cut ↑		cut ↓	– – –	– – –	cut ↓
102	105	198	148	61	316	73	336	273	257

⇓

109	5	336	273	257	119	275	278	255	26
				+					
102	105	198	148	61	316	73	180	287	346

FIGURE 5.22
Standard two-point crossover.

5.8.3.3 Wraparound Crossover

Two-point crossover is more versatile than one-point crossover because a much wider variety of child strings can be created from a given pair of

---

* A closer analysis suggests that the situation is rather more involved than this argument implies. Suppose that two genes in a string are correlated, so that if they have similar values this improves the fitness of the string, but, if they have very different values, the fitness of the string is lowered. If the two genes are neighbors, thus are very close in the string, one-point crossover is unlikely to separate them, so a good combination of genes will probably survive crossover. By contrast, uniform crossover will separate them and, therefore, reduce fitness of the string 50 percent of the time. On the other hand, if the genes are very far apart in the string the probability that they will be separated by one-point crossover becomes high, while uniform crossover will still separate them just 50% of the time; the latter is then *less* disruptive.

parents. Earlier we noted that, if no string in the initial population has the value of 90° for the first dipole, one-point crossover could not move it there. Two-point crossover, because it shuffles angles around in the strings, is able to manage this, provided that the value of 90° exists somewhere in the population.

As a result of these advantages, two-point crossover is more valuable than its one-point cousin, but, in the form described above, it hides a subtle problem. When one-point crossover is applied, every gene has an equal chance of being in the segment that is moved. (There is no distinction between the segment that is swapped between strings and the segment that is left behind; thus it is inevitable that the swapping probability is the same at all positions along the string.) However, in two-point crossover, genes in the middle of the string are more likely than genes at the extremities to be included within the segment that is selected for swapping; this difference becomes more pronounced as the size of the string increases (Table 5.10).

As genes in the center of the strings are frequently swapped by two-point crossover, the algorithm is constantly creating new combinations of genes in this part of the strings. Through this shuffling of genes and retesting as the generations pass, the algorithm may make good progress optimizing the middle of the string, but leave the ends of long strings relatively untouched over many generations; as a consequence, the outer regions of the strings will be slower to improve than the midsection.

A simple solution to this unequal treatment of different sections of the string exists. The order in which the cut points for crossover are generated is noted. If the second cut is "earlier" in the string than the first cut, the crossover wraps around at the end of the string. Suppose that the two strings shown at the top of Figure 5.23 are to be crossed. The two cut points selected at random in the first string are between genes 2 and 3 and between genes 5 and 6. This defines a segment of length 4 that will be swapped into the other string. A cut point is chosen in the second string, say between genes 8 and 9. The second cut point in this string is chosen with equal probability either four genes to the left of this point or four genes to the right. If the second cut is to the right, this takes us beyond the end of the string, so the cut will be between genes 2 and 3 and the cut wraps around. This will yield the two strings shown in Figure 5.23.

TABLE 5.10

Probability That a Gene at the Midpoint of a String, p_{mid}, or at the End of a String, p_{end}, Will Be Swapped during Two-Point Crossover

String Length	5	11	17	51
p_{mid}	0.6	0.545	0.529	0.510
p_{end}	0.2	0.091	0.059	0.020

109	257	102	105	198	119	275	278	255	26
	cut↕		*cut* ↓	*cut* ↑			*cut* ↓		
180	287	346	148	61	316	73	336	273	5

⇓

109	257	273	5	180	287	275	278	255	26
				+					
198	119	346	148	61	316	73	336	102	105

FIGURE 5.23
Wraparound crossover.

Figure 5.24 illustrates wraparound in the first string. The cut points in the first string are chosen as 8/9 and 2/3 (in that order). The segment that is swapped from this string is still four genes in length, but this time comprises genes 9, 10, 1, and 2.

109	257	102	105	198	119	275	278	255	26
	cut ↑		*cut* ↓				*cut*↕		
180	287	346	148	61	316	73	336	273	5

⇓

73	336	102	105	198	119	275	278	61	316
				+					
180	287	346	148	255	26	109	257	273	5

FIGURE 5.24
Wraparound crossover: a second example.

The wraparound operator gives every position in the string an equal chance of being selected because, on average, half of the time the second cut point will be to the left of the right cut point, thus wraparound will apply and it will be the outer parts of the string rather than an interior segment that will be swapped. This is the equivalent of joining the two ends of the string and choosing cut points at random from within the circular string thus formed (Figure 5.25). This demonstrates that wraparound treats all genes equally.

Wraparound is not limited to one-dimensional strings. In this chapter, we will not be dealing in detail with two-dimensional GA strings, which are strings that take the form of an array rather than a simple list of values.

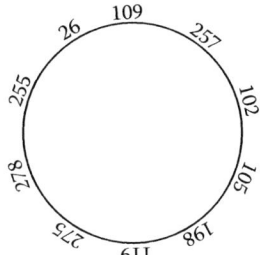

FIGURE 5.25
Interpretation of a genetic algorithm (GA) string as a circle.

$$S = \begin{bmatrix} 45 & 7 & 388 & \dots \\ 71 & -11 & 46 & \dots \\ 0 & 71 & 0 & \dots \\ 34 & -100 & \dots & \dots \end{bmatrix} \tag{5.7}$$

Such a string may be useful when two types of variables independently affect the quality of a solution. Suppose that an analytical laboratory monitors the air quality at a downtown site, measuring the level of fifteen different pollutants. A theoretical model is established that links the amount of these pollutants measured at the monitoring point with the amounts of each pollutant released at twenty different sites around the city. Because there are fifteen pollutants and twenty pollution sources, a 15×20 matrix is required to model the problem and GA strings can be constructed in this form.

Two-dimensional (or indeed n-dimensional) GA strings can be handled in the usual way, with an n-dimensional chunk of the strings being swapped by the crossover operator rather than a linear segment. However, care must then be taken to ensure that wraparound is applied in all n dimensions, not just one. (See also Problem 2 at the end of this chapter.)

5.8.3.4 Other Types of Crossover

Other mechanisms for crossover also have been described. One of the more interesting assumes that crossover is more effective in creating high-quality strings when it is applied at some positions in the string than at others, so the crossover cuts should be made with greater probability in certain regions of the string. Since the best places for the cuts are not likely to be known in advance, the normal GA string is expanded by adding a *punctuation string* that specifies which sites in the string should be favored in crossover. The punctuation string, as it is part of the overall string, also evolves; therefore, the string gradually learns where it is most productive to slice itself up.

5.8.4 Population Size

The success of a GA calculation is never critically dependent on the exact size of the population, provided that the population is not unreasonably small. A run that uses a population of forty-eight strings is about as likely to find a good solution to a problem as one that uses fifty strings. However, very large or very small populations are less effective than populations of intermediate size. If the level of diversity in the population is extremely large or is negligibly small, the chance that the algorithm will discover high-quality solutions is reduced. If the population becomes so large that the processing time per generation prevents the evaluation of more than a small number of generations, proper evolution becomes impossible.

The GA requires population diversity. If a situation is reached when every individual in a population of five hundred strings is identical, the population contains no more information than a population of just one string, evolution ceases unless mutation can start generating new strings, and the advantages of working with a population are lost. On the other hand, when the number of strings is very small, the population will converge quickly toward a single string and its mutated variants. The search then becomes similar to a random search under the influence of the mutation operator alone, starting from some modestly fit starting point.

The dangers of using too small a population are evident in Figure 5.26, in which we see how the fitness of the best string in the dipoles problem varies with generation number when a population of four strings, which is unrealistically small, is used. String fitness does improve from generation 0, but the uncertain progress made by the algorithm does not give one confidence that the optimum string (of fitness 2.0) will ever be found. In a very small population, the best string at any time is in a perilous position; in the run shown in the figure, there will be at most four copies of the best string; often there will

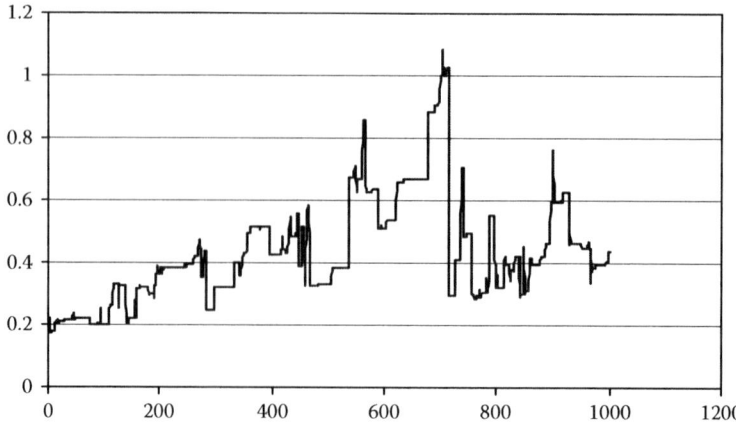

FIGURE 5.26
The fitness of the best string in an attempt to solve the dipoles problem using a population of four strings: $p_m = 0.2$, $p_c = 1.0$.

be just one. Crossover or mutation may easily destroy it and, with no backup copy available, the algorithm will repeatedly need to retrace its steps to try to rediscover information in a string that has been lost.

By contrast, when the algorithm is run with a population of one hundred strings (Figure 5.27), the fitness of the best string increases with almost no setbacks, and a high-quality solution emerges in fewer than two hundred generations. The optimum solution in this run is found (and then lost) twice around generation 770, but the algorithm settles permanently on the optimum solution at generation 800.

If a modest increase in the number of strings, say, from ten to one hundred, is helpful, perhaps a more dramatic increase, to fifty thousand or one hundred thousand, would be even better. However, it is rarely beneficial to use huge populations. Duplicates of a string provide protection against its possible destruction by the genetic operators, but a very large population may contain hundreds or thousands of identical strings whose presence does nothing to increase diversity; it merely slows the calculation.

The success of the algorithm relies on the GA processing many generations; solutions can only evolve if evolution is possible. Excessively large populations are especially problematic if evaluation of the fitness function is computationally expensive as it is, for example, in the chemical flowshop problem (section 5.10) in which evaluating the total time required for all chemicals to be processed by the flowshop in a defined order may take far more time than the execution of all other parts of the GA combined.

When the fitness function is challenging to evaluate, the use of a very large population will restrict operation of the algorithm to just a few generations. During these generations, the fitness of the best string will improve, which may give the impression that the algorithm is effective, but the improvement occurs because in each generation very many essentially random new

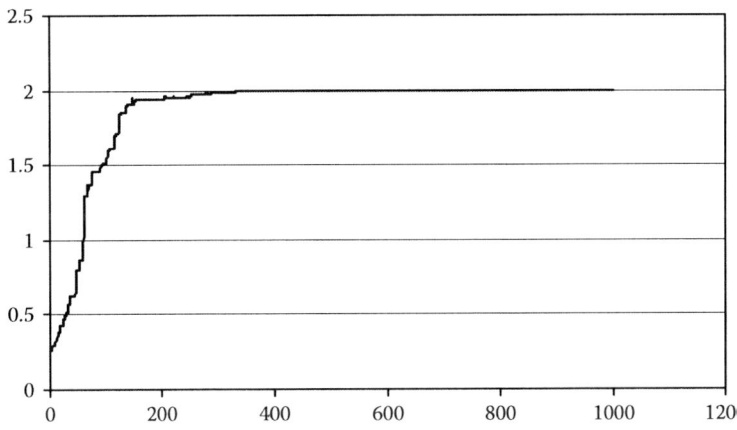

FIGURE 5.27
The fitness of the best string in an attempt to solve the dipoles problem with a population of one hundred strings: $p_m = 0.2$; $p_c = 1.0$.

strings are created, not because of evolution; the result of running a GA on one hundred thousand strings for five generations is not very different from creating a single generation of half a million random strings.

5.8.5 Focused Mutation

Figure 5.27 shows how the fitness of the best string improves rapidly near the start of a run, while the rate of progress becomes more leisurely later. It might seem that the nearer the best string is to the optimum solution, the easier the task of finding that solution should become, but this is not so.

The reason is not hard to find. In generation 1000, string 0 in the dipoles problem is (Table 5.7)

$$96 \quad 92 \quad 90 \quad 93 \quad 93 \quad 97 \quad 95 \quad 88 \quad 88 \quad 88$$

In this string one of the dipoles has the 90° alignment required for all the dipoles in the perfect solution, but for each of the remaining dipoles, the required value is present neither in this string nor in any of the other strings in the population. We can estimate how many generations must pass on average before an angle of 90° is created by the mutation operator for the first dipole in string 0, assuming that the string survives the selection process each generation.

The probability of mutation per string per generation is 0.2; on average, therefore, two strings are mutated each generation. Taking all strings together, there is a 2/10 probability that a particular position is chosen for mutation,* and, if it is, the probability that the required value of dipole orientation (90) will be generated is 1/360. The probability that the required value will be generated in the desired position in the string in which it required, thus, is about 2/10 × 1/360, substantially less than one chance in a thousand per generation.

It is frustrating that the angle of 96° is nearly correct, and yet the probability that the algorithm can move it by just a few degrees to the optimum value seems to be so low, but there is a solution. When only small changes need to be made to the angles to arrive at the optimum solution, the mutation operator is a blunt tool with which to make those changes. The values generated by mutation span the entire range of angles, so the operator may replace a near-optimum value such as 96° with 211° or some other unsuitable angle, which is a move in the wrong direction. One way of proceeding is to adjust the mutation operator so that, at the start of the calculation, it generates any angle with equal probability, while in the later stages of a run the angle generated is not picked at random from across the entire range of possible values, but is instead related to the value already in place. Rather

---

* Roughly. Two strings are mutated each generation on average, but since the process is stochastic fewer or more strings may be mutated during a given cycle. Furthermore, there is a chance that the same position will be chosen for mutation in each string.

than generating a completely new value for the angle, mutation generates an *offset*, which adjusts the current value by a small amount. The size of the offset is chosen randomly from within a normal distribution whose width diminishes slowly with generation number; the effect is to focus the search increasingly on a region around the current solution.

This is *focused mutation*, also known as *creep*, a name that is suggestive of the way in which the mutated gene investigates the region around its current value. Focused mutation is one of a number of ways in which we can adapt crossover and mutation. In crossover, for example, the genes in the child strings might be created not by swapping values between parents, but by taking an arithmetic or geometric average of those values (provided that the genes are real numbers). Such operators are not widely used, but the fact that they have been proposed indicates the extent to which one has the freedom to "invent" operators that might be of value in solving a problem while still remaining within the realm of the GA.

5.8.6 Local Search

An alternative, or an addition, to focused mutation is local search. In a local search, the algorithm occasionally takes a brief diversion, during which small nonevolutionary changes are made to a randomly selected string to see whether its quality can be improved. A string of dipoles is picked and a random angle within it changed by a few degrees. If the fitness of the string improves, a second small change is made to the same angle to see if a further improvement could be achieved, and the process continues until a change fails to improve the fitness of the string. If during the course of this process a string is created that is better than the original, the revised string replaces the original in the population. If there was no improvement, the string is returned unchanged and the GA then continues in the normal way.

Local searches may be helpful when the efficiency of mutation falls away in the later stages of a run. However, local search must not be used too enthusiastically by applying it in succession to every angle in a string or at a very early stage in the calculation, perhaps. If this is done, strings will be forced up to a local fitness maximum that will make it harder for the algorithm to discover the global maximum.

5.9 Encoding

5.9.1 Real or Binary Coding?

In any genetic algorithm application, the physical problem must be translated into a form suitable for manipulation by the evolutionary operators. Choice of coding is an important part of this process.

In early work, GA strings were binary coded. Computer scientists are comfortable with binary representations and the problems tackled at that time could be easily expressed using this type of coding. Binary coding is sometimes appropriate in scientific applications, but it is less easy to interpret than alternative forms, as most scientific problems are naturally expressed using real numbers.

When deciding what type of coding to use, one has to consider the effect that binary coding may have on the effectiveness of the algorithm as well as whether one type of coding is more difficult to interpret than another. For example, suppose that we decided to use binary coding for the dipoles problem. String 1 from the original population was

103   176   266   239   180   215   69   217   85   296

In binary form this string is

{001100111 010110000 100001010 011101111 010110100 011010111 001000101 011011001 001010101 100101000}

where the gaps have been introduced to show the breaks between successive angles, but do not form part of the string itself. We notice immediately that, whereas in the real-valued format it was simple to spot whether two dipoles pointed in approximately the same direction, in binary format this is a good deal harder. The situation is worse still if the gaps between the angles are omitted:

{001100111010110000100001010011101111010110100011010111001000101011011100 1001010101100101000}

The GA operates upon binary strings using selection, mating, and mutation operators in much the same way as when real number coding is used, but a binary representation brings with it several difficulties. Consider a second string with which we can cross the string given above:

74   1   111   64   217   2   31   99   351   70

In binary, this string is

{001001010 000000001 001101111 001000000 011011001 000000010 000011111 001100011 101011111 001000110}

Applying one-point crossover at a randomly chosen position, the thirty-second binary digit in each string, we get the two new strings:

{001100111 010110000 100001010 011100000 011011001 000000010 000011111 001100011 101011111 001000110} and

{001001010 000000001 001101111 001001111 010110100 011010111 001000101
011011001 001010101 100101000}

When these new binary strings are translated back into real values so that
we can check what has happened, the result is

102   176   266   224   217   2   31   99   351   70

and

74   1   111   79   180   215   69   217   85   296

We see that crossover has swapped some values between the strings, just
as happened when real-valued strings were used, but at the crossing point
the values of two angles have not been swapped but changed because the cut
fell within the binary coded value for a single number; where before the real-
valued strings contained 239 and 64, they now contain 224 and 79. This alteration of the values in the string could be avoided by allowing the crossover
operator to cut the binary string only at the position where one real-valued
number is divided from another, but if cuts are only permitted at those points,
the algorithm now resembles closely the situation when real value coding is
used and the procedure is reduced to a disguised real-value algorithm.

The evolution of the algorithm using binary strings will not be identical to
the evolution when real-valued strings are used because the effect of mutation on an angle represented in binary will be different from the effect of
mutation on a real-valued angle. A binary string is mutated by selecting a
gene at random and flipping the bit, turning $0 \rightarrow 1$ or $1 \rightarrow 0$. Suppose we flip
the eighth bit from the left in this string:

{001100111 010110000 100001010 011100000 011011001 000000010 000011111
001100011 101011111 001000110}

In the mutated string, the first angle, in real values, has changed from 103
to 101, so the value of the angle has changed only slightly. At the start of the
GA run, large changes in the values of the genes are desirable because many
genes will be far from their optimum values, but if the groups of binary bits
that encode for a real number are long, most random changes to a binary
representation will alter the real-number value of the gene by only a small
fraction of the total possible range. In the present example, if any one of the
five rightmost bits is changed by mutation, the greatest possible change in the
angle is just 16°. Binary coding thus biases mutation toward small changes
in angle, in contrast to the mutation of real-valued angles in which mutation
changes one angle to any other with equal probability.

We could apparently get around this difficulty by allowing changes to be made to not just one but every bit in the mutation of the 9-bit binary string, but this then again returns us to a situation in which we are, in essence, manipulating real numbers disguised in a binary cloak.

Further difficulty arises if the string represented in binary format does not have the same range as the string represented in a real number format. Nine binary bits are needed to represent all the values of angles between 1° and 360°, but further binary numbers that correspond to the real numbers from 361 to 511 could be created using these 9 bits. There is no physical difficulty associated with using these angles, but since there will be two binary representations of some angles (for example, 000011000 is the same as 24°, while 1100000 is 384°, which corresponds to the same dipole orientation), but only one that corresponds to other angles, such as 349°, the calculation will now be biased toward certain angles. The use of real-valued strings avoids these difficulties and this type of representation is much more common in physical and life science applications than in computer science.

One well-known but not widely adopted alternative to conventional binary coding is to use Gray codes. The binary representation of adjacent integer values differs by one binary digit using Gray codes. The Gray codes for a few integers are given in Table 5.11. The particular advantage of Gray codes is that successive real-valued numbers differ by only one binary digit, so a small change in the Gray code may translate into a small change in the real value of a parameter. Despite this minor advantage, Gray codes have not yet attracted a wide following in science.

TABLE 5.11

Gray Codes of the Real Numbers 0 to 7

Integer Value	Gray Code
0	000
1	001
2	011
3	010
4	110
5	111
6	101
7	100

5.10 Repairing String Damage

In the alignment of dipoles discussed earlier, no matter what modifications were made to the strings by crossover and mutation, every new string was still a valid solution. In some types of problems, this no longer applies. In a permutation problem, the strings consist of the integers from 1 to n in some order, and strings that no longer represent feasible solutions can readily be created by crossover or mutation. An example is provided by the widely studied traveling salesman problem (TSP) or, because Europe is full of good restaurants, the European traveling gourmet problem (Figure 5.28).

In a grueling challenge, a gourmet must visit each of the n locations in the British Isles in which a two- or three-starred Michelin restaurant is to be

FIGURE 5.28
A possible (nonoptimum) tour of some two- and three-starred Michelin restaurants in the British Isles.

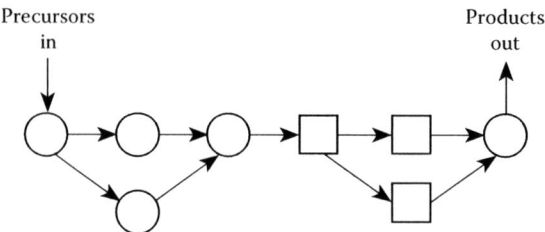

Precursors in

Products out

FIGURE 5.29
A schematic of a small chemical flowshop. The circles and squares represent units, such as reactors, dryers, or centrifuges.

found, eating in each restaurant exactly once before returning to the starting point. In order to save enough money to pay for the meals, the shortest possible route that includes every establishment must be taken. The route can be expressed as a vector listing the restaurants in the order to be visited. The number of possible orders is $n!$; thus, when n is large, an enumerative search (in which all the routes are inspected one by one to see which is the shortest) is not possible.

The traveling gourmet problem is in itself not of much interest to physical and life scientists (or, more accurately, is probably of acute interest to many of us, but well beyond our means to investigate in any practical way), but problems that bear a formal resemblance to this problem do arise in science. The chemical flowshop is an example (Figure 5.29).*

Chemical flowshops are widely used in industry for the medium-scale synthesis of chemicals. In a flowshop, a number of different physical units, such as reactor vessels, ovens, dryers, and distillation columns are placed in a serial, or largely serial, arrangement. Rather than being devoted to the preparation of a single chemical, many different chemicals are synthesized one after another in a flowshop by feeding in the appropriate precursors and reagents as the intermediates pass through most or all of the units. This mode of operation is an effective use of the physical plant because no part

---

* Scheduling problems are extraordinarily common within and beyond science, so much so that the TSP has been used as a test problem for almost every form of AI algorithm. The problem itself can be formulated in many different ways; the most obvious is to require, as here, that the sum of the distances traveled be minimized. An equivalent requirement, but one that hints at the many other existing ways of tackling the problem, is to regard the complete tour as being defined by an enormous rubber band stretched from city to city. If the tour is long, the band is stretched so its energy is high; as the length of the tour decreases so does the energy, so the optimum tour is that of lowest energy. An advantage of this interpretation is that by introducing "friction" at the position where the rubber band is wrapped around an imaginary post in the center of each city, the tension in the band need not be the same in each segment of the tour, so different portions of the tour can be partially optimized by minimizing their energy without having to worry about what is happening in other parts of the tour.

of the plant should stand idle for long, but it creates scheduling problems. When making different products, the residence times of chemicals in each unit may be very different. The synthesis of one material might require that a particular unit be used for a period of several hours, while the preparation of a second material may require use of the immediately preceding unit for only a matter of minutes, thus the long-residence chemical blocks the progress of the chemical that follows it through the line.

Because the residence times of different chemicals in the units vary, the order in which chemicals are made in a flowshop has a profound effect on the efficiency of its operation. If the optimum production order can be identified and used, chemicals will move through the flowshop in something that approaches lock-step fashion, with the contents of every unit in the flowshop being ready to move on to the next unit at about the same time. Such perfect matching of the movement of chemicals throughout the flowshop will almost never be achievable in practice, but if something approaching this sort of coordinated movement can be managed, all units would be used for a high proportion of the available time and the flowshop would be running at high efficiency.

Choice of the optimum production order is a scheduling problem. Like the problem of the traveling gourmet, a GA solution for the flowshop consists of an ordered list. In the flowshop, this list specifies the order in which chemicals are to be made:

$$\{6, 1, 9, 10, 7, 3, 2, 5, 4, 8\}$$

This string tells us that product 6 is to be made first, followed by product 1, then by 9, and so on; no product is made more than once. The GA is a powerful means of finding the optimum order, but it is easy to see that the crossover and mutation operators may cause problems. Suppose that the strings

$$\{6, 1, 9, 10, 7, 3, 2, 5, 4, 8\} \text{ and } \{7, 3, 9, 1, 2, 10, 5, 4, 8, 6\}$$

were subjected to two-point crossover at genes 3/4 and 7/8 in both strings to give the two new strings

$$\{6, 1, 9, 1, 2, 10, 5, 5, 4, 8\} \text{ and } \{7, 3, 9, 10, 7, 3, 2, 4, 8, 6\}$$

After crossover, neither string is valid because both specify that some products must be synthesized twice and other products not at all. It is tempting to allow the GA itself to sort out this problem. A large penalty could be imposed on the fitness of any string that contains duplicate products, thus ensuring that the string would soon be removed from the population by the selection operator. However, in a flowshop of industrial scale, operating a schedule that calls for the synthesis of fifteen to twenty products, a large majority of the strings created by the evolutionary operators will be invalid, therefore, the algorithm will spend much of its time identifying, processing, and destroying strings of no value. This flooding of the population with invalid strings will significantly slow evolution.

Instead of relying on the selection process within the GA to filter out invalid strings, it is more efficient to inspect new strings as they are created by crossover and to repair any damage before the strings are fed back into the GA. This can be done by sweeping through pairs of strings as they are formed by crossover and checking for duplicate entries. When the first repeated chemical is found in one string, it is swapped with the first duplicate in the second string (Figure 5.30), and the procedure is continued until all duplicate entries have been repaired.

This procedure must always succeed because the same number of duplicates will appear in both strings and, if a chemical appears twice in one string, it must be missing entirely from the other string.

Similar repair work following mutation is rendered unnecessary if a small change is made to the way in which mutation is applied. Instead of changing a single entry in the string to a random new value, the positions of two randomly chosen entries are swapped, thus automatically leaving a valid string (Figure 5.31).

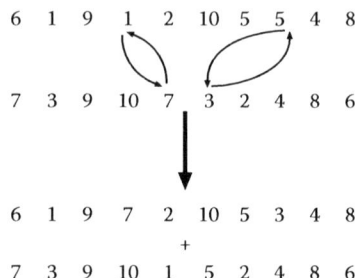

FIGURE 5.30
String repair after crossover.

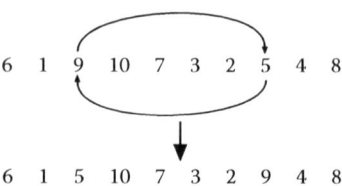

FIGURE 5.31
Mutation in a permutation string.

5.11 Fine Tuning

5.11.1 Elitism

Roulette wheel or tournament-based selection can fail to select the best string in a population as a parent, even if no other string is of comparable fitness. The effect of this failure, if it occurs, is most severe in the middle stages of a run, when the best string may have only a slim advantage in fitness over its fellows, but there may be no other copies of that string to replace it if lost. Stochastic remainder selection provides some protection for the best string, as it guarantees selection of the best string as a parent. However, the crossover or mutation operators might still damage the child strings subsequently created from it. A method that guarantees that the best string is never lost is elitism, which is independent of the type of selection operator.

In an elitist strategy, the first step in the selection process is to identify the best string in the current generation and copy this into the parent pool without it having to compete against any other strings for selection. This certain selection as a parent is combined with protection from crossover and mutation; thus the best string in any generation is sure to reach the next one unscathed, until eventually displaced by a string of superior fitness in some future generation. The effect of elitism on the progress of the fitness of a population can be seen in Figure 5.32, where the noise accompanying the best fitness string seen in Figure 5.16 has been replaced by a steady rise. Elitism is of greatest value in the middle of a GA run. In the early stages promising

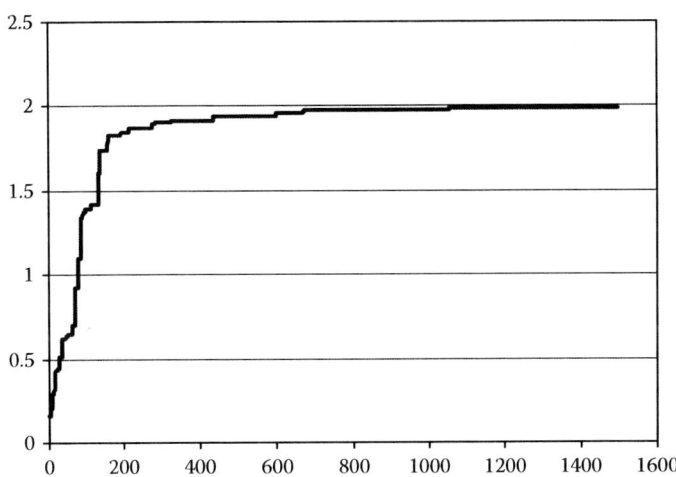

FIGURE 5.32
The effect of elitism on the fitness as a function of generation number.

new solutions are generated readily, so that even without elitism the fitness rises rapidly, while toward the end of the run, the population contains many similar strings of high fitness, therefore, it is not detrimental to lose one of them.

Since elitism guarantees that the best string in any generation cannot be lost, it can be safely combined with an increased mutation rate, especially in the later stages of a run. A higher mutation rate toward the end of a GA run helps to promote diversity and encourages a wider exploration of the search space at a stage when many strings in the population will be similar, thus the disruption caused by mutation is less of a concern than it would be earlier in the run.

5.11.2 String Fridge

The string fridge contains a group of strings that is isolated from all the GA operators; these strings comprise a seed bank of information from earlier generations (Figure 5.33). At the start of a run, the fridge is filled with random strings; evolution then begins. Periodically a string from the current GA population is copied into the fridge and swaps places with a random string already there.

Once the GA has been running for a while, the fridge, which may be several times the size of the main GA population, will contain snapshots of strings that evolved at many different stages of the calculation. This store of old strings can be valuable if it is relatively easy to find good building

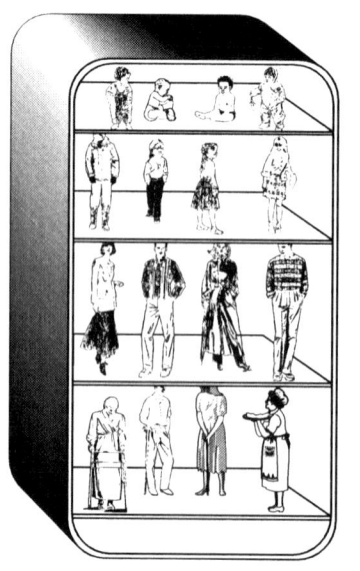

FIGURE 5.33
The string fridge: Strings of various ages are stored in the fridge to provide a random selection of information from previous generations.

blocks, but difficult to retain them while at the same time discovering other good building blocks. A string stored in the fridge at an early point in the algorithm that contains a nugget of valuable information may be resurrected thousands of generations later and brought out to inoculate the GA population with information that, having been discovered near the start of the run, could subsequently have been lost.

String fridges contain much outdated information, and most strings that are defrosted find themselves back in a population whose average fitness far exceeds that of the population when the string was shifted into the fridge, so are culled without delay. However, because the fridge door is rarely opened and its use only involves copying a single string in and out, its computational cost is almost zero, and for a complex problem this approach can be of value.

5.12 Traps

As in the application of any nondeterministic Artificial Intelligence method, it is not safe to assume that all that is needed for a successful GA calculation is to open the jaws of the algorithm, drop in the problem, and then wait. Various factors must be considered to determine whether the GA might be effective.

Most fundamental of all is the need for evolution, which requires that the population must be permitted to run for a significant number of generations during which it can change. In some applications it is difficult to run the GA for long enough for this requirement to be satisfied. Typical of situations when evolution is difficult to bring about are attempts to combine the GA with experimental measurements, for example, by using the GA to propose a formulation of materials that might be fused together at high temperature to create a solid possessing potentially interesting or valuable properties. The number of generations may be severely restricted by the need to evaluate the fitness of each potential solution by performing an experiment in the laboratory. Although in principle an experimental approach is productive, if it is possible to run the algorithm only for a few generations, such problems may be better tackled by other methods, and there may be little justification for regarding the calculation as having converged.

Equally, very short GA strings that contain only two or three genes are not suitable for a GA approach. Better methods than the GA for the optimization of a function of two or three variables are usually available. It should also be clear that converting a very short real-valued string into its much longer binary-coded equivalent merely disguises the underlying problem of a short string. Writing it in binary form so that it appears more complex does not alter the inherent simplicity of the string.

5.13 Other Evolutionary Algorithms

Several other types of evolutionary algorithms (EAs) exist, differing in the selection method for parents and the way in which one population is formed from the previous one or in the interpretation of the GA strings. In addition, evolutionary agent-based models are starting to appear. Although these methods show promise, they are currently less widely used in science than the GA, so are covered here briefly.

5.13.1 Evolutionary Strategies

In evolutionary strategies, a parent string produces λ offspring; the fittest of the $1 + \lambda$ individuals is selected to be the single parent for the next generation of offspring. There is no crossover operator in evolutionary strategies, only mutation.

Before the algorithm is allowed to start, a reasonable range for each parameter in the string is established, such that we are confident that the optimum solution lies within the range covered by the parameters:

$$S = \{x_{1,min} \leftrightarrow x_{1,max}, \ x_{2,min} \leftrightarrow x_{2,max}, \ ..., \ x_{n,min} \leftrightarrow x_{n,max}\} \tag{5.8}$$

A standard deviation, which expresses our estimate of how likely it is that the correct value of x_i lies in the middle of the range and how much variance we expect the parameter to show, is chosen for each x_i.

Values for each x_i are then chosen at random from within the allowed range and the solution to which this set of values corresponds is calculated:

$$X = f(x_1, x_2, x_3, ... x_n) \tag{5.9}$$

A new solution is now created by modifying each x_i by an amount chosen at random from a normal distribution with mean 0 and standard deviation σ_i:

$$x_{i,new} = x_i + a(0, \sigma_i) \tag{5.10}$$

This reflects the observation that genetic change among living species is based on the current chromosome (hence, the mean of zero) and that small changes to it are more likely than large ones (hence, the dependence of the size of the change on a normal distribution with a limited standard deviation). The solution, which corresponds to the new set of parameters, is then calculated.

$$X_{new} = f(x_{1,new}, x_{2,new}, x_{3,new}, ... x_{n,new}) \tag{5.11}$$

In a (1,1) algorithm, the two solutions, X and X_{new}, are compared, the better solution is kept, and the process repeats until a solution of sufficient quality emerges.

The primary evolutionary operator in the algorithm is mutation, but this is used in a very different way from the GA, since *every* gene in the string is mutated. This process, in which all genes in a candidate solution are manipulated simultaneously, implicitly recognizes that the quality of the solution may depend on not just the values of individual genes, but also how the different genes may act together to determine the quality of the solution. This is a contrast with the more block-like approach of the GA, in which substantial portions of a string remain unchanged from one generation to the next.

5.13.2 Genetic Programming

Genetic programming (GP) should be the Holy Grail of scientific computing, and indeed of many other sorts of problem solving. The goal of GP is not to evolve the solution to a problem, but to evolve a *computer program* that when executed will yield that solution.

The potential of such a method is very considerable since, if it can be made to work, there would no longer be a need to write an explicit program to solve a problem. Instead a description of the problem would be fed into the GP program, the user would retire from the scene, and in due course a complete program to solve the problem would emerge.

A computer program can be written as a sequence of operations to be applied to a number of arguments. A GP algorithm manipulates strings just as the GA does, but in a GP the strings are composed of fragments of computer code or, more accurately, a sequence of instructions to prepare those fragments. Each instruction codes for an operation which is applied to zero or more variables that also form part of the string, so by constructing a GP string as a sequence of operations and arguments, the entire string may be interpreted as a (possibly) fully functioning program. This program is often shown in the GP literature as a *tree diagram*. Generally the program is understood to be in the LISP language, although, in principle, a GP could be used to construct a program in any computer language. However, LISP programs have special abilities, such as being able to operate on themselves to make changes or to generate further LISP programs, which makes them well suited to a GP application.

Unlike the GA, GP strings do not have a fixed length, so can grow or shrink under the influence of the genetic operators. The quality of each string is measured by unraveling it, turning it into a computer program, running the program, and comparing the output of the program with a solution to the problem set by the scientist. Notice that this process is almost the reverse of the procedure that we use to run a GA. In the GA, we have available a method by which the quality of any arbitrary solution can be calculated, but usually have no way of knowing what the optimum value for the solution to a problem is. In the GP, the optimum value that solves the problem is known, but the means by which that value should be calculated is unknown.

The GP is initialized with strings that code for random sequences of instructions and the fitness of the strings is assessed. Those strings that generate a program whose execution gives a value closest to the correct one receive the highest fitness; programs that offer poor solutions, or generate no output at all, are allocated a low fitness. The genetic operators of selection, crossover, and mutation are then applied to generate a new population of candidate solutions.

In constructing a GP program, it is necessary not only to define GA-like parameters, such as the population size, the type of crossover, and the number of generations, but also parameters that relate specifically to the components of the candidate programs that the GP will build. These include:

- The *terminals*: These correspond to the inputs that the program will need. Even though each candidate program will be manipulated by the evolutionary operators, if we know in advance that pieces of input data are certain to be required for any GP-generated program to function correctly, we must at least ensure that the program has access to them.

- The *primitive functions*: These comprise the various mathematical and logical operations that the program may need. They will usually include mathematical functions such as $+$ $-$, $/$, and $*$, logical functions, programming constructs, such as loops, and possibly other mathematical functions, such as trigonometric, exponential, and power functions.

- The *fitness function*: This is defined by the problem to be solved. The function is determined not by how closely the GP-generated program gets to the correct solution when run once, but how close the program comes to the correct answer when tackling many examples of the same problem. The reason for this will become clear from the example below.

Example 3: Genetic Programming

Suppose that GP was required to evolve a program that, given values for four input variables a, b, c, and d, correctly computed some prespecified output value. A particular set of the input values might be $a = 1$, $b = 2$, $c = 3$, and $d = 4$, and the desired output from the program might be 10. The GP could create a program in which the problem was solved through the following function:

Program 1: $a + b + c + d = 10$

This program would be allocated maximum fitness as it exactly matches the required output. However, the same GP run might create a second program that used the function:

Program 2: $c \times d - (a \times b) = 10$

or even a third that used:

Program 3: $b \times d + c - a = 10$

If only one correct function exists, at most one of these equations can be right, yet each program that returned the value of 10 for its output would be allocated the same fitness. By providing a second set of input data, say $a = 2$, $b = 4$, $c = 7$, $d = 0$, and the associated solution (–8), the fitness of those programs that had found an incorrect function would be diminished when the fitness was calculated over a range of examples.

This area of evolutionary algorithms, which owes much to the work of John Koza,[1] is as challenging as it is potentially productive. The most serious difficulty (of several) is that of deception, mentioned earlier. A computer program created by GP might be just one or two instructions away from being perfect and yet generate no output at all, or output that is useless and provides no hint to the algorithm that the program that it has made is almost correct. Without some indication from the fitness function that the correct solution is close at hand, there is little information available to guide the GP search toward better solutions.

Example 4: Deception

A fundamental equation in biosensor analysis is

$$R_{eq} = R_{max} \left\{ \frac{a_o K}{a_o K + 1} \right\} \tag{5.12}$$

R is a measure of the energy absorbed by a surface plasmon, K is the equilibrium constant for a reaction between two polymers, A and B

$$A + B \leftrightarrows AB$$

and a_o is the (constant) concentration of A. Suppose that a program created by the GP implemented the following equation:

$$R_{eq} = R_{max} \left\{ \frac{K^2}{a_o K - 1} \right\} \tag{5.13}$$

This has a form similar in some respects to the correct expression, but several changes are required to reach that expression. Depending on the values of the parameters, making one of these changes (e.g., changing the minus to a plus in the denominator) might improve or worsen the fitness. There is often very little information that indicates to the algorithm whether a proposed change will be beneficial in moving toward the optimum solution.

Analogies to this exist in nature: Human and chimp genomes are approximately 99 percent identical, so the change in the genome required to turn a chimp into a human is 1 percent. Suppose that we were trying to evolve the chromosome of an object that would perform consistently well when set

FIGURE 5.34
Potential differential equation solvers.

the problem of solving differential equations. We carry out an experiment in which the task is to solve some typical differential equations, measuring success by the quality of the solution generated after twenty minutes. The participants in this challenge are a chimp, a snail, and a piece of cheese. We can be confident that all would be equally poor at completing the task. Therefore, there is nothing in the performance of the chimp (or the snail or the cheese) that tells us that a solution to the problem of finding a living differential equation solver is just 1 percent away (Figure 5.34).

Finding a way to create and identify useful subprograms that can reliably be determined to be components of an optimum program is thus very challenging and has restricted the use of GP to small problems. Its application to large-scale scientific problems seems a long way off.

5.13.3 Particle Swarm Optimization

Swarm intelligence is a term that is applied to two rather different techniques — ant colony or pheromone trail optimization and particle swarm optimization. We deal here briefly with the latter.

The ability of birds to flock is notable. Although it seems that quite subtle and complex navigation and communication skills among birds might be required to explain the V-shaped formations of migrating flocks and the democratic rotation among the flock of the bird that leads the V, Eberhart and Kennedy[2] showed in the 1990s that flocking behavior need not involve complicated calculations by the birds, but could be the result of the application of a few simple rules.*

Imagine a flock of starlings searching for food, which lies in a small area of the local countryside. None of the birds knows in advance where the food is located, so each of them will search for food independently, but will not move so far from the rest of the flock that they lose touch with it. By acting

---

* Realistic flocking behavior can be generated by constructing some simulated birds (often known as "boids"), each of which obeys the rules: (1) move toward other members of the flock, (2) don't crash into anything, and (3) match your speed with the boids near you.

in this way, the entire search covers a large area, but once one member of the flock discovers the food, the message will quickly spread throughout the entire flock. Each bird constitutes what is known as an *agent*, carrying out local searches that are loosely correlated.

A similar approach is used computationally in particle swarm optimization (PSO). Each particle or agent within a large group moves semirandomly across the search space, but with a bias in the direction of the best solution found so far by that agent, p_{best}, and a bias also in the direction of the best global solution found by the entire swarm, g_{best}. As in the GA, we need a means by which the solution for which an agent encodes can be assessed. Each move made by the agent is calculated from its current position and velocity, modified by a randomly chosen amount of movement towards both g_{best} and p_{best}, together with a small element of movement in an entirely random direction.

$$v(k+1) = v(k) + c_p \times rand_p \times (p_{best} - x(k)) + c_g \times rand_g \times (g_{best} - x(k)) + c_r \times rand_r \quad (5.14)$$

$$x(k+1) = x(k) + v(k+1) \quad (5.15)$$

In these equations, $v(k)$ and $v(k+1)$ are the particle's velocity in step k and step $(k+1)$, respectively; c_p is the learning factor for movement toward the personal best solution; c_g is the learning factor for movement toward the global solution; c_r is a random learning rate; and $x(k)$ and $x(k+1)$ are the current and next positions of the particle. c_p and c_r typically have values in the range 1 to 2.5, while c_r is in the range 0 to 0.2. To prevent particles oscillating with large amplitude in the search space (a potential problem if c_p or c_g are > 2), a maximum velocity is imposed on all particles.

Each agent therefore moves in a way that is determined both by the success of its own search and the success of the entire flock. When a good solution is found, the flock moves toward it and, because many agents are present, the general area is then investigated thoroughly.

Particle Swarm Optimization

1. Choose parameters for the optimization.
2. Give particles random positions and velocities in the search space.
3. Calculate the current fitness for each particle; if this exceeds p_{best}, update p_{best}.
4. Set g_{best} equal to the best fitness of all particles.
5. If the value of g_{best} is not yet good enough, calculate new velocities according to equation (5.14), determine the new positions through equation (5.15) and return to step 3.

The effect of this process is rather similar to evolution in a GA. The initial search covers a wide area, but is unfocused. The information that is passed between agents causes the search to become focused around the better solutions. However, the mechanism by which evolution occurs is quite different from the GA. There is no selection or crossover, and mutation exists only inasmuch as each particle can adjust the direction and speed with which it moves. There is also no removal of particles, the population remains of constant size and always features the same particles, but the particles do have an individual memory of where they have been and where the most promising positions visited by the swarm are located, thus each functions as a semi-independent entity.

5.14 Applications

The GA is now widely used in science, and this discussion of typical applications can just touch upon a few representative examples of its use. No attempt is made to provide a comprehensive review of published applications, but this section will give a flavor of the wide range of areas in which the GA is now used.

There has been a notable change in the way that the GA has been used as it has become more popular in science. Some of the earliest applications of the GA in chemistry (for example, the work of Hugh Cartwright and Robert Long, and Andrew Tuson and Hugh Cartwright on chemical flowshops) used the GA as the sole optimization tool; many more recent applications combine the GA with a second technique, such as an artificial neural network.

The use of hyphenated algorithms has been encouraged by the rapid growth in computer power. For example, Hemmateenejad and co-workers have combined a binary-coded GA with *ab initio* calculations in a study of blood–brain partitioning of solvents.[3] It is now feasible to use the GA to optimize molecular structures by interpreting each GA string as a recipe for building a molecule, generating multiple strings corresponding to possible structures, running a quantum mechanical calculation on each structure to determine a fitness for each molecule, and then using the normal evolutionary operators to select and change the structures. Numerous workers have used the GA in combination with some means of calculating molecular energies as a route to the prediction of molecular structure. For clusters of large numbers of atoms, the quantum mechanical calculations are demanding, which makes the use of the GA to guide these searches only barely feasible; this has led some workers to run the GA for only a few generations, which renders it less effective. Among the more interesting applications are those by Djurdjevic and Briggs,[4] who have used the GA in protein fold prediction, as have Cox and Johnston,[5] while Wei and Jensen[6] have addressed the question of how to find optimal motifs in DNA sequences. Abraham and Probert[7] have used the GA for polymorph prediction. Johnston's group has been particularly active in this area, especially in the study of metal clusters.[8]

A number of workers have used the GA to find values for the set of parameters in some predefined model. Hartfield's work on the interpretation of the spectrum of molecular iodine is typical.[9] A particular difficulty with such an application is that the problem may well be deceptive, but because this may not be obvious from the form of the problem, the fit may be of lower quality than anticipated.

Computer screening allows pharmaceutical companies to test virtual drugs against known protein structures to decide whether synthesis is financially worthwhile. This screening takes place at all stages of drug development because the later a drug is abandoned the greater the cost. GP has been used to select a small group of molecular descriptors from a large set that best explains the propensity of a drug to degradation by P450 (which is responsible for the degradation of most drugs in the body). This is significant as an important cause of the withdrawal of a drug from trials is the prevalance of adverse side effects. One advantage of the GP approach over a neural network is that the model that the GP creates is open to inspection, so it is possible to determine whether the molecular descriptors that the GP believes to be important are in agreement with what human experts would expect.

Applications of swarm models are just starting to appear in science. In a recent application, each agent codes for a molecular structure, initially randomly generated, which calculates its own nuclear magnetic resonance (NMR) spectrum. This spectrum is compared with the experimentally determined spectrum of a protein, and the degree of match indicates how close the agent is to the correct protein structure. The movement of one agent toward another is accomplished through the sharing of structural information, which ensures that agents that satisfactorily reflect at least part of the structure can spread this information through the population.

5.15 Where Do I Go Now?

Several readable texts exist that cover the area of genetic algorithms, although there is a preponderance of edited multiauthor texts available that dive straight into the subject at the research level. One of the best texts for a new user is one of the oldest: *Genetic Algorithms in Search, Optimization and Machine Learning* by Goldberg[10] was published in 1989 and, although it is now difficult to find, it provides a lucid introduction to the field. *Handbook of Genetic Algorithms* is a rather more advanced, wider-ranging text, edited by Davis,[11] more suitable once some background in the area has been gained.

A couple of more recent texts provide gentle introductions to the subject. *Evolutionary Computation* by De Jong[12] looks beyond the basic genetic algorithm, but does so at a studied pace, thus is suitable for those who do not yet have much experience in the field. Finally, *Swarm Intelligence*, by Kennedy and

Eberhart[2] is one of the few recent introductions to that field, but also covers genetic algorithms and genetic programming by way of setting the scene.

5.16 Problems

1. Aligning quadrupoles

   Consider a quadrupole, in which positive charges and negative charges are arranged at the ends of a cross. In the dipoles problems discussed in this chapter, it was immediately clear that a head-to-tail arrangement of dipoles would give the optimum energy, but the lowest energy orientation for a line of quadrupoles is less obvious. They might orient themselves as a set of pluses so that positive and negative charges lie as close as possible to each other along the x-axis, or as a series of multiplication signs, which allows a greater number of favorable interactions, but of a lower energy as the charges are farther apart (Figure 5.35).

   Write a GA to investigate which arrangement is the more stable and investigate whether there is some critical ratio of the dimension of the quadrupole to the distance apart that determines which geometry is of lower energy.

2. A field of dipoles

   Use two-dimensional strings to construct a GA that can find the lowest energy arrangement of 100 identical dipoles positioned evenly across a square.

3. Universal Indicator

   Consider the formulation of a universal indicator; this is a solution whose color is pH-dependent across a wide pH range. Universal indicators usually contain several weak acids, in each of which

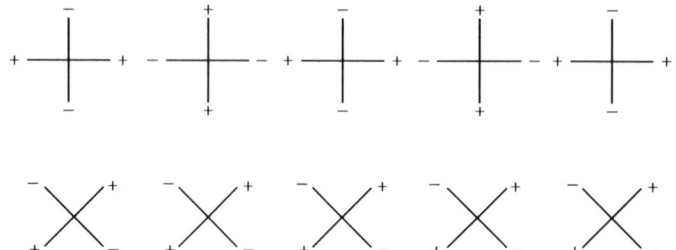

FIGURE 5.35
Two possible ways in which quadrupoles might align when positioned at equal distances along a straight line.

either the protonated or the unprotonated form (or both) are colored. The solution must have the following two characteristics:

a. The indicator solution must be colored at all pHs and change color to a noticeable extent when the pH changes by a small amount, say 0.5 units.

b. Every pH must correspond to a different color.

Consider what information you would need to evolve a recipe for a good Universal indicator. The visible spectra of many suitable weak acids and their pK_a values can be found on the Web if you want to test your algorithm. The formulations of some typical commercial Universal indicators can be found on the Web sites of chemical suppliers.

References

1. Koza, J.R., Keane, M.A., and Streeter, M.J., What's AI done for me recently? Genetic Programming's human-competitive results, *IEEE Intell. Systs.* 18, 25, 2003.
2. Kennedy, J. and Eberhart, R.C., *Swarm Intelligence*, Morgan Kaufmann, San Francisco, 2001.
3. Hemmateenejad, B., et al., Accurate prediction of the bloodbrain partitioning of a large set of solutes using *ab intio* calculations and genetic neural network modeling, *J. Comp. Chem.*, 27, 1125, 2006.
4. Djurdjevic, D.P., and Biggs, N.J., *Ab initio* protein fold prediction using evolutionary algorithms: Influence of design and control parameters on performance, *J. Comp. Chem.*, 27, 1177, 2006.
5. Cox, G.A. and Johnston, R.L., Analyzing energy landscapes for folding model proteins, *J. Chem. Phys.*, 124, 204714, 2006.
6. Wei, Z. and Jensen, S.T., GAME: Detecting cis-regulatory elements using a genetic algorithm, *Bioinformatics*, 22, 1577, 2006.
7. Abraham, N.L. and Probert, N.J., A periodic genetic algorithm with real space representation for crystal structure and polymorph prediction, *Phys. Rev. B.*, 73, 224106, 2006.
8. Curley, B.C. et al., Theoretical studies of structure and segregation in 38-atom Ag-Au nanoalloys, *Eur. Phys. J. D. Atom. Mol., Opt. Plasma Phys.*, 43, 53, 2007.
9. Hartfield, R.J., Interpretation of spectroscopic data from the iodine molecule using a genetic algorithm, *Appl. Math. Comp.*, 177, 597, 2006.
10. Goldberg, D.E., *Genetic Algorithms in Search, Optimization and Machine Learning*, Addison Wesley, Reading, MA, 1989.
11. Davis, L. (ed.), *Handbook of Genetic Algorithms*, Van Nostrand Reinhold, New York, 1991.
12. De Jong, K.A., *Evolutionary computation: A unified approach*, MIT Press, Cambridge, MA, 2006.

6

Cellular Automata

Numerous systems in science change with time or in space: plants and bacterial colonies grow, chemicals react, gases diffuse. The conventional way to model time-dependent processes is through sets of differential equations, but if no analytical solution to the equations is known, so that it is necessary to use numerical integration, these may be computationally expensive to solve.

In the cellular automata (CA) algorithm, many small cells replace the differential equations by discrete approximations; in suitable applications the approximate description that the cellular automata method provides can be every bit as effective as a more abstract equation-based approach. In addition, while differential equations rarely help a novice user to develop an intuitive understanding of a process, the visualization provided by CA can be very informative.

Cellular automata models can be used to model many systems that vary in both space and time, for example, the growth of a bacterial colony (Figure 6.1),

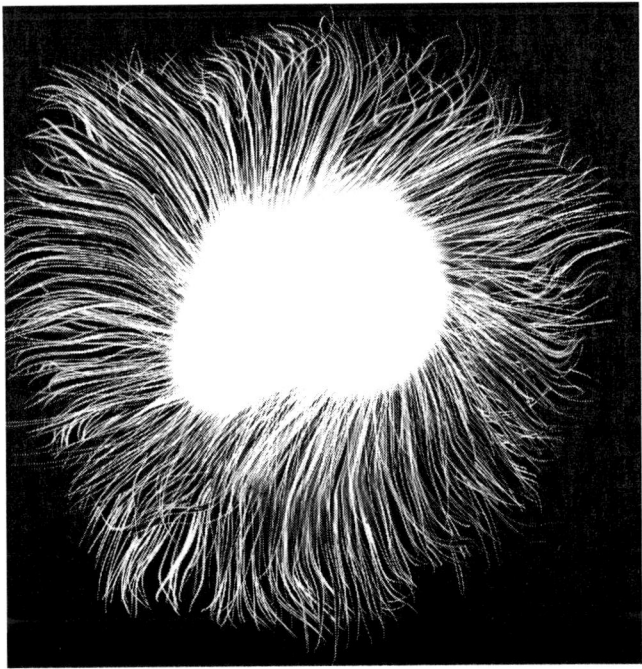

FIGURE 6.1
A simulation of the filamentary growth of a bacterial colony.

FIGURE 6.2
A computer model of the development of chemical waves in the Zaikin–Zhabotinsky reaction.[1]

or the spreading spirals of concentration waves in the Zaikin–Zhabotinsky reaction (Figure 6.2). The method has been widely used in biochemistry and biology; indeed, one of its earliest applications was in the description of biological reproduction. By modeling self-organization in an effective way, the use of CA has been extended to chemistry, physics, fluid dynamics, and numerous other areas. We shall meet several scientific examples in this chapter.

6.1 Introduction

Cellular automata model scientific phenomena by breaking up the region of simulation into many cells, each of which describes a small part of the overall system. The *state* of each cell is defined by a limited number of variables whose values change according to a set of rules that are applied repeatedly to every cell. Even though the rules that are used within the algorithm are usually simple, complex behavior may still emerge as cells alter their state in a way that depends both on the current state of the cells and that of their neighbors. CA are conceptually the most straightforward of the methods covered in this book, yet, despite their simplicity, they are versatile and powerful. In fact, it is possible to demonstrate that a CA model is capable of universal computation; that is, with a suitable choice of starting conditions and rules, the model can emulate a general-purpose computer.

The method is different in nature from the techniques discussed elsewhere in this book because, despite the fact that the cells that comprise a CA model evolve, they do not learn. The lack of a learning mechanism places the method at the boundaries of Artificial Intelligence, but the algorithm still has features in common with AI methods and offers an intriguing way to solve scientific problems.

6.2 Principles of Cellular Automata

Although CA lie at the edge of AI, their range of applicability is broad. In addition, although this is hardly a recommendation for their use in science, CA generate intriguing and beautiful patterns that are sometimes presented as works of modern art. A CA model is formed by constructing an array of *finite state automata* (Figure 6.3). A finite state automaton* is a simple computational unit, comparable in complexity to a node in an artificial neural network.

Although there is the possibility of confusion in terminology when dealing with the applications of CA, since they are often applied to the simulation of processes that involve biological cells, we shall follow standard practice and refer to both the finite state automaton and the place in which it is located as a "cell."

Many individual finite state automata are joined together to form a regular array in one, two, or more dimensions; this entire array is the cellular automaton. The CA evolve, as all cells in this array update their state synchronously. Into each cell is fed a small amount of input provided by its neighbors. Taking account of this input, the cell then generates some output, which determines the next state of the cell; in deciding what its output should be, each cell consults its *state*, which consists of one piece, or a few pieces, of information stored within it. In the most elementary of automata, the state of the cell that comprises this finite state automaton is very simple, perhaps just

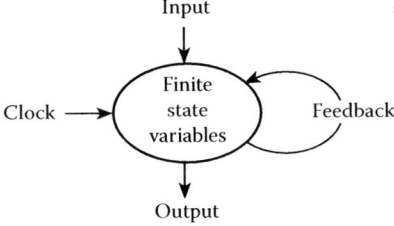

FIGURE 6.3
A finite state automaton.

---

* A finite state automaton is also known as a *finite state machine*.

"on" or "off." The cell also incorporates feedback, thus some of the input is generated by the cell reading its own output from the previous cycle.

The cells in a CA use *transition rules* to update their states. Every cell uses an identical set of rules, each of which is easy to express and is computationally simple. Even though transition rules are not difficult either to understand or to implement, it does not follow that they give rise to dull or pointless behavior, as the examples in this chapter illustrate.

Example 1: Pascal's Triangle

Suppose that several cells are placed in a line to form a one-dimensional CA. Every unit has an initial state of 0, apart from the central cell, which has a state of 1.

$$0 \quad 0 \quad 0 \quad 1 \quad 0 \quad 0 \quad 0$$

At regular intervals as determined by a virtual clock, every cell updates its state, using a rule that depends both on its current state and on the current state of its neighbors. In this first example, in which successive states of the CA are written one below the other, the transition rule is

1. The state of the cell in the next cycle is determined by adding together the current values of the states of the two cells immediately above it.

Thus, in the second cycle, the CA has progressed to:

$$0 \quad 0 \quad 1 \quad 1 \quad 0 \quad 0$$

So, as to make clear the evolution of the CA, we will write out all of the cycles, one after the other, so the first two cycles are

```
0     0     0     1     0     0     0
   0     0     1     1     0     0
```

And, after a further cycle:

```
0     0     0     1     0     0     0
   0     0     1     1     0     0
   0     0     1     2     1     0     0
```

Continuing in this way, the familiar sequence of numbers in Pascal's triangle emerges:

```
0     0     0     1     0     0     0
   0     0     1     1     0     0
   0     0     1     2     1     0     0
      0     1     3     3     1     0
   0     1     4     6     4     1     0
      1     5    10    10     5     1
      1     6    15    20    15     6     1
```

Something of scientific interest (among numerous other applications, Pascal's triangle is used to predict nuclear magnetic resonance [NMR] multiplet splitting patterns) has been created from the repeated application of a simple rule.

This evolution of a complex set of numbers from something very simple is rather like a recursion rule. For example, the wave function for a harmonic oscillator contains the Hermite polynomial, $H_v(y)$, which satisfies the recursion relation:

$$H_{v+1} - 2yH_v + 2vH_{v-1} = 0$$

The recursion rule itself is straightforward, yet it generates a complex sequence of functions, even though each function in the sequence is defined only by the recursion rule and the value of the preceding function in the sequence.*

In the example given above, the generation of Pascal's triangle has been brought to a halt once the nonzero values reached the outer limits of the small set of cells that are available. We cannot proceed farther until we decide what to do at these boundaries. (This is dealt with in section 6.5.)

6.3 Components of a Cellular Automata Model

The sequence of operations in the CA is straightforward and is described by the flow diagram in Figure 6.4.

The components of a CA are

1. A set of equivalent *cells* arranged in a regular lattice and a set of *states* that characterize the cells.

2. *Transition rules* that define how the state of the cell in the next cycle is to be determined from the state of the cell and the states of its neighbors in the current cycle.

3. A function that defines the *neighborhood* around each cell. A cell can interact only with other cells that are within its neighborhood.

We shall consider each of these components in turn.

6.3.1 The Cells and Their States

The cells of the CA lie at the vertices of a regular lattice that homogenously covers a region of space. Although it is possible in principle to use an irregular lattice, we need to be able to define unambiguously the "neighborhood"

---

* The first few Hermite polynomials are
$H_0(x) = 1$, $H_1(x) = 2x$, $H_2(x) = 4x^2 - 2$, $H_3(x) = 8x^3 - 12x$, $H_4(x) = 16x^4 - 48x^2 + 12$

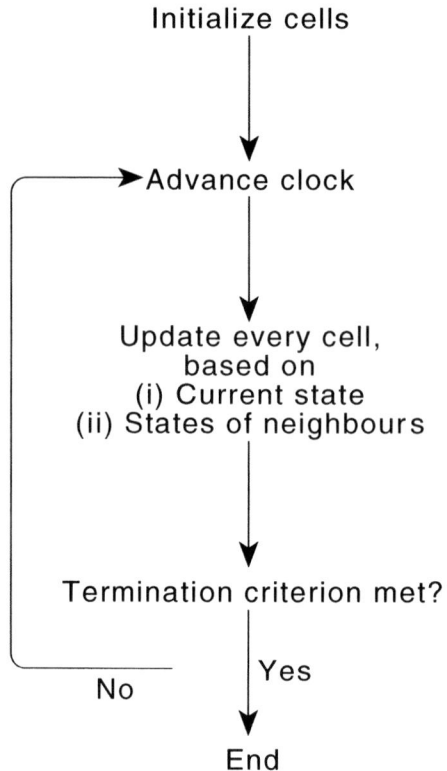

FIGURE 6.4
Cellular automata: the algorithm.

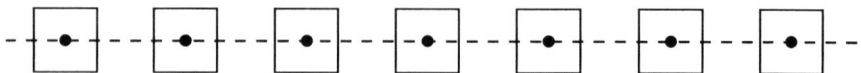

FIGURE 6.5
The arrangement of cells in a one-dimensional cellular automaton.

around each cell. Since all cells in a regular lattice are equivalent, the neighborhood in such a lattice must always include the same number of cells. In an irregular lattice, this requirement is difficult to meet.

In the simplest CA, cells are evenly spaced along a straight line (Figure 6.5) and can adopt one of only two possible states, 0 and 1. More flexibility is gained if the cells are arranged in a two-dimensional rectangular or hexagonal lattice and the number of permissible states is increased. We could further increase flexibility by including in the cell's state parameters whose values can vary continuously. This is a natural extension to the model to consider in scientific applications, in which the data very often can take any value within a real number range. The number of possible states for each cell is then infinitely large. A model in which there is no limit to the number of possible states that a cell may access is not a true CA, but is nevertheless often still referred to, and can be treated as, a CA.

6.3.2 Transition Rules

The state of every cell is updated once each cycle, according to transition rules, which take into account the cell's current state and the state of cells in the neighborhood. Transition rules may be deterministic, so that the next state of a cell is determined unambiguously by the current state of the CA, or partly or wholly stochastic.

Transition rules are applied in parallel to all cells, so the state to be adopted by each cell in the next cycle is determined before any of them changes its current state. Because of this feature, CA models are appropriate for the description of processes in which events occur simultaneously and largely independently in different parts of the system and, like several other AI algorithms, are well-suited to implementation on parallel processors. Although the states of the cells change as the model runs, the rules themselves remain unchanged throughout a simulation.

Example 2: Sierpinski Triangles

If the state of a cell in the next cycle is determined by its current state and those of its neighbors, organized — even attractive — behavior may emerge as the entire CA evolve. Suppose that the transition rules for a one-dimensional, two-state (0 and 1) CA model are

1. Sum the current state of a cell and that of its two neighbors;
2. If the sum equals 0 or 3 the state of the cell in the next cycle will be 0, otherwise it will be 1.

Figure 6.6 shows the first few cycles in the evolution of this CA. Each successive cycle is drawn higher up the page, thus the starting state is at

FIGURE 6.6
The evolution of a one-dimensional, two-state cellular automaton. The starting state is at the bottom of the figure and successive generations are drawn one above the other. The transition rules are given in the text.

the bottom of the figure and "time" (the cycle number) increases from the bottom upward. Cells that are "on" are shown in black, while those that are "off" are shown in white.

From a single starting cell, an intricate pattern emerges.*

As Example 1 and Example 2 suggest, CA rules are easy to apply using a computer. An early example of a CA with which most computer users are familiar is the Game of Life,[2] which is a two-state CA with simple transition rules. This became one of the earliest dynamic screen savers.**

The introduction of the Game of Life, developed by John Conway, ignited a burst of interest in CA models in the 1970s. At that stage, they were regarded mainly as a curiosity and few researchers anticipated that there would be much direct application for them in the physical and life sciences. Steven Wolfram's work in the 1980s on the use of CA in physics[3] established the method as one with potential in many scientific problems and since then numerous scientific applications have appeared.

Transition rules may be made as complicated as one might wish, though, paradoxically, very involved rules may give rise to behavior that is so disorganized that it is of little predictive value, while simpler rules can lead to correlated, yet still complex patterns, as Figure 6.6 shows. In order to simulate scientifically interesting behavior, we can expect that the rules will have some link to the physical problem to which the model relates. There may also be global variables, such as the concentration of a chemical or a gravitational field, that will be taken into account. However, the emergence of Pascal's triangle from a rule that has no obvious connection with NMR shows that links between rule and application might not always be foreseen.

6.3.3 Neighborhoods

6.3.3.1 The Neighborhood in a One-Dimensional Cellular Automata

The future state of any cell upon which we focus our attention, known as the *target cell*, is determined from its current state and that of the cells that lie in its neighborhood, hence, we must define what is meant by the neighborhood. In one dimension, it is easy to identify neighbors: If we assume that the lattice is infinite, each cell in a one-dimensional CA has both a left and a right neighbor, so the number of immediate neighbors is two. The neighborhood

---

* The pattern is a set of Sierpinski triangles; these are triangles in which an inner triangle, formed by connecting the midpoints of each side, has been removed.

** The Game of Life is run on a two-dimensional square lattice; each cell is either "dead" or "alive." The transition rules are:

1. If a live cell has one neighbor or none at all, it dies of isolation; if it has more than three neighbors, it dies of overcrowding ("becomes unoccupied" might be a better description than "dies," as the cell may be reoccupied subsequently).
2. If an occupied cell has two or three neighbors, it survives to the next generation.
3. If an unoccupied cell has exactly three neighbors, it becomes a birth cell and is occupied in the next generation.

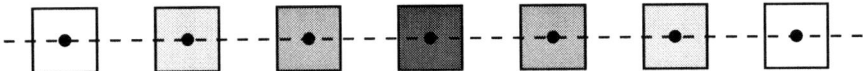

FIGURE 6.7
The neighborhood in a one-dimensional cellular automaton. Usually this includes only the immediate neighbors, but it can extend farther out to include more distant cells.

includes the target cell itself, so that the state of the target cell is fed back to help determine its state in the next cycle (Figure 6.7).

The neighborhood may be expanded to include cells that do not actually touch the target cell, so that the states of more distant cells can influence the future state of the target. The farther away a cell is from the target cell, the less will be its influence on the target. Therefore, some function, which might be linear, Gaussian, or exponential, can be used to define how much influence cells have depending on their distance away. In the majority of applications, only neighbors that actually touch the target cell appear in the transition rules.

6.3.3.2 The Neighborhood in a Two-Dimensional Cellular Automata

One-dimensional CA have found some application in science; the pattern on the shells of the cone shell, *conus textile*, is strikingly similar to one that can be generated using a one-dimensional CA. However, two-dimensional CA are more powerful, thus they are more able to model complex behavior. Since the CA lattice is space-filling, all cells in the CA are geometrically equivalent (almost — an exception is noted in section 6.5 below). As the only regular polyhedra that form a tessellation in two dimensions are triangles, squares, and hexagons, these are the only types of regular lattices used for two-dimensional CA.

On a two-dimensional lattice, there is more flexibility in the definition of a neighborhood than exists in one dimension. The von Neumann and Moore neighborhoods are both widely used on a rectangular lattice (Figure 6.8). The von Neumann neighborhood consists of the four cells that share an edge with the target cell; the centers of these cells are nearer to the center of the target cell than are any other nearby cells.

The Moore neighborhood (Figure 6.8b) includes all cells that touch any part of the target cell, either a face or a corner. The extended von Neumann or axial neighborhood, (Figure 6.8c) is simple computationally, but includes as neighbors some cells whose centers are farther away than the centers of other cells that are not included as neighbors, so this has only limited applicability. The radial neighborhood includes all cells whose centers lie within a defined distance from the center of the target cell (Figure 6.8d). In each of the four cases shown, the neighborhoods of nearby cells overlap.

No more theory is required to start investigating CA applications, so we turn now to some further examples.

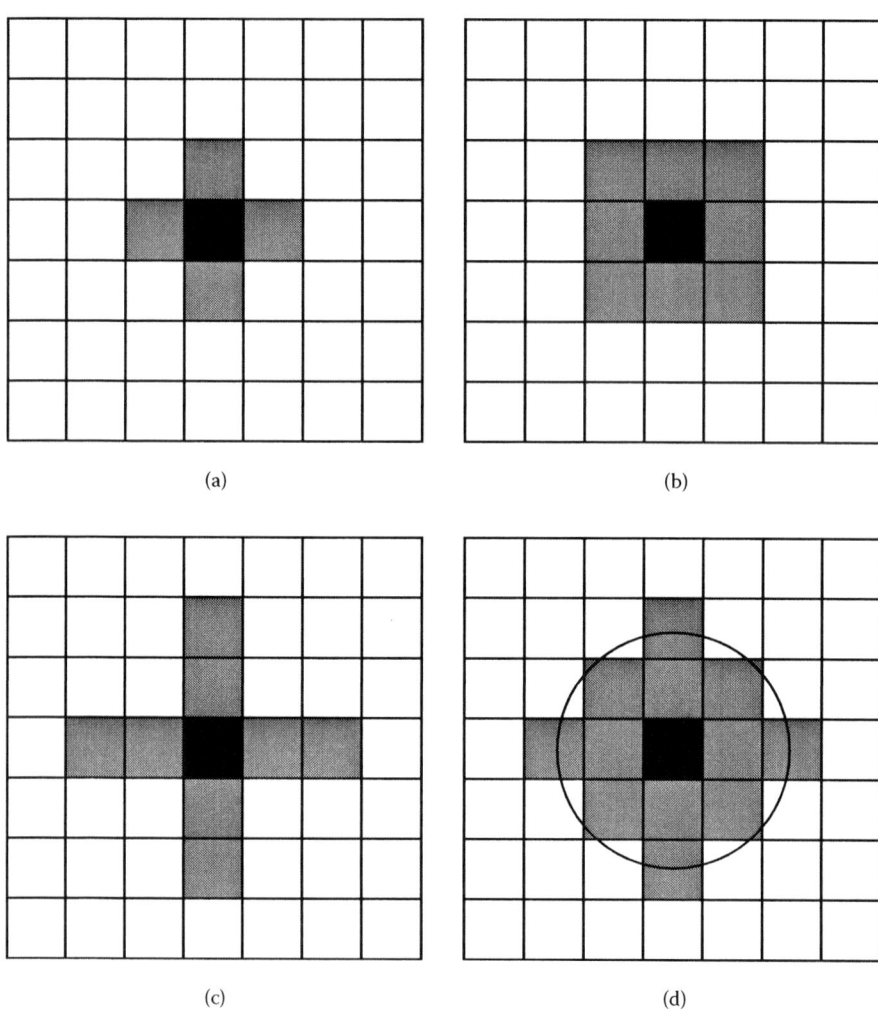

FIGURE 6.8
Possible neighborhoods in a two-dimensional cellular automaton: (a) von Neumann, (b) Moore, (c) axial, and (d) radial.

6.4 Theoretical Applications

6.4.1 Random Transition Rules

The simplest possible, and perhaps the least useful, transition rule is:

> The state of the cell in the next cycle is selected at random from all possible states.

FIGURE 6.9
Successive states of a two-dimensional cellular automaton in which the state of every cell varies randomly.

When such a transition rule is applied, the state of each cell and, therefore, of the entire system varies completely unpredictably from one cycle to the next (Figure 6.9), which is unlikely to be of much scientific interest. No information is stored in the model about the values of the random numbers used to determine the next state of a cell, thus once a new pattern has been created using this rule there is no turning back: All knowledge of what has gone before has been destroyed. This *irreversibility*, when it is impossible to determine what the states of the CA were in the last cycle by inspecting the current state of all cells, is a common feature if the transition rules are partly stochastic. It also arises when deterministic rules are used if two different starting patterns can create the same pattern in the next cycle.

The behavior of CA is linked to the geometry of the lattice, though the difference between running a simulation on a lattice of one geometry and a different geometry may be computational speed, rather than effectiveness. There has been some work on CA of dimensionality greater than two, but the behavior of three-dimensional CA is difficult to visualize because of the need for semitransparency in the display of the cells. The problem is, understandably, even more severe in four dimensions. If we concentrate on rectangular lattices, the factors that determine the way that the system evolves are the permissible states for the cells and the transition rules between those states.

It might seem that transition rules that are predominantly random would not give rise to interesting behavior, but this is not entirely true. Semirandom rules have a role in adding noise to deterministic simulations and, thus, leading to a simulation that is closer to reality, but even without this role such rules can be of interest.

Example 3: Random Rules and the Speed of Light

Suppose in a two-dimensional CA the rules were

1. If the current state of the target cell and that of all of its neighbors is zero, the state of the cell in the next cycle is zero.
2. If the current state of the target cell or any of its neighbors is not zero, the state of the cell in the next cycle is either 0 or 1 with equal probability.

The behavior of the CA is random at the level of individual cells, but nonrandom on the bulk scale (Figure 6.10).

In a pattern that mimics the movement of an expanding gas, the "filled" region of the simulation spreads to previously empty areas. With the transition rules given, the maximum possible speed at which the front of this disturbance moves is one cell per cycle. This is known as the *speed of light* because no disturbance can propagate across the CA lattice more rapidly.

Example 4: A Random Walk

Suppose that a single cell in a rectangular lattice has the state ON in an environment in which all other cells are OFF. The transition rule is

1. If the current state of a cell is ON, turn the cell OFF and turn ON a randomly selected cell in the von Neumann neighborhood.

This rule will lead to a cell executing a random walk across the grid (Figure 6.11). A random walk is the behavior shown by particles, such as smoke or pollen, undergoing Brownian motion; indeed, observation of the extent of Brownian motion provides a way to estimate the value of Planck's constant.

Probabilistic rules can be introduced into the CA in several ways. In "the speed of light," a rule that contains a random element leads to the propagation of a wavefront. When updating each cell, we could make a random choice between several different rules to introduce stochastic behavior, but we could also determine the future state of the cell by reference not to the

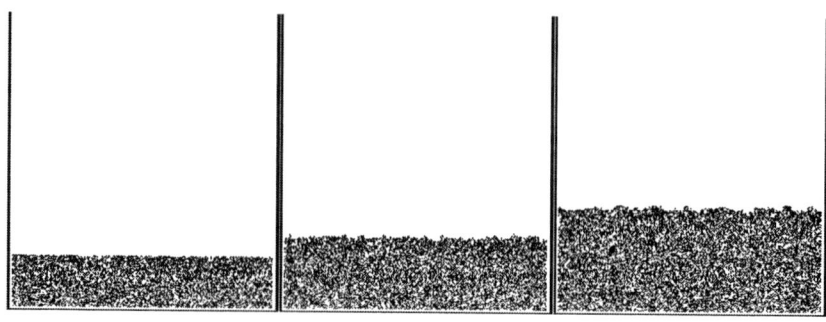

FIGURE 6.10
Random rules giving rise to a propagating wave front.

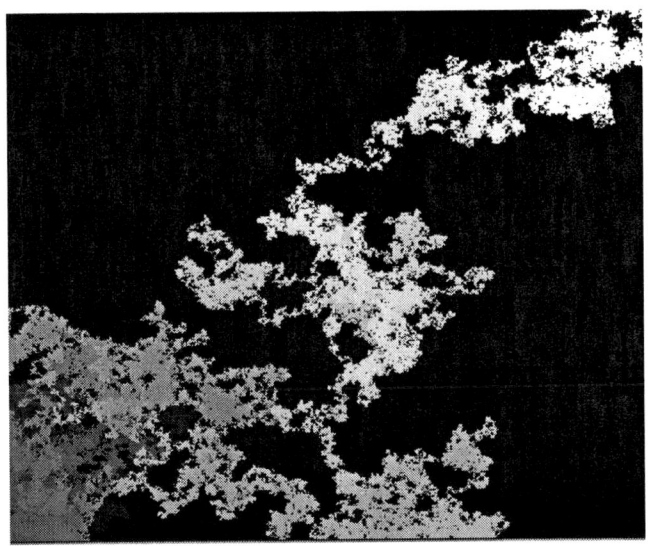

FIGURE 6.11
A random walk by a particle across a square grid. The most recent movements are shown in white.

states of all its equivalent neighbors, but to the state of a randomly chosen neighbor. An alternative would be to move away from an entirely synchronous update of every cell and instead update each cell after some random time step, chosen separately for each cell, has passed.

6.4.2 Deterministic Transition Rules

Deterministic rules, or a combination of deterministic and random rules, are of more value in science than rules that rely completely on chance. From a particular starting arrangement of cells and states, purely deterministic rules, such as those used by the Game of Life, will always result in exactly the same behavior. Although evolution in the forward direction always takes the same course, the CA is not necessarily reversible because there may be some patterns of cells that could be created by the transition rules from two different precursor patterns.

6.4.2.1 Voting Rules

When voting rules are used, the future state of a cell is determined by the number of its neighbors that share a particular state. The target cell is forced to adopt the state shared by the majority of its neighbors, or to adopt a state that is determined by the largest number of neighbors voting for it. The Game of Life is an example of a simulation that, while it does not precisely follow the voting rules mechanism, uses something close to it.

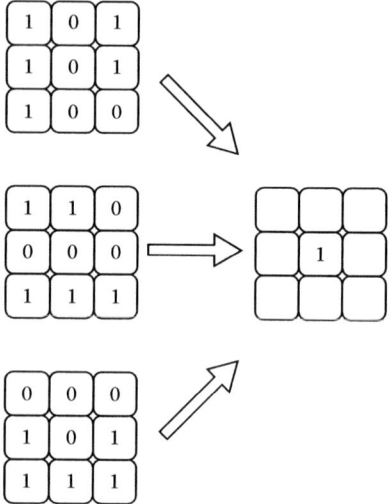

FIGURE 6.12

The many-to-one mapping that may arise in voting rules. Several different combinations of states may give rise to the same state in the next cycle.

Voting rules are irreversible as several different configurations of cells in the neighborhood can lead to the same set of states in the following cycle. In other words, there is a many-to-one mapping of states (Figure 6.12).

Example 5: Separation in a Multiphase System

The application of voting rules leads unavoidably to pattern formation, since its effect is to delete cells in a state that is poorly represented in the local environment and replace them by cells in a state that is common; therefore, cells of the same state collect in neighboring regions of the simulation space. Figure 6.13 shows an example.

Figure 6.13 illustrates a five-state simulation in which the rule is that in every cycle a cell will adopt the state that is most common in its Moore neighborhood. The effect observed is similar to the separation of mutually insoluble liquids on a surface, as small homogeneous regions start to aggregate and grow. Unlike physical systems, though, the boundaries of the phases that form are initially jagged. There is only a slow evolution toward phases in which surface energy is reduced by minimizing the length of the boundary.

We can correct the appearance of these fragmented boundaries by introducing a simple additional rule, which we shall apply in a two-state (0,1) system: If the voting within the neighborhood of a two-state system is split 5/4 or 4/5, the choice of the new state is 1 or 0 with equal probability. The effect of this is illustrated with a two-state CA in Figure 6.14. Coagulation into larger regions that try to minimize their surface area is apparent.

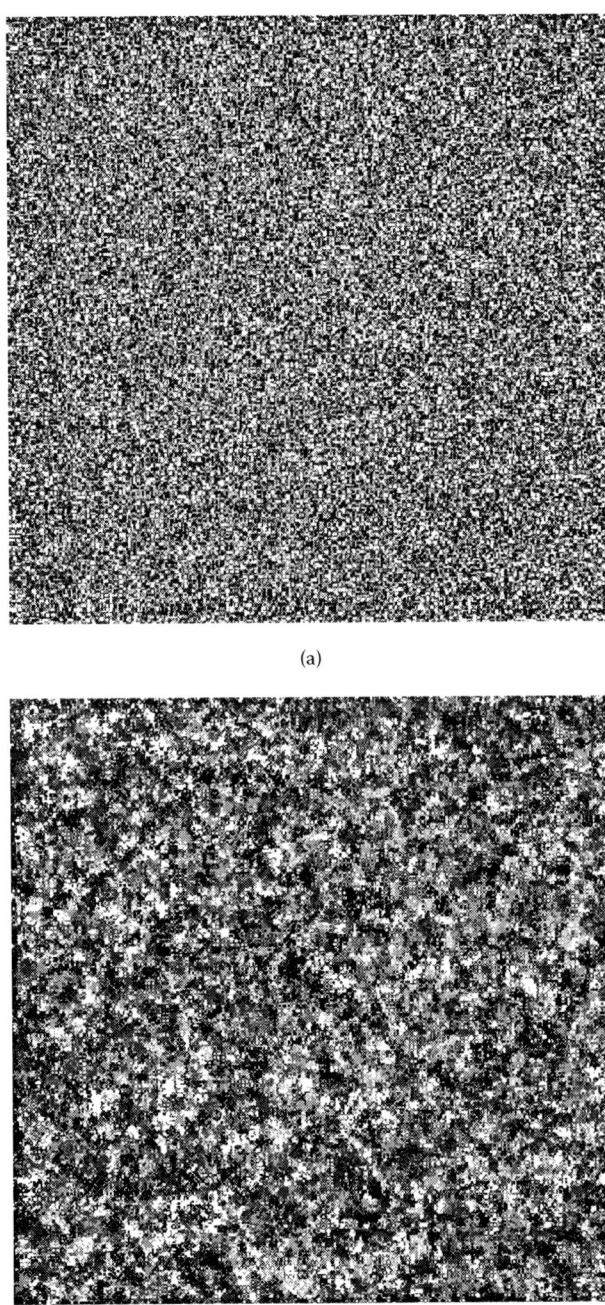

(a)

(b)

FIGURE 6.13
The coagulation of cells in a five-state cellular automata model as a result of the use of voting rules.

(c)

FIGURE 6.13 (CONTINUED)
The coagulation of cells in a five-state cellular automata model as a result of the use of voting rules.

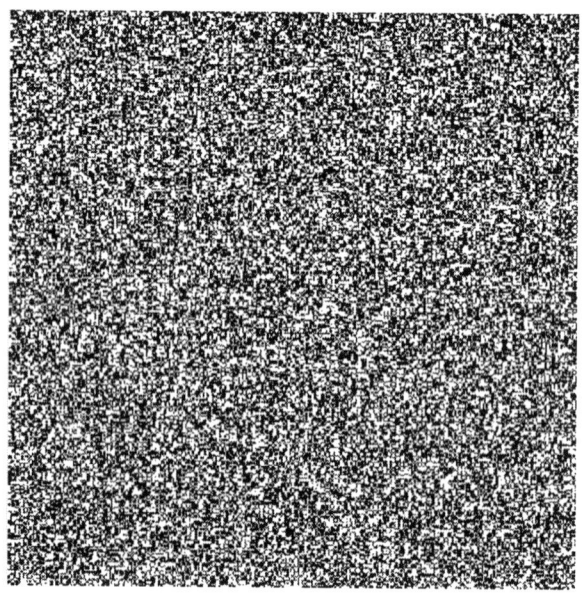

(a)

FIGURE 6.14
Coagulation in a two-state system with random voting.

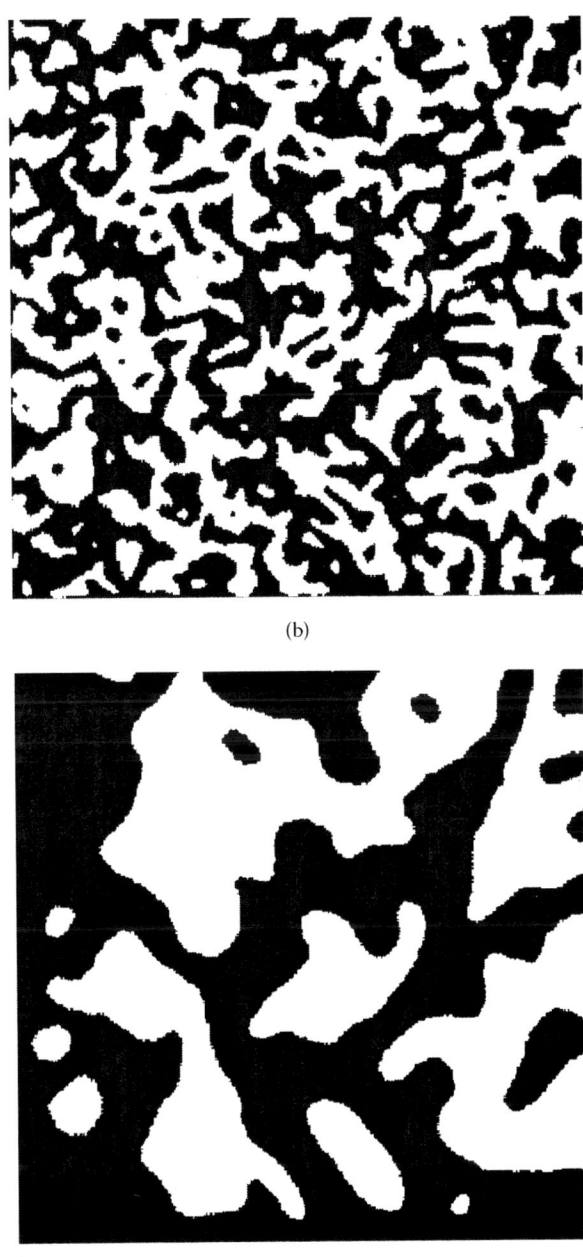

(b)

(c)

FIGURE 6.14 (CONTINUED)
Coagulation in a two-state system with random voting.

This change in the behavior of the simulation is counterintuitive. One would expect the introduction of greater uncertainty in the transition rules to increase the entropy of the system, but it has the opposite effect, leading to a reduction of randomness in the pattern. The reason is that, in the absence of the rule, the state of some cells oscillates between 1 and 0 over large numbers of cycles; hence, the average local pattern changes little with time. The introduction of the random element breaks this oscillation and introduces a surface tension, in which pools of cells of the same state tend to aggregate into larger regions.

CA in which many filled cells execute a random walk but never interact with one another, cannot give rise to stable pattern formation since the cells will move at random forever. However, if cells can interact when they meet, so that one diffusing cell is allowed to stick to another, stable structures can be created. These structures illustrate the modeling of diffusion-limited aggregation (DLA), which is of interest in studies of crystal formation, precipitation, and the electrochemical formation of solids.

In DLA, occupied cells that play the role of particles in solution or in colloidal suspension are allowed to travel across the CA lattice in a random walk. They have a certain sticking probability if they meet a fixed object, which may be another particle that has been immobilized or simply an object that is larger than the moving particle so is traveling at a much smaller rate. If they do stick, they can provide a nucleus for further particles to stick. A CA model may give rise to many different types of growth, depending on the density of occupied cells, the sticking probability, the extent of the random walk, and whether, once stuck, the cells can escape into the body of the liquid again.

Example 6: Diffusion-Limited Aggregation (DLA)

Colloidal aggregation is an application naturally suited to a CA approach because each cell can take on the role of a small volume element in a liquid, which may or may not contain a colloidal particle. Appropriate rules might be of the form

1. A colloidal particle will transfer from one cell to a neighboring cell that contains no particle with a probability q.
2. If both the target cell and a neighboring cell represent colloidal particles, then the two particles aggregate with probability p.

Under the first rule each colloidal particle executes a random walk through the liquid, while the second rule permits aggregation. The effect of running a CA that uses these rules is shown in Figure 6.15.

Extension of diffusion to three dimensions is computationally straightforward, though more difficult to visualize on the pages of a book. In real systems, we would expect that aggregates of particles would begin to settle under the influence of gravity once they reach a size that is too great for them to remain in suspension as a result of Brownian motion. We can introduce this effect by amending rule 1 so that an aggregate of, say, more than two particles has a slightly higher chance of moving in the "down" direction, thus illustrating the effects of precipitation. Similar rules to DLA can be applied to electrochemical deposition, in which metal ions in solution are attracted to, and deposit on, a negatively charged electrode.

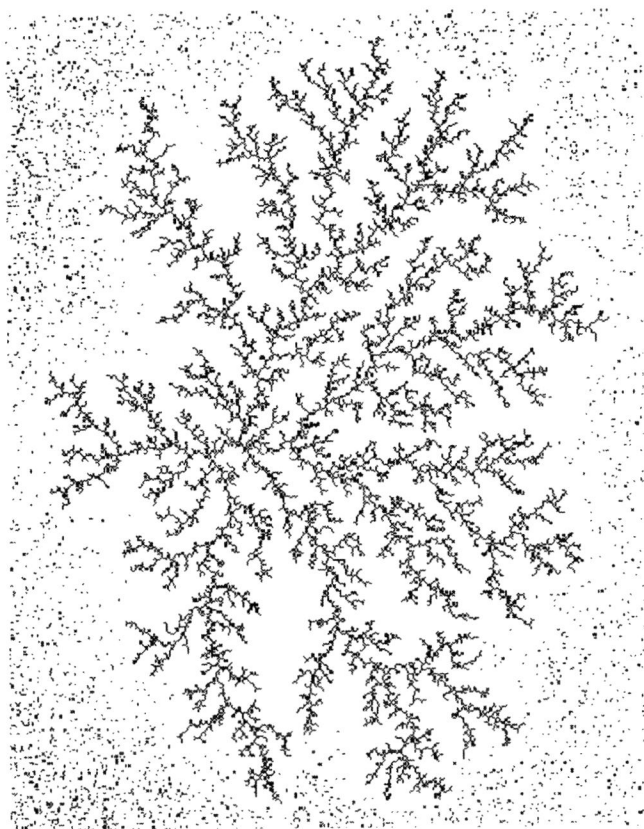

FIGURE 6.15
Diffusion-limited aggregation (DLA).

6.5 Practical Considerations

6.5.1 Boundaries

To be computationally tractable, the CA lattice must be of finite size. Boundaries, therefore, exist at the edge of the lattice.

A cell at the boundary has fewer neighbors than cells in the body of the lattice. Since transition rules depend on the states of neighboring cells, the behavior of cells at the lattice edges might differ from the behavior of cells in the body of the lattice. Figure 6.16 illustrates what may happen in the generation of Sierpinski triangles if we forget to take into account the boundaries in this system.

In some systems the boundaries may be ignored, as the focus of the simulation is static. If a prototype bacterial colony is placed at the center of a large CA simulation, it may grow, but will not move as a whole. If the CA is run for

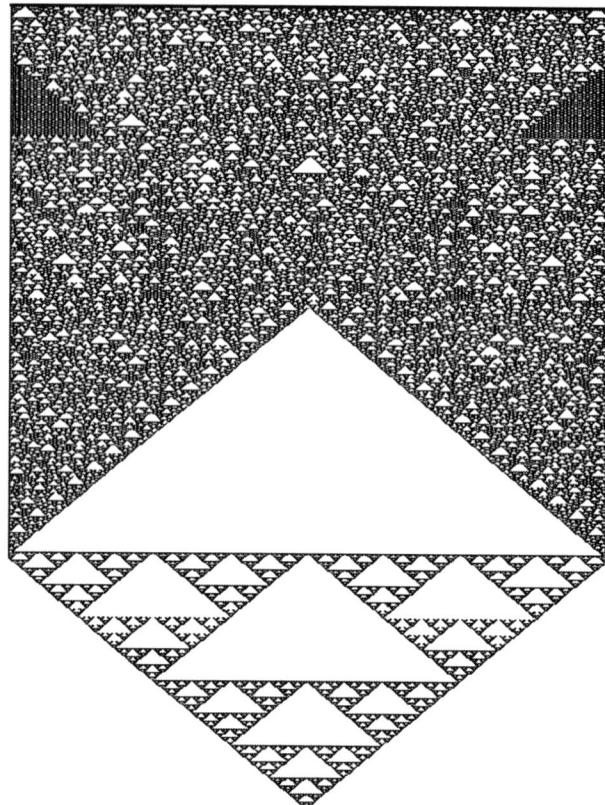

FIGURE 6.16
Cellular automata (CA) using the transition rules given in Example 2, but without special steps being taken at the boundaries to permit development of the pattern to continue.

a very large number of generations, the colony may grow to such an extent that it eventually comes into contact with a boundary, but by starting with a sufficiently large grid, the simulation can be run without the boundaries being of consequence.

If the boundaries might affect the simulation, several approaches are possible. In the first, "brute force" method, we simply try to move the boundaries still farther away, accepting as we do so that this increase in the scale of the CA will slow the execution. This is cheating because we are dodging the issue of how to deal properly with the boundaries, but may be feasible if the key region of the simulation can be positioned close to the center of the lattice so that any anomalies that could arise at the boundaries will not affect the global behavior of the CA. The growth of bacterial colonies falls into this category.

A second way to deal with boundaries is to eliminate them. This can be achieved by wrapping the lattice around on itself as we did with the self organizing map, generating a ring in one dimension or a torus or a sphere in two dimensions. This is an initially attractive approach, but potentially problematical because it may create anomalies when separate patterns that are moving independently toward the left- and right-hand edges, or toward the top and bottom edges of the CA, collide.

Suppose that a CA is used to set up a "predator and prey" simulation. In such a simulation, two populations of animals occupy the same area of land. One population, such as gazelles, multiplies readily and acts as a prey for a population of lions, whose numbers grow more slowly. The lions are entirely dependent on the gazelles for food. If the number of gazelles increases because lush grazing is available, the lions, finding that food is abundant for them, too, will also multiply, though the growth in lion population will lag behind that of the gazelles. Eventually the number of lions becomes so great the gazelle population is decimated, which then leads to a collapse in the lion population. Now, with few lions around, the gazelle population will recover, to be followed later by the lion population. This is illustrated by the periodic fluctuations shown in Figure 6.17.

In a real population, the dynamics are complicated by territorial disputes between predators. A pride of lions may establish and defend a territory against other lions by driving away intruders. A "repulsion" term that keeps members of different packs apart can be included in a CA simulation by allowing cells that contain lions to interact over large neighborhoods. However, if the two-dimensional simulation "wraps around" on itself, the region that a pack of lions controls may become mirrored, so that what appear to be intruders encroaching into one side of the territory are, in fact, members of the same pack positioned at the other side of the territory that have been

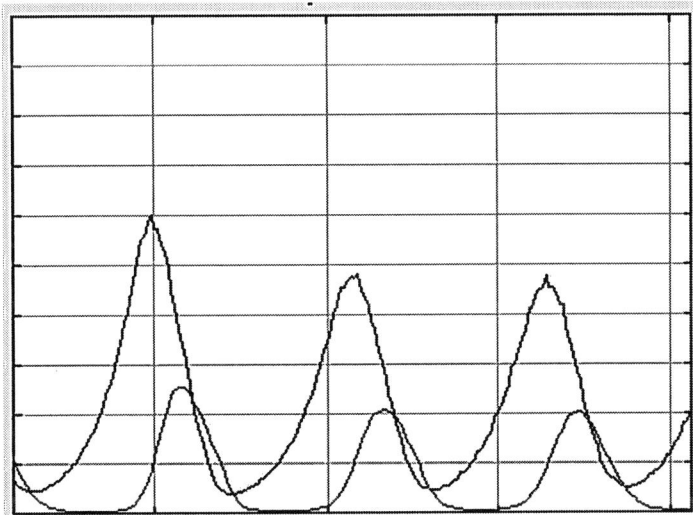

FIGURE 6.17
A predator and prey simulation. The population of gazelles (the higher line) has been scaled down since, in reality, it would be many times that of the lions.

wrapped around by the algorithm. Such effects need to be anticipated so that they can be dealt with appropriately.

The wrapping around is similar to, but subtly different from, the use of periodic boundary conditions. Such conditions are used in molecular dynamics simulations, in which a finite number of particles are allowed to move under the influence of classical forces within a "simulation box" (Figure 6.18). If the movement of a particle leads to it escaping from the box through one of the faces, a copy of it is generated moving into the box through the opposite face. The incoming particle is identical to the one lost from the box in all respects, other than its position. The central box in the simulation, therefore, is surrounded by eight copies of itself, thus no particle in that central box can find itself at an "edge."

Much the same sort of procedure can be used in the CA in an approximation to an infinite grid. If the disturbance due to an evolving CA reaches the boundary, a unit possessing the same state as the one at the boundary is created just outside the opposite boundary.

A third way of dealing with boundaries is to use *infinity pool rules*. These are rules whose purpose is to absorb silently any CA cells that become active at the edges of the simulation. The aim is that these cells will drop over the horizon and their removal will have no effect on the global behavior of the

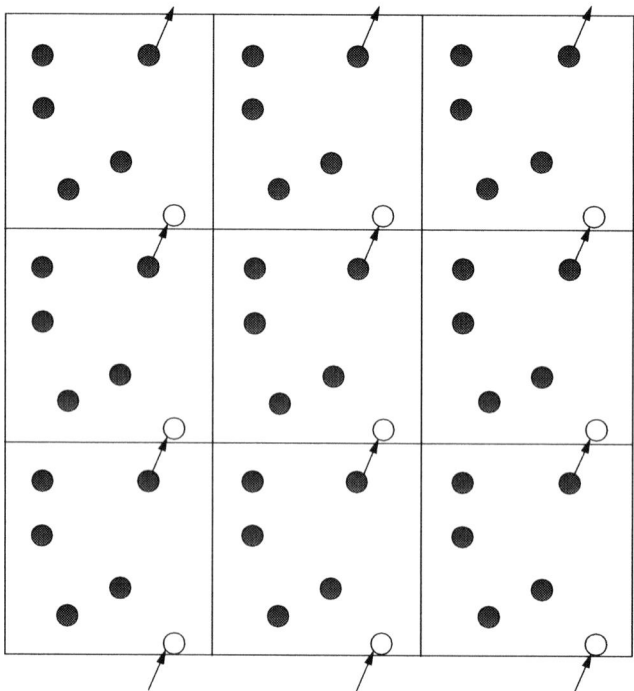

FIGURE 6.18

A cellular automata simulation box using periodic boundaries.

simulation. This can be accomplished successfully only with certain types of transition rules.

A further possibility is that all cells on the boundary itself are given a fixed state, which cannot be changed. This is somewhat analogous to the boundary conditions that may have to be satisfied when a set of differential equations is solved. This is the approach taken in Figure 6.16 in which all cells at the boundary are forced to maintain a state of 0 at all times.

6.5.2 Additional Parameters

In "pure" CA, each cell can adopt one of a small number of discrete states. However, it is possible to loosen this limitation on the number of states and permit the state of a cell to include the values of some continuous variables. If the simulation was of a reacting liquid, the state of a cell could contain details of the temperature of the liquid in the cell, its direction of motion, the concentration of all chemicals within it, and so on. The state of the cell may also be subject to universal rules that apply equally to every cell, e.g., gravity that pulls cells downward; real time, which ages the contents of the cells, moving them toward a dying state; or a level of illumination, which affects the chance that they will be photochemically excited, or to local rules, such as a local electric field.

In a similar vein, we may relax the requirement that the same rule must apply to every cell. We, in fact, have already done this, for example, in the simulation shown in Figure 6.10, in which cells at the walls of the container were unable to move into or through the wall, so were treated differently from those in the body of the liquid.

If not all cells are to be treated in the same way, this can be accomplished in a simple fashion by including an additional state for each cell that specifies what combination of rules will apply to the cell.

6.6 Extensions to the Model: Excited States

In an interesting variant of the CA model, cells can adopt "excited states." This has been used to model the spatial waves observed in the Zaikin–Zhabotinsky reaction.

In this model, run on a square grid, each unit can adopt one of N states, 0, 1, 2, 3,..., (N − 1); states other than 0 are "excited states." The neighbors are those cells that share an edge with the target cell (a von Neumann neighborhood). The transition rules can be divided into two types: first we have reaction rules:

1. If the current state of a cell is k, where $0 < k < N - 2$, the state of the cell in the next cycle is $k + 1$.

2. If the current state of a cell is $N - 1$, its state in the next cycle is 0.

These rules mimic the spontaneous first order reaction of a chemical (a cell in state x) that transforms into a different chemical (the cell in state $x + 1$).

Diffusion rules relate the future state of the target cell to those of neighboring cells. This mimics a second order reaction:

3. If the current state of a cell is 0 and one of its neighbors is in an excited state, the next state of the cell is 1.

4. If the current state of a cell and that of all its neighbors is 0, the cell remains in that state.

Figure 6.19 shows an example of the sort of pattern that results from the application of these rules, with a disturbance moving out from a central region in a sequence of waves with constant wavelength (equal to the number of accessible states). In a lattice in which most cells were given an initial state of 0, a small number in a block at the center were given an initial state of 1. A small amount of random noise has been added to the rules by applying them at each cell with a probability of 0.98 and the effect of this can be seen as the waves gradually become more disorganized and diffuse; eventually they will dissolve into the background noise. In this simulation, infinity pool boundary conditions mean that cells at the edge of the simulation are lost, thus the simulation has the appearance of being run on a lattice of infinite size.

With careful choice of rules, excited state CA can produce convincing simulations of the spirals and waves that can be produced by oscillating reactions.

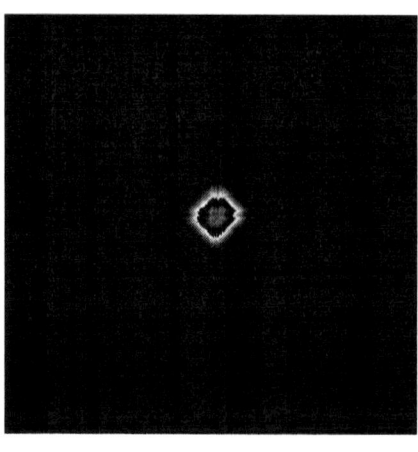

(a)

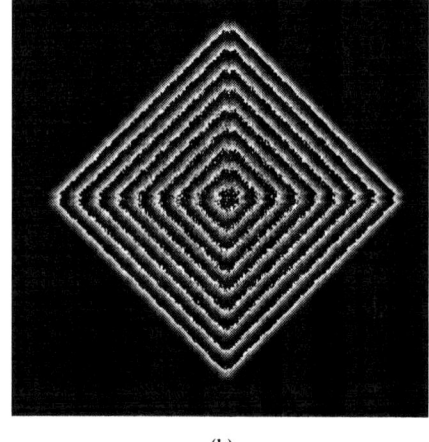

(b)

FIGURE 6.19
A disturbance in an excited state cellular automata (CA).

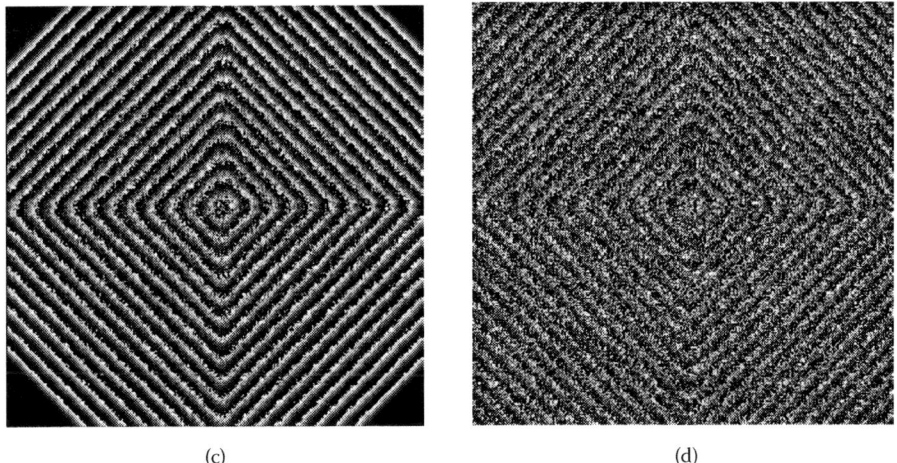

(c)   (d)

FIGURE 6.19 (CONTINUED)
A disturbance in an excited state cellular automata.

6.7 Lattice Gases

A lattice gas model is a macro-level simulation of a fluid that uses micro-level rules. The dynamics of a CA model are determined by a local transition rule that defines the state of the cell in the next cycle in terms of its current state and that of its neighbors. When the local rule is deterministic for a given starting state, only one evolutionary path is possible. By contrast, a Lattice gas model is probabilistic. A second difference between a lattice gas and the cellular automata model is that in the CA the number of "empty" cells is rarely constant. In the Game of Life, certain arrangements of live cells generate an unending stream of "gliders," which drift away from the simulation at constant speed, thereby increasing the number of live cells indefinitely. In a lattice gas, however, lattice points are occupied by particles that have a physical meaning — they might be the molecules that constitute a fluid perhaps — so the number of particles is fixed for the duration of the simulation.

The principle of a lattice gas is to reproduce macroscopic behavior by modeling the underlying microscopic dynamics. In order to successfully predict the macro-level behavior of a fluid from micro-level rules, three requirements must be satisfied. First, the number of particles must be conserved and, in most cases, so is the particle momentum. States of all the cells in the neighborhood depend on the states of all the others, but neighborhoods do not overlap. This makes application of conservation laws simple because if they apply to one neighborhood they apply to the whole lattice.

Second, there must be local thermodynamic equilibrium so that the energy of the system does not fluctuate. Finally, the scale of the micro and macro simulations must be significantly different.

A lattice gas consists of a number of particles of equal mass moving through a space that contains a fixed imaginary two- or three-dimensional lattice. Each site on the lattice may be occupied or empty. Every particle has either zero velocity or unit velocity in one of the directions specified by the von Neumann neighborhood. Thus, in two dimensions, every particle will be moving independently in the $+x$, $-x$, $+y$, or $-y$ direction. Each lattice site may contain, at most, one particle with a given velocity; since each particle may be stationary or moving, the maximum occupancy for each two-dimensional lattice site is five. In the cycle updating, every particle moves in the direction specified by its velocity. If a collision occurs because too many particles are trying to enter the same lattice site, the velocities of the colliding particles are deflected by 90°.

In the macroscopic limit, this model running on a square lattice tends to the Navier–Stokes equation, but a hexagonal lattice rather than a square lattice gives a simulation that is more scientifically justifiable and permits the determination of a range of parameters, such as transport coefficients.

Lattice gas models are simple to construct, but the gross approximations that they involve mean that their predictions must be treated with care. There are no long-range interactions in the model, which is unrealistic for real molecules; the short-range interactions are effectively hard-sphere, and the assumption that collisions lead to a 90° deflection in the direction of movement of both particles is very drastic. At the level of the individual molecule then, such a simulation can probably tell us nothing. However, at the macroscopic level such models have value, especially if a triangular or hexagonal lattice is used so that three-body collisions are allowed.

6.8 Applications

Because the CA are well suited to the description of temporal phenomena, it will be no surprise that an early application of CA in science was in hydrodynamic flow. Wolfram, who has been an enthusiastic advocate of CA, has used them successfully in the simulation of turbulent flow — a demanding area for simulations of any type. Indeed, Wolfram's work covers a range of topics in physics and uptake of the method among scientists owes much to his work.[4]

The CA is a powerful paradigm for pattern formation and self-organization, an area of increasing importance in nanotechnology. CA have not yet been extensively used in nanotechnology applications, though their use in quantum dot applications is growing.

A further area of interest is bacterial growth. The form that a bacterial colony adopts when growing is affected by several factors: the life cycle of the bacterium, the level of nutrient in the immediate environment, the rate at which nutrient diffuses into that environment to replenish that used by the colony, the temperature and the presence of pollutant, among others. Studies in this area are of practical interest because of the increasing use of bacterial remediation, in which industrial sites polluted with toxic organic matter can be cleaned in a "green" process by injecting a mix of bacteria and nutrient into the soil. With a suitable choice of bacteria, pollution can be brought down to levels low enough that the land can be used for residential housing, comparatively rapidly and cheaply. At present, most use of bioremediation is based on empirical evidence, but CA models show potential in this area and are well suited to such studies.

A natural progression from using CA to model bacterial growth is to model tumor growth and the development of abnormal cells. There has been considerable work on this topic. Features such as cell mutation, adhesion, layered growth, and chemotaxis can readily be incorporated into a CA model.[5] Deutsch and Dormann's book provides a useful introduction to this area.[6]

Until recently, there have been few applications of three-dimensional CA models because of the substantial computational demands that these make.[7] For even fairly simple processes, one might want to use a lattice whose dimensions are at least $100 \times 100 \times 100$, but even this may be too small for meaningful results if the behavior of the phenomenon being simulated is complicated. Simulation of a lattice in which there are more than one million cells has not been feasible for long, so only now are some really large simulations starting to appear.

Reaction-diffusion systems can readily be modeled in thin layers using CA. Since the transition rules are simple, increases in computational power allow one to add another dimension and run simulations at a speed that should permit the simulation of meaningful behavior in three dimensions. The Zaikin–Zhabotinsky reaction is normally followed in the laboratory by studying thin films. It is difficult to determine experimentally the processes occurring in all regions of a three-dimensional segment of excitable media, but three-dimensional simulations will offer an interesting window into the behavior of such systems in the bulk.

6.9 Where Do I Go Now?

Kier, Seybold, and Cheng[8] have described the application of CA to a number of chemical systems, including the formation of interfaces and chemical kinetics. The models they describe are comparatively simple and their aim principally is to introduce the use of CA into the undergraduate practical course.

By contrast, Deutsch and Dormann[5] cover the use of CA in much greater depth and with a good deal of relevant theory. Containing a large number of examples, this book provides a very helpful introduction to those who wish to study the topic more deeply.

6.10 Problems

1. Bacterial remediation

   CA models are well suited to the modeling of bacterial remediation. The first step is to build a model of a colony of bacterial cells that can reproduce and grow. In doing so, you will need to take into account several factors such as:

   1. The need of each cell to consume a certain amount of nutrient per time step to stay alive.

   2. Diffusion of nutrient from the surrounding environment into the region of the bacterial cells to replenish that consumed by the bacteria.

   3. The time required before a cell is mature enough to divide and create a new cell.

   4. Any dependence of the rate of aging of the cells on the level of nutrient; in particular, the possibility that a bacterial cell may enter a state of "suspended animation" if the nutrient concentration falls below a certain level.

   Construct a two-dimensional CA to model bacterial growth and investigate the conditions under which your model will give rise to:

   1. Rapid, regular growth leading to a circular colony.

   2. Irregular growth leading to a fractal colony.

   3. Sparse growth leading eventually to the colony dying.

   Also investigate the behavior of the colony in the presence of a non-uniform concentration of pollutant. Assume that the colony can metabolize the pollutant and that the presence of the pollutant (1) inhibits or (2) promotes growth of the colony.

   You may wish to compare the results of your simulation with Ben-Jacob's experimental results.[9] Virtually all of the behavior described by Ben-Jacob and co-workers can be effectively modeled using CA.

References

1. Field, R.J., Noyes, R.M., and Koros, E., Oscillations in chemical systems. 2. Thorough analysis of temporal oscillation in bromate-cerium-malonic acid system, *JACS*, 94, 8649, 1972.
2. Gardner, M., *Wheels, Life and Other Mathematical Amusements*, Freeman, New York, 1993.
3. Wolfram, S., Statistical mechanics of cellular automata, *Revs. Mod. Phys.*, 55, 601, 1983.
4. Wolfram, S., Computation theory of cellular automata, *Commun. Math. Phys.* 96, 15, 1984.
5. Chuong, C.M., et al., What is the biological basis of pattern formation of skin lesions? *Exp. Dermatol.*, 15, 547, 2006.
6. Deutsch, A. and Dormann, S., *Cellular Automaton Modeling of Biological Pattern Formation*, Birkhauser, Boston, 2005.
7. Wu, P.F., Wu, X.P., and Wainer, G., Applying cell-DEVS in 3D free-form shape modeling, *Cellular Automata, Proceedings Lecture Notes in Computer Science (LNCS)*, Springer, Berlin, 3305, 81, 2004.
8. Kier, L.B., Seybold, P.G., and Cheng, C-K, *Modeling Chemical Systems Using Cellular Automata*, Springer, Dordrecht, 2005.
9. Ben-Jacob, E., Cohen, I., and Levine, H., Cooperative self-organization of microorganisms. *Adv. Phys.* 49, 395, 2000.

7

Expert Systems

Expert systems appear in some ways to be the "classic" application of Artificial Intelligence. An expert system is a kind of personal advisor that engages the user in a conversation with the aim of providing help on a specialist topic. In sophisticated systems, the interaction between the expert system (ES) software and the user may be so natural that the user almost forgets that it is a computer rather than a human that is holding up the other end of the conversation. As developers of an ES usually do their best to create software that can participate in "intelligent" conversations, in order to enhance the user's confidence in the system, expert systems can seem the most human and friendly side of AI.

Many of the earliest AI programs were primitive expert systems. An example was Winograd's SHRDLU system, which was developed in the early 1970s. SHRDLU knew basic facts about a virtual world that was populated with children's blocks of different shapes and colors. It also possessed a rudimentary understanding of relative position and of simple grammatical rules. SHRDLU was able to generate instructions to a user (who might have a real set of children's blocks to play with, which could be used to test the behavior of the system) to rearrange the blocks, such as:

```
Place the green cube under the red pyramid.
```

SHRDLU knew enough about the properties of blocks and what could be done with them, including the meaning of phrases such as "on top of" and "pick up," that it was able to determine not only whether the move of a block that it or a user might propose was feasible, given the current positions of all the blocks, but was also able to respond correctly to queries from the user about the current location and relative position of the blocks. This led some researchers to argue that in a sense the program showed "understanding." However, this view suggests that understanding can be measured by observing behavior, which is rather a weak requirement for intelligence* and few in the AI community would now argue that SHRDLU understood at all.

---

* The way in which birds flock together seems to imply some sophistication in both decision making and communication, but in fact, as we saw in Chapter 5, flocking can arise from the application of a few very simple rules and requires no intelligence. This is not to suggest that the mechanism by which real birds flock is through application of the three rules mentioned in that chapter, but does tell us that what we perceive as intelligent behavior need not be evidence of deep thinking.

Programs such as SHRDLU were written at a time when some people felt that the long-term goal for AI-based software should be the development of a tool that could solve a problem that was completely new to it, working virtually "from scratch." This is a hugely ambitious goal, and even now, several decades after the emergence of the first primitive expert systems, computer scientists are far from achieving it. Though researchers have made rapid advances in building multipurpose AI-driven robots equipped with sensors that provide good vision, and endowing the computers that run them with advanced reasoning, no silicon being comes near to matching the performance of humans.

In fact, expert systems are almost the exact opposite of a machine that could solve a problem in some arbitrary area from scratch. They focus on a single specialist topic and know almost nothing outside it. They make up for this blinkered outlook on life by possessing a knowledge of the topic that few humans could match and then complement it with a reasoning ability that mimics that of a human expert, allowing them to make deductions from data supplied by the user, so that their naming as an "expert" system is appropriate.

7.1 Introduction

Expert systems are problem-solving tools, so are very practical in nature. Each ES is written to meet the need for advice in a specific area. Software whose role is to provide practical advice, be it advising a worker on the choice of a pension plan or helping a scientist to interpret nuclear magnetic resonance spectra, must contain at least two components: a database that contains the relevant background information, and rules that define the context of the problem and provide a framework that allows the software to manipulate data. Given both components, computers can manage many tasks that require specialized knowledge, such as providing investment advice or playing a competent game of chess.

Chess intriguingly combines simplicity in its rules and goal with complexity in its tactics and play, so it is a favorite topic of many in the AI community. An average Computer Science undergraduate could write an ES whose role was to suggest the next move to a novice chess player during a game. As both the rules of chess and its aim are simply stated and unambiguous, it is straightforward from the current state of the playing pieces on a chessboard to work out all the possible moves. For each legal move, the rules of chess can then be applied a second time to work out an opponent's next move. Each combination of a user move followed by an opponent's response can be assessed to see whether the user's move is advantageous, perhaps as judged by whether the user gains an opponent's piece.

A chess-playing ES could go a little way beyond this simple reliance on knowledge of the legal moves in the game. Some of the simpler strategies that players adopt can also be expressed as rules, so these too could be incorporated into the software, but an approach that relies on an exhaustive evaluation of every possible move cannot be pushed far. The number of possible sequences of moves from a given starting position grows very rapidly; this is an example of *combinatorial explosion*, which is very common in science. An important scientific example is the extremely rapid growth in the number of possible molecules that might act as drugs as the number of atoms in the molecule grows.*

A computer that hoped to compete effectively against international Grand Masters could not possibly rely on an exhaustive search strategy, since no Grand Master plays chess just by applying simple strategies and rules. An ES could, and often does, form the basis of recreational chess-playing software, but any chess-playing software that has ambitions to play at a higher level requires far more advanced software programming.

Fortunately, not all problem-solving tasks in science require the degree of lateral and inventive thinking needed to play championship-level chess. If the area of knowledge in which the computer is required to make deductions and provide advice is not unmanageably large, computers can not only compete with human performance, they can surpass it.

Expert systems are a widely used application of AI. In science, suitable domains for deploying an ES might be the methods used to analyze aqueous samples for heavy metals or the interpretation of images of abnormal cells under a microscope. In both applications, a degree of expert knowledge is required for competent work. More widely, they are a crucial component within many search engines, both general and scientific, and are a part of most text-based adventure games, where a sophisticated ability to interpret natural text responses is essential. It is an open question whether expert systems really are intelligent, but a well-designed system may appear to be so to the user, and they are specifically constructed to follow lines of reasoning similar to those used by human experts. They remain an area of active research within AI.

7.2 An Interaction with a Simple Expert System

Experts possess specialized and often valuable knowledge. In their area of expertise, they would expect to be able to solve problems that would defeat most nonexperts and to be able to solve problems faster than the average

---

* It is estimated that the total number of compounds that might in theory have potential as drugs is of the order of 10^{40}. Although extremely large, this number is not effectively infinite as almost all drugs of therapeutic interest are quite small molecules.

person in the field. The knowledge that a human expert possesses may be sufficiently specialized that it cannot readily be transferred to other people. In addition, experts in some fields may be so rare that it is prudent to take precautions against the possibility that their knowledge might be lost if the expert retires, dies, or runs off to join one's competitors. Therefore, it is not surprising that computer scientists long ago spotted an opportunity; if a human expert is not always available when needed, it might be possible instead to access expert knowledge through a computer keyboard if a silicon equivalent could be constructed.

Expert systems are the most important application within a field known as *knowledge-based systems.* This broad area includes tools that manipulate symbolic knowledge as well as those that can handle only statistical or mathematical data. At the dumb end of the spectrum, the simplest knowledge-based systems have a limited reasoning ability, or none at all, and consist of little more than a large database of information about some particular topic, accessible through a simple query system. They behave like an electronic card index that can instantly reveal which team won the Stanley Cup in 1980.* A true expert system has reasoning ability; this allows it to interact with the user in what appears to be an intelligent manner within the domain of its expertise, and take advantage of rules and relationships in its database to make deductions and provide advice in response to user input.

Example 1: Holidays

Here is an imagined conversation between an ES and a scientist:

```
Biologist: I've been working too hard. I need a break.
Expert System: Have you chosen where to go?
B:  No
ES: Do you like warm weather?
B:  Yes
ES: Do you like to lie on a beach?
B:  No
ES: Do you like good food and wine?
B:  Definitely
ES: France or Italy would be suitable. Do you like art?
B:  Yes
ES: I suggest Florence for your holiday.
```

This is a fairly mundane exchange. You would not need years of experience before you were able to decide for yourself where you should go on a holiday, so there is little commercial potential for an expert system advising overworked biologists on how to spend their summer break. Nevertheless, this simple interaction gives us an early insight into how an expert system appears to the user. The system has provided one half of a conversation, eliciting information from the user through a sequence of focused questions as

---

* New York Islanders.

the options of a holiday location are narrowed down. The user is eventually offered a recommendation based on the system's analysis of the information that the user has provided, combined with whatever information the software already contained.

The interaction between a user and the ES resembles the kind of conversation that might take place between a nonspecialist and a human expert.* Within the task domain, not only should the conversation proceed along much the same lines as it would if two humans were talking, but an ES should also approximately mirror the decision-making processes of a human expert and at least match their level of competence. Some researchers have taken the view that it is not sufficient that the ES should only give the *appearance* of working like a human; they argue that emulation requires that the software should as far as possible actually follow the same sort of deductive pathway that a person would take. This is a far stiffer requirement than that an ES should appear from the outside to be reasoning "along human lines" and is rather difficult to implement, given that our knowledge of how the human brain operates is still fragmentary, thus making impossible a proper judgment of the degree to which the two types of reasoning match. Nevertheless, if we at least try to reproduce human reasoning in the ES, this may make it more likely that the reasoning of the system will be robust and predictable.

Notice how the conversation above reveals that the system has made an attempt to reason logically by finding out what the user considers to be important before it makes any recommendations. This approach is much more productive than a random search in which the interaction might run as follows:

```
Biologist: I need a break
Expert System: Do you want to go to Orlando?
B:   No
ES: Do you want to go to Mindanao?
B:   No.
ES: Do you want to go to Novosibirsk?
B:   !!!
```

with possible holiday destinations extracted by the "expert" system from its database and thrown out at random. An exhaustive search is no better. In this approach, the system might first propose all the locations it knows about that start with A, then suggest all those that start with B, and so on. It is clear

---

* Although most expert systems are designed to advise fairly unsophisticated users, some act instead as assistants to established human experts; DENDRAL, which we introduce shortly, falls in this category. Rather than replacing the expert, these systems enhance their productivity. Assistants of this sort have particular value in fields, such as law or medicine, in which the amount of potentially relevant background material may be so large that not even a human expert may fully be cognizant with it.

that neither a random nor an exhaustive approach would provide a practical way to provide advice.

7.3 Applicability of Expert Systems

Scientists are pragmatists, so an ES that provides advice on chemical or biological problems will be of much greater value than one that can balance the merits of Florida against those of the Costa del Sol. Expert systems address real-world problems in a well-defined, generally narrow *task domain*; this defines the field in which the system has knowledge of value. Typical task domains include:

- The selection of analytical procedures for the determination of toxic metals in polluted river water.
- The design of multivariate experiments.
- The analysis of near infrared reflectance spectra of foods.
- The diagnosis of illness from patient symptoms.
- The choice of the most appropriate glove materials to provide protection when handling toxic chemicals.

SHRDLU was an example of a system that operated in a well-defined task domain and this software was an important steppingstone in the development of AI programs, enabling computer scientists to better understand how to construct expert systems and how to handle the user–computer interaction. However, the market for software that can rearrange children's blocks is limited and the development of the first ES in chemistry was a far more significant milestone in science.

DENDRAL was developed at Stanford University from the mid-1960s onward. Provided with a knowledge of mass spectrometry fragmentation rules, DENDRAL took as input the mass spectra of organic compounds and used that input to help the user derive plausible structures for the samples. The fragmentation pattern of an organic molecule is often complex, since a molecule with energy well above its dissociation limit may fragment in many different ways. Furthermore, the route that fragmentation takes may depend on the conditions under which the spectrum is measured, for example, the type of ionization used, so the area is a fruitful one for the provision of expert advice.

Rather than a single program, DENDRAL was a group of interrelated programs that took turns to participate in a discussion with the user, proposing possible molecular structures for the user to consider and removing from consideration any structure that the user deemed not to be chemically stable, incompatible with the observed fragmentation pattern, or otherwise unreasonable. Though limited in the complexity of the spectra that it could analyze, DENDRAL had a knowledge of mass spectrometry fragmentation patterns superior to that of many of its users, so was widely used by chemists to help in ascribing structure to molecules. On its own, DENDRAL did much to persuade scientists of the potential of expert systems.

Like all early expert systems, DENDRAL and SHRDLU required exact knowledge to function. The way that expert systems work depends on whether the knowledge that they manipulate is exact ("The temperature is 86°C") or vague ("The temperature is high"). We shall first consider how an ES can use exact knowledge to provide advice. Methods for dealing with ill-defined information form the topic for the next chapter, which covers fuzzy logic.

7.4 Goals of an Expert System

In its field of expertise, an expert system should be able to meet the following objectives:

- Provide advice that at least matches in reliability the advice that would be provided upon analysis of the same data by a human expert.
- Interact with the user through a natural language interface.
- When required, provide justification for, and an explanation of, its decisions and advice.
- Deal effectively with uncertainty and incomplete information.

Unlike methods, such as neural networks, which can tackle any one of a wide range of tasks provided that they are given suitable training, each ES can work in just one field, but the range of possible fields is large. Working expert systems provide instruction for car mechanics, remote medical diagnoses for nurses, and help in the diagnosis of faults in machinery. They are valuable in control, interpretation, prediction, and other types of applications; in each case though, an ES must be built specifically for the desired application.

As Table 7.1 shows, the advantages of a well-constructed expert system are considerable, but anyone choosing to build an ES faces significant challenges. The goal is to construct a software tool that emulates human reasoning, is fast in operation, reliable, and at least matches the performance of human experts in an area where few humans are experts. This is asking a lot. No ES is yet able to match humans across a large and random selection of areas, but a well-constructed ES can offer a passable impression of a human expert in a particular area.

TABLE 7.1

Some Advantages and Disadvantages of Silicon-Based Experts

Advantages of Expert Systems	Disadvantages of Expert Systems
Low operational cost, once built	Expensive and time-consuming to create
Reliable and reproducible; will not make mistakes if correctly programmed	Narrow focus; can only provide advice on a single area or a small number of closely related areas
Almost unlimited memory	Require programming, so quality of the advice is related to quality of programming
Not affected by external factors, such as dislike of user or having had a bad night; always available; expert systems do not have strops	Boring
Able to explain reasoning	Unable to use hunches or intuition
Readily reproduced; unlimited number of copies of an expert system can be made once the first one has been constructed	Hard for an ES to learn on its own
Consistency — all identical queries will be handled in precisely the same way.	ES has only a limited ability to recognize when a query lies outside its area of expertise and a poor ability to handle this
Breadth — by combining the knowledge of many experts, a broad knowledge base may be assembled	No creative ability

7.5 The Components of an Expert System

The components of an ES are shown in Figure 7.1. The key ingredient is factual knowledge; the system is constructed around a repository of information and rules relevant to its area of expertise known as the *knowledge base*. The knowledge base is not accessed directly by the user. This task is handled by an *inference engine*, which is responsible for retrieving, organizing, and making deductions from data that are relevant to the problem. Within the knowledge base there may be a library of *case-based data* that offers a window onto relevant examples from the past. The *user interface* is the route through which user and software converse. Finally, if the system is still growing or being tested, a *knowledge-base editor* will be available to provide an opening into the system through which a human expert can update and improve it as needed.

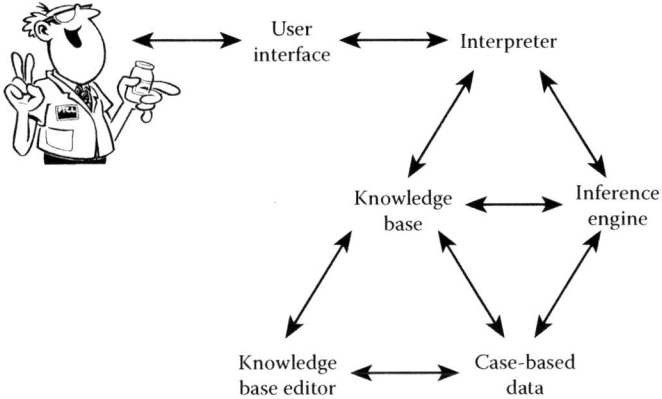

FIGURE 7.1
The components of an expert system.

7.5.1 The Knowledge Base

The knowledge base is a storehouse of facts, rules, and heuristics that relate to the domain in which the system provides advice. As far as possible, it contains everything that might be useful to the ES in solving problems.

The *knowledge domain* to which all this information relates is a subset of a larger body of information known as the *problem domain*. A problem domain might consist of the procedures required to perform a complete analysis of samples of river water, including determination of their heavy metal content, level of pesticide residues, the biochemical oxygen demand, and the amount of dissolved oxygen. Because a problem domain that covers all aspects of the analysis of river water might be too unwieldy or complex for a single ES to cover adequately, a practical system in this area might focus on only one aspect of this subject, such as the analysis of heavy metals (Figure 7.2).

7.5.1.1 Factual Knowledge

Much of the information in the knowledge base consists of *factual knowledge*. This is information that is widely accepted to be correct by those with expertise in a particular field, but it need not consist only of expert knowledge. Particularly in scientific applications, some of the information that an expert will use to make deductions may be of far broader use than just the area of expertise of the ES, but is nevertheless required for the system to operate. The factual data in the knowledge base, therefore, typically ranges from rather general, widely-accepted "textbook" knowledge to highly specialized information.

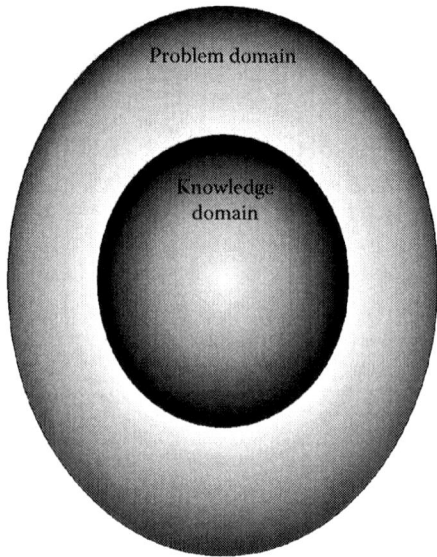

FIGURE 7.2
The knowledge domain is an area within what is usually a much broader problem domain.

```
pV = nRT
The American Civil War lasted from 1861 until 1865
The world is round
π = 3.14159
Streetlights show d-line emission from excited sodium atoms at
wavelengths of 589.76 nm and 589.16 nm.
```

In most expert systems, only small sections of the information in the knowledge base are *causal*. Causal knowledge consists of statements of facts and relationships, linked to information that explains *why* these facts are true. The "why" part is often not required for the ES to function effectively and, therefore, not included in a working ES.

```
If the battery is old and flat, the car won't start.
```

This is useful, albeit basic, information for an ES working in the domain of sick cars. If you are employed as a car mechanic, you are not likely to benefit from, or even be interested in, the presence within the knowledge base of details about how lead sulfate batteries degrade with age and how any diminution in concentration of the active materials in the battery can be related through the Nernst equation to a reduction in its voltage. In the context of car repair, all that you need to know is that the battery is dead and the car will not start, so it will need replacing or recharging.

Even though the domain in which an ES is applied may be clearly defined, the contents of the knowledge base may extend a little beyond the immediate

TABLE 7.2

Typical Nondomain Data in an Expert System Based in a Laboratory

Type of Information	Data
Instrument availability	Types of analytical instruments within the laboratory and their current status
Instrument specification	Operating ranges of instruments and their maximum throughput in samples per hour
Instrument configuration	Types of solvents that are available for running HPLC, sampling accessories on hand for IR, detectors available for GC, and the sensitivity toward particular analytes of each detector and instrument
Regulatory procedures	Information relating to standards and procedures for analytical measurements that determine how analyses must be performed to meet legal requirements
Scheduling data	Time required for each analysis

boundaries of that domain. The reason is not only the need to include the general knowledge mentioned above, but also that the ES may need to tell the user how to *use* the expert advice that it will provide. An ES that advises technicians in an analytical laboratory in which river water samples are analyzed for their metal ion content might contain a variety of operational data that relate not to the general problem of determining heavy metal content, but to the specific problem of doing so in the laboratory in which the ES is operating. This might include detailed data on instrumentation and procedures (Table 7.2).

7.5.1.2 Heuristic Knowledge

As well as factual information, the knowledge base may contain *heuristic knowledge*. This is judgmental knowledge, which is more difficult to justify through logical reasoning than is factual knowledge and is also more difficult to encode in the database. What is thought by one person to be "true" or "obvious" may be regarded as mere speculation by someone else, thus heuristic knowledge is sometimes described as the "art of good guessing." It consists of information about what "normally" happens in particular circumstances, what we "expect," or what "seems likely":

```
Days in spring are often breezy.
It can be dangerous to get close to a swan.
The Mets won't win the World Series.
```

Because of the uncertainty inherent in such statements, an ES that uses some heuristics is not guaranteed to succeed in answering the user's questions, and may propose a solution with some degree of confidence. Heuristics are closely linked with fuzzy data, which we shall cover in the next chapter.

7.5.1.3 Rules

An expert system does much more than extract information from a database, format it, and offer it up to the user; it analyzes and processes the information to make deductions and generate recommendations. Because an ES may be required to present alternative strategies and give an estimate of the potential value of different courses of action, it must contain a reasoning capacity, which relies on some sort of general problem-solving method.

One of the simplest and most powerful ways to manipulate information in the knowledge base is through *rule-based reasoning*, which relies on *production rules*. A production rule links two or more items of information in a structure of the form:

```
IF <condition> THEN <conclusion>
```

For example,

```
IF    <the activation energy for the reaction is high>
THEN  <the rate of reaction at room temperature will be low>

IF    <the sample is labile in acid>
AND   <the pH is below 4>
THEN  <the risk of sample degradation is high>
```

The information to which the rule is applied might be extracted from the knowledge base, it might be provided by the user in response to questions from the ES, or it may be provided by combining the two. An expert system that uses rule-based reasoning is, quite reasonably, known as a *rule-based system*. This is the most widely used form of expert system in science, and it is on this type of system that this chapter concentrates.

7.5.1.4 Case-Based Library

As well as purely factual data and production rules, the knowledge base may contain a section devoted to *case-based information*. This is a library of specific examples that in the past have proved informative and that relate to the area in which the system has expertise.

For example, the factual component of a knowledge base in an ES whose role is to advise on measures that could be taken to limit the global spread of disease might contain data on the effect of air travel on the movement of disease between countries, rules that describe how the rate of spread of disease depends on the local climate, and so on. The knowledge base might also include case-specific data, such as detailed information about particular instances when a disease has spread unusually quickly from one country to another and the reasons for this atypical behavior. Such examples could be used for comparison with current data to determine whether a similar situation has arisen in the past. If similarities are found, the case library can provide information about the action that was taken in the past example and the result of that action.

7.5.2 Inference Engine and Scheduler

To provide suitable responses to user queries, the ES must not only contain the data needed to reach a valid conclusion, but also be able to find that data and recognize its relevance. It, therefore, must make intelligent choices about what data in the knowledge base are pertinent to the user's query, and then apply rules in the knowledge base to make deductions. Both tasks are the responsibility of the *inference engine*, which is the brain of the expert system. It trawls through the knowledge base, assessing the facts and applying the rules using roughly the same methods that a human would.

In a complex environment the knowledge base might contain scores or hundreds of rules and hundreds or thousands of facts. The order in which production rules are processed may influence the conclusion that the ES reaches; thus, within the inference engine is a scheduler that decides in what order rules should be processed and also, with the help of the inference engine, sorts out any conflicts that arise. Because the scheduler is responsible for organization and timetabling rather than deduction, it could be regarded as a component of minor importance, but without an effective scheduler, rules would be processed in an unpredictable order or in the order in which they appear in the knowledge base, leading, at best, to a reduction in efficiency and, at worst, to faulty reasoning.

It might seem, if the knowledge base contains all the rules that are needed to reason about some topic and these rules are internally consistent, that conflicts could not arise, but conflicts can appear as readily within an expert system as they can in the laboratory. For example, consider the two instructions:

1. There is a backlog of samples for analysis. To reduce the backlog, first process those samples whose analysis can be completed in the shortest time.
2. All samples sent by the hospital should be processed without delay.

The potential conflict is obvious.

7.5.3 The User Interface

An *interpreter*, or *explanation system*, forms the interface between the user and the core software. It interprets queries from the user so that the inference engine can make sense of them and presents the conclusions of the inference engine in language appropriate to the needs and level of knowledge of the user. It is also responsible for generating explanations of the reasoning that the ES has used if the user is suspicious of, or baffled by, the advice that the ES has offered and wants to discover the logic behind it.

The dialogue in which the ES engages is not preprogrammed; there is no set sequence of questions that the system is sure to ask, no matter how the user responds. Instead, after each response, the system analyzes the information that it now possesses and then decides what to do next.

Although the *user interface* just acts as the messenger for the interpreter, sound and logical design of the interface is crucial in giving the user confidence in the system. The user interface employs a conversational format, so the interpreter and user interface must between them have some knowledge about how information should be presented to humans. In the most effective expert systems, the interactions between software and user take into account the level of expertise of the user.

The interface may be self-adaptive, noting the identity of the user each time they log in to use the system and tracking the interactions that follow, learning about the level of expertise of the user, and building up a personal profile of them so that the approach of the ES can be tailored to them individually. This is sometimes apparent in the Help system on a PC, which is usually based around a type of ES. Since the level of knowledge, experience, and confidence among PC users varies widely, if a Help system is to be of the greatest value, it is wise to construct it so that it can adjust its approach and presentation to match the level of expertise of the user.

Together, the user interface and the explanation system, therefore, are important components in the ES. A user who finds the software difficult to use, or confusing, may become irritated with the system or be misled by it and stop using it. Worse still, they may misinterpret the advice that it provides.

7.5.4 Knowledge-Base Editor

The final component in the system is a *knowledge-base editor*, which is required during the construction of the ES. This provides a tool that the human expert, who is responsible for building the system, can use to enter information during the development phase and later update the knowledge base by adding new rules and data. We consider some of the challenges that arise in the construction of the knowledge base in section 7.10.

We turn now to the central question in a discussion of expert systems: How does the software manage to make deductions?

7.6 Inference Engine

In order to make decisions on which a user can depend, the ES must have facts and rules. As we have seen above, production rules contain both a condition and an action part.

```
IF    <the patient's Body Mass Index is well above recommended
      levels>
THEN  <recommend weight loss to reduce the risk of a heart
      attack>
```

The condition part is a testable statement whose truth may be determined by inspection of data within the knowledge base or provided by the user in response to questions from the ES. The data consist of clear statements of fact:

```
The patient's BMI is 31.
More than four inches of rain fell in Boston during April.
The sample is a liquid.
```

The action in a production rule may take various forms; it might be an instruction to the user to bring about some event:

```
Lose weight.
Buy an umbrella.
Use GC-MS (gas chromatography-mass spectrometry) to analyze the
sample.
```

Or a conclusion:

```
Bicycles go rusty during the spring in Boston.
The sample will require pretreatment to remove phosphates.
```

The conclusion could be that a certain piece of data can now be demonstrated to be true ("The sample has a melting point below $10°C$") and this might just be added to the ES's own database. Alternatively, the conclusion could be an instruction to the scheduler to change the order in which it will assess rules in the database. It follows that not every conclusion will be reported. Most will form just one step in a chain of reasoning, only the end of which will be seen by the user.

Using its pool of facts, the ES can attempt to reason. Humans use several different methods to reach a conclusion. We may reason logically:

```
If the temperature outside is below 0°C, one should dress
warmly when walking to the shops.
```

Or, we may make a decision based as much on convention as on logic:

```
Eating fish tonight? You should have white wine with the meal.
```

Heuristics are widely used by experts:

```
The sample of river water comes from an area that once housed
a sheet metal works. Heavy metal contamination is common in
```

the leachate from such sites, so we should suspect the presence
of heavy metals in this sample.

And case-based reasoning, in which the decision on how to act in one situation is guided by what action was successful when a similar situation was encountered in the past, is also of potential value:

These patient symptoms are similar to those met in an outbreak
of Legionnaire's disease last year in which effective isolation
of patients proved essential, so our first action must be to
provide suitable isolation facilities.

An ES may use all of these techniques to deal with a query from a user, but is likely to rely primarily on production rules. These rules are a type of knowledge, so are stored in the knowledge base, often using a format that makes their meaning obvious.* Each rule is "standalone," hence, although its execution may affect the conclusion reached by applying it in combination with other rules, a rule can be added, deleted, or modified without this action directly affecting any other rule in the database. Correction and updating of the expert system, therefore, is much more straightforward than would be the case if different rules were linked together using program code.

7.6.1 Rule Chaining

No sensible expert system, when it receives a query from the user, blindly executes every rule in the database in the hope that the sum total of all its conclusions might provide an appropriate response to the user. Even assuming that the output of large numbers of rules executed simultaneously could be massaged into some meaningful format (which is optimistic), a plan of action is needed. The *paradigm* or *problem-solving model* is this plan. It sets out the steps that the inference engine should take to solve the query posed by the user. In a rule-based system, the paradigm often relies on identifying a group of connected production rules that, when applied in sequence, will yield a line of reasoning that leads to the desired conclusion.

7.6.1.1 Forward Chaining

The system can do this by applying several rules in succession to construct a logical sequence of deduction, which is known as *chaining*.

There are two ways in which chaining may be performed. The dialogue at the start of this chapter illustrates the application of *forward chaining*. The ES starts from a broad set of conditions and data and uses its interaction with the user to move from this very general starting point to a conclusion. The ES interrogated the biologist to gain information and narrow down the number

---

* Computer code, which is descriptive, so it can be read and understood without having to run
  the program to find out what its purpose is, is known as *declarative* code.

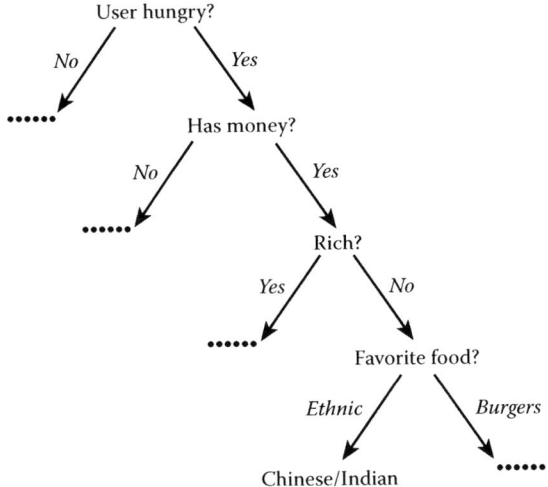

FIGURE 7.3
A tree diagram that illustrates forward chaining.

of possible holiday destinations from a large initial set to a manageable number and eventually to one.

In forward chaining, given a response from the user to some question, the inference engine searches the knowledge base for any conditions that will be satisfied by this piece of information. If it finds a rule which is satisfied, the THEN portion of the rule is triggered, which provides the inference engine with further information and prompts further questions to the user. The user's positive answer to the question: "Do you like warm weather?" leads the inference engine to search through its knowledge base for data on warm countries, using knowledge of the local climate. It will be able to rule out Greenland, Antarctica, Patagonia, and several other areas as possible holiday destinations.

Forward chaining can conveniently be represented as a tree structure. To illustrate how transparent this process is, Figure 7.3 introduces a second interaction between an ES and a user without any further explanation. The logic of the reasoning should be clear. This type of reasoning is also known as *data-driven reasoning* because the order in which rules are fired is determined by the data that are at that stage available to the inference engine.

7.6.1.2 Backward Chaining

In *backward chaining,* the conclusion is known in advance, but the reasoning that is needed to reach that conclusion is not. This sounds a little odd. How can we know what we are going to conclude and yet not know how to reach that conclusion? In fact, it is common for us to know where we want to go, but not how to get there; we have a hypothesis, which is to be proved or disproved, or a goal that is to be attained, but are unclear about what steps are needed to reach that goal. Just like forward chaining, backward chaining

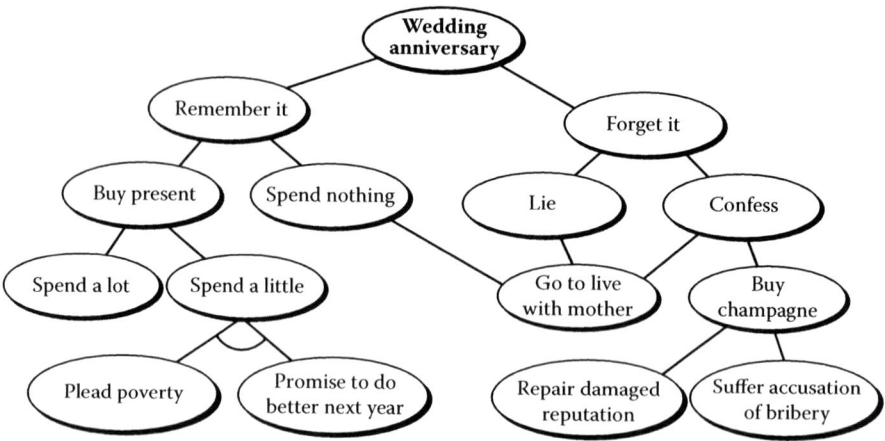

FIGURE 7.4
Backward chaining.

can be conveniently displayed as a tree, in this instance, an AND/OR tree (Figure 7.4).

In a backward-chaining tree, routes from an oval node, working down from the top, are drawn as unconnected lines if either route can be taken; this corresponds to an OR condition:

```
IF <wedding anniversary> THEN <remember it> OR <forget it>
```

The two lines are joined by an arc if both steps must be taken.

```
IF <spend a little> THEN <plead poverty> AND <promise to do
better next year>
```

We can conclude from Figure 7.4 that, if we want the marriage to last, we should not lie when we forget our wedding anniversary, or, if we confess to our lapse of memory, we may be able to recover from the disaster by buying champagne (note that no guarantee is provided for any advice contained within this book).

Suppose that, as our wedding anniversary approaches, our goal is to avoid having to return home to live with mother. Assuming that we are spineless enough that we need to consult an expert system for advice on how this catastrophe can be avoided, the goal is defined as:

```
NOT (go to live with mother)
```

In backward chaining, the inference engine would search through the knowledge base until it found a rule, the conclusion of which was "Go to live with mother." It would find three such rules:

```
IF <spend nothing> THEN <go to live with mother>
IF <lie> THEN <go to live with mother>
IF <confess> THEN <go to live with mother> OR <buy champagne>
```

Since several rules have the same conclusion, the ES would take the first rule it encounters and test the condition (<spend nothing>) to see whether it is true. If it is not known at this point whether the condition is true, the condition provides a new goal for the ES. It will now attempt to find out whether the amount spent on a present was different from zero. The system will again search the knowledge base to see if it contains a rule whose conclusion is the amount spent. If it finds no such rule, it will ask the user whether she/he can provide that information. The ES thus starts at the final conclusion to be reached and works backward through a sequence of goals until it either proves the entire set of conclusions or runs out of data. Because the route that the ES takes is at each stage determined by the need to meet some goal, this is also known as *goal-driven reasoning*.

A prime advantage of backward chaining or goal-driven reasoning is that the questioning in which the ES engages is focused from the very start because it begins at the final step of a possibly long chain of reasoning. The questions that are asked, therefore, should all be relevant. Contrast this with forward chaining, in which the early stages of the interaction may comprise vague, unfocused questions because the user might be interested in anything across the entire range in which the system has expertise:

```
"Well, Hal, how are you today?"
```

From such a woolly starting point, time may be wasted while the ES tries to find out what the user really wants. On the other hand, if there are many different rules that lead to the action <go to live with mother>, it is possible that an ES using backward chaining will pursue many fruitless chains before reaching a satisfactory conclusion.

Whether it follows forward or backward chaining, the system will pose a series of questions that allow it to gather more information about the problem until it successfully answers the user's query or runs out of alternatives and admits defeat. Just like the way that the system chooses to respond to a query, the exact sequence of steps that it takes is not preprogrammed into the knowledge base, but is determined "on the fly" by the scheduler, which reassesses the options after each response that the user provides.

Although the ES can be constructed using rules with only one action and one condition, rules need not be restricted to a single condition or action. They may contain several parts joined by logical operators, such as AND, OR, and NOT, so that the conclusions depend on more than one premise or test.

```
IF <remember wedding anniversary> THEN <buy present> OR <spend
nothing>
```

```
IF <the analysis requires more than three hours>
AND IF <the analysis cannot be interrupted once it has been
started>
AND IF {<it is Friday afternoon> OR <someone has turned off the
air-conditioning>}
THEN <advise the user to go home> AND <to start the barbecue>
```

Example 2

Suppose that we have the following rules that are to be used to work out what might be a suitable solvent for some material.

```
R1: IF the sample has a low molecular weight AND IF
    the sample is polar, THEN try water as the solvent.
R2: IF the sample has a low molecular weight AND IF
    the sample is not polar, THEN try toluene as the
    solvent.
R3: IF the sample has a moderate molecular weight AND
    IF the sample is ionic, THEN try water as the sol-
    vent ELSE suspect insoluble in water and toluene.
R4: IF the sample is volatile, THEN the sample has low
    molecular weight.
R5: IF the sample boiling point is <50°C, THEN the
    sample is volatile.
```

In a goal-driven system, the inference engine would start with the first rule because the conclusion of that rule is the recommendation of a solvent. This forms its first goal. In other words, it would try to demonstrate that water is a suitable solvent for the sample. In order to test this rule, the ES needs to know whether the molecular weight is low.* If this information is not already available within the database, the next step is to check whether the knowledge base contains a rule, the conclusion of which is that the sample has a low molecular weight. In this instance, R4 provides this information, but this rule, in turn, depends on whether or not the sample is volatile, so this gives the ES a secondary goal: to show that the sample is volatile. It would then find a further rule (R5) that states that a sample is volatile if its boiling point is less than 50°C. If the required boiling point is not in the knowledge base, the ES would turn to the user.

```
ES: What is the boiling point of the sample?
U:  43°C
```

and the ES can now turn to the next step of determining whether the sample is polar.

---

* Whether a molecular weight is "low" or "moderate" involves personal judgment. To make use of these rules, the ES will need to have a precise definition of these words, in terms of some range of molecular weights. We shall see in the next chapter what options exist if an exact definition is not available.

7.7 Explanation and Limits to Knowledge

Some types of knowledge-based systems may make judgments based on data that contain uncertainty; we shall learn more of this in the next chapter when we encounter fuzzy systems. Even when the information that the system reasons with is unambiguous, the system's conclusions may come as a surprise to a nonexpert. If the user doubts whether the ES has reasoned correctly, it is natural for them to seek reassurance that the line of reasoning used is robust, so the ES must be able to do more than merely provide advice, it should be able to explain how it has reached a particular conclusion.

Explanations can be triggered by a user response of "Why?" or "How?" We can revisit the example from the start of the chapter to see this in action:

```
Biologist: I've been working too hard. I need a break.
Expert System: Have you chosen where to go?
User: No.
ES:    Do you like warm weather?
B:     Yes.
ES:    Do you like to lie on a beach?
B:     No.
ES:    Do you like good food and wine?
B:     Definitely. Why?
ES:    If you like good food and wine, I will suggest only coun-
       tries in which my knowledge base tells me that the food
       and wine are good. France or Italy would be suitable. Do
       you like art?
B:     Yes.
ES:    I suggest Florence.
B:     How?
ES:    Florence is in Italy. In Italy, the weather is hot and the
       food and wine are good. The Uffizi Gallery, which contains
       world-famous art, is in Florence.
```

Comments from the system help to give the user confidence that the ES knows what it is doing. They are particularly valuable if the ES has engaged in some complicated reasoning or if its conclusions are unexpected (though, in the example given above, the replies from the ES hardly fall in that category).

The explanatory dialogue that the system offers is limited, however, for several reasons. First, the evidence that the ES draws on to support its conclusions will consist of the logic that it used to derive those conclusions. Consequently, any explanation that it provides will be a restatement or rephrasing of the rules and facts that it has already used in generating its advice. It is true that much of the reasoning may have been hidden from the user — the holiday-advising expert did not explain its reasoning until prompted — but, if within the chain of reasoning is a doubtful step, there may be no way that the user can test it. Recall that the rules that an ES uses are unlikely to be causal; thus, if the conclusions of an ES appear strange and there is a rule about which the user has concerns, it is unlikely that the ES will contain sufficient information to independently justify the rule.

Furthermore, an ES is unable to step outside the boundaries of what is contained in its knowledge base This knowledge base may, knowingly or otherwise, incorporate the biases of the person who built the system and any bias or prejudice will color its view of how to interpret responses from a user. For example, the ES interprets the positive answer to the question: "Do you like good food and wine?" in a particular way, that is, that the user will appreciate French or Italian cooking. However, for the user, a visit to McDonald's® or Kentucky Fried Chicken® may be high on their list of culinary treats, so for them the expert system is going off on the wrong track if it proposes a holiday in Florence when the biologist would really be happier in South Dakota.

In addition, most expert systems are unable to rephrase an explanation. If the user does not understand the reasoning of the system after being given an initial explanation, there is no Plan B that the ES could use to explain its conclusions in another way.

Paradoxically, the explanation system is actually of the greatest use to the person who might be expected to least need it (the expert who built the system) before the end-user even gets a look in. This is because the flow of control in an expert system is not linear — rules are not executed in the order in which they are entered into the database, but in an order determined by the scheduler, so following the flow of control and thus understanding the system's reasoning may be difficult. During building and testing, if the ES provides an invalid or unexpected response, it can be asked to explain itself. This can be a significant aid in tracking any faults and ambiguities.

As well as being able to provide some explanation of its reasoning, it is important also that an ES should know and report its limitations. This is routine for a human:

```
I suspect that the reason your car won't start is a flat bat-
tery, but until I check it, I can't be sure.
```

Not only does this human know the limits of his knowledge ("I suspect that...," "I can't be sure..."), but he makes it clear to the listener that his conclusion about the stranded car is only provisional ("...until I check it...."), so the listener can judge how much weight to attach to the opinion. Expert systems are far less able in both respects. Expert systems report every conclusion with the same degree of confidence unless they are explicitly programmed to deal with uncertainty, so it is important that, as an ES bumps up against the boundaries of its knowledge, it should be able to give some hints that it is venturing into unreliable territory. This requires "knowledge about knowledge." Knowing what it knows, its limitations and the circumstances within which it is applicable, is an important attribute of the more flexible expert system. This type of knowledge is *meta-knowledge* and is an active area of research across many AI fields.

7.8 Case-Based Reasoning

If the ES does not have sufficient information to be able to answer a user's query using only information from its rule base and repository of facts, it may refer to the contents of its case library.

The aim of case-based reasoning is to provide advice based on a set of known examples that are judged to be relevant to the user's query. Files within the library contain data about past cases relevant to the area of expertise, how they were tackled, what the results of this approach were, and whether the action taken was appropriate and successful. Each case is tagged with a set of attributes that describe the case, so that when the library is searched for relevant material, it can quickly be identified through some form of similarity metric.

If a case is found that matches the user's query closely, this is used to provide appropriate advice. If, on the other hand, there is no case that matches the problem presented by the user sufficiently closely, the system can try to modify a case that is present in the library to bring it into line with the user's query in what is known as *structural adaptation*, or it may be able to create a new solution using as a starting point a similar case from the past (*derivational adaptation*).

If neither approach works, but the problem is eventually solved through discussion with the user, the system can add the current case and its solution to the library, thus expanding the system's understanding of relevant cases.

7.9 An Expert System for All?

Computers are much cheaper than experts, so it might seem that an ES should be considered for any situation in which advice from a human expert might be helpful. However, many factors limit their use: the two most crucial are cost and the difficulties of the extraction of knowledge.

The principle cost of creating an ES is not the hardware on which the system runs or the software that is used to create it, but the cost of employing a human expert to create the system. Not only are human experts expensive creatures, but they also may be understandably reluctant to spend time developing a computerized copy of themselves, knowing that, the more perfectly they manage to distil their expertise into a computer system, the greater the chance that they will make themselves redundant.

It may be fairly simple in principle to build an ES in areas such as analytical chemistry, in which most conditions and actions are clear-cut, but matters are not always so straightforward. A doctor uses a variety of tests and questions to determine what illness might be responsible for a patient's symptoms. While some of the data on which a doctor relies will be unambiguous (e.g., the result of a test for blood sugar level or a measurement of body temperature), other forms of knowledge may be hard to encode in a silicon expert. A doctor might feel that a patient consulting her is looking "unwell" and, on the basis of their clinical experience, be able to determine what potency of drug to prescribe to best help the patient. A lay person might also recognize that the person is unwell, but be unable to draw reliable conclusions about the level of medication required. Thus, the question of how to extract expert knowledge and how readily it can be converted into production rules is central to any decision on whether to try to create a new ES. Tools to make this less painful and more efficient are discussed in the next section.

7.10 Expert System Shells

Although the knowledge base that forms the heart of the ES varies from one application to another, the core software components, such as the inference engine and scheduler that are responsible for the way that information is manipulated, are similar. There is also a clear division within an ES between the general domain knowledge, which is its long-term memory, much like the contents of an encyclopedia, and the ephemeral data provided by a user, as part of an attempt to solve a specific problem. It is the former that is filled during the construction of the system. Furthermore, in contrast to the almost limitless range of topics that might form the focus of an expert system, only a few ways are known to represent knowledge so that it can readily be manipulated in an expert system, to support inferences and provide explanations.

Taken together, all of these points suggest that it might be possible to prepare a toolkit consisting of the essential components of an ES, apart from the knowledge base, and then fill it with application-specific data. This is such a useful way to work that in all expert systems there is a clean separation between the information that the ES manipulates and the tools that are required to perform that manipulation. This division between the part of an ES that changes between applications and that which is constant has led to the development of the *expert system shell*.

The user interface, the interpreter, and the inference engine together comprise the ES shell or a *skeletal system*. As these components are largely independent of the specific application, all that is needed to create a working ES from the shell is to feed it with rules and facts.

Shells greatly simplify the programming of an ES, but they do not directly diminish the effort required in *knowledge acquisition*, which is the key task in its development. During the knowledge acquisition process, the relevant information is collated so that it can be inserted into the ES. This process obviously requires a human expert and, in addition, frequently needs the guidance of a knowledge engineer. In developing an effective ES, the choice of reasoning method is important, but less crucial than the accumulation of high-quality knowledge.

An ES shell resembles a good kitchen. All the utensils and ingredients needed to prepare a meal are present. What is required to make enticing food is the specialized knowledge that cooks possess. The process of entering the recipes, the domain-specific knowledge needed to turn a shell into a working system, is known as *knowledge engineering*.* In the more straightforward of expert systems, the shell can be used by an expert working on his or her own to build up a working system. It is possible to buy software shells that will run on a PC and can be used by any computer-literate user. If the ES is to be large and complex, however, the process of building a working system will probably require specialist understanding of how expert systems work and will be overseen and driven by a *knowledge engineer* in cooperation with the human expert. The process is shown in Figure 7.5.

Figure 7.5 makes the construction of an ES look agreeably simple, but this is misleading. The most notable difficulty, among several, in building an effective ES is that of determining what constitutes a complete set of rules so the ES can deal successfully with all queries that the end-user might pose. A further difficulty is finding a way to express as a rule that the inference engine can use something that may be as vague as an "expert hunch." Hunches and general background knowledge are important: some rules that a human expert uses to make deductions may be so much a part of his everyday understanding that he might not even realize that they are actually required until prompted to explain how he reached a particular conclusion.

Before any information can be entered into the knowledge base, decisions must be made about how the rules and data will be represented. The question of knowledge representation is a matter for the knowledge engineer who, from a preliminary discussion with the expert, and meetings with potential

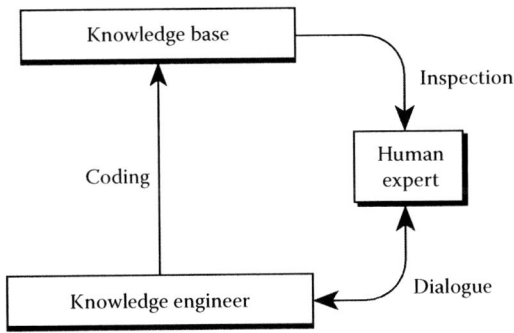

FIGURE 7.5
The iterative process by which the first version of an expert system is built.

---

* In broad terms, the aim of knowledge engineering is to integrate human knowledge with computer systems.

users of the finished system, will have a feel for what type of representation will be most effective. Once a suitable representation has been selected, the next step in building the knowledge base is a detailed conversation between the knowledge engineer and the expert, so that the knowledge engineer can gain an overview of the subject area and develop an understanding of the role that the ES is expected to fill. Once some initial data have been fed into the system, the conversation between expert and software engineer becomes more wide-ranging, and often is full of open-ended questions from the knowledge engineer such as:

```
Suppose this situation arose; what would you conclude?
If this happens, what will happen next?
I don't understand your logic; why do you think that?
```

These questions allow the engineer to elicit relevant information, which is then coded and placed into the knowledge base. As the core of data in the knowledge base grows, testing can begin. The expert starts to interrogate the database, asking the sort of questions that a user might pose, identifying inconsistencies, faults, and gaps in the responses of the software. This refinement of the knowledge base is an important and lengthy process, but even when errors are discovered, substantial reconstruction of the database should not be necessary because each fact or rule is independent; instead, individual items of knowledge and rules are deleted, modified, or inserted.

After some preliminary testing, the expert returns to a conversation with the engineer in order to resolve any problems. There may be many such cycles before a prototype system is judged ready for testing by users, so the process, often known as the *knowledge acquisition bottleneck*, is long. It might seem strange to regard as a bottleneck what appears to be just a matter of some conversations between the expert and the knowledge engineer, followed by a period during which the resulting information is fed into a waiting system. However, when one realizes that, in a knowledge base that may eventually contain several hundred rules, the rate at which useful rules are added may be as low as five or six per day, the picture of this stage as a bottleneck becomes more understandable.

There is a further reason for this slow rate of progress; experts use a special language. Stock market traders talk of an "aggressive" stance with regard to a share, or "going long on the stock," expressions, which to an outsider, make little sense (Figure 7.6 and Figure 7.7). Organic chemists describe the process of product workup or of performing a reaction under mild conditions, terminology that is meaningless to a nonchemist.

Any such "insider" terms must first be explained to the knowledge engineer, then massaged into a form that can be fed into, and used by, the expert system. In addition to being obscure, this specialist language may be imprecise. Even after an organic chemist explains to the knowledge engineer what "mild reaction conditions" means, it may not be possible to pin down these

FIGURE 7.6
An "aggressive stance."

FIGURE 7.7
Going long on stock.

conditions precisely. Mild conditions do not indicate an exact temperature or pH, but rather a range of conditions with fuzzy boundaries.

It is also common when building the knowledge base to find that the insertion of one piece of knowledge generates the need for further material. One step in a chemical analysis may require strongly acidic conditions, which are achieved by using a low pH. This implies that the ES needs to know about pH, which in turn demands an understanding of what is meant by the concentration of H^+ ion. This may then require that the ES contains some information that relates to acid dissociation constants.

An ES generally learns by being told what it needs to know by the expert, with some input from the knowledge engineer, but this is not the only way to learn. Instead, or as well, the system may learn by querying the expert in the relevant area or by the inspection of examples in datasets. Learning through the inspection of examples, known as *automatic knowledge elicitation*, is a far quicker way of building an ES than querying a human expert, provided that a suitable dataset of examples is already available. Such a database can be generated by asking the human expert to assess examples that have been chosen to cover the entire range of problems that the user might reasonably be expected to want to know about. However, as well as being quicker, learning by example is much more difficult to implement, since the shell is required not just to examine the results of the assessment by the human expert, but also to derive production rules from those results (a task which one might rightly suppose could benefit from the services of an expert system). As a result, this type of system generation is at present used only in the building of a minority of systems.

Most expert systems are built incrementally — a small prototype system is built and tested before further rules are added. This reduces the risk that the full system will contain errors that are not found until practical completion, when tracking down their origin can be difficult, and also gives the expert an early chance to test the look-and-feel of the system, so that any changes required in its appearance can be incorporated before development has proceeded too far.

7.11 Can I Build an Expert System?

The user of an ES needs to know little of what goes on behind the scenes. Provided that the system is robust, easy to use, and fully covers a suitable range of topics, the user need not worry about the form in which information is held in the ES or the type of reasoning on which it relies.

However, if a potential need for an ES is identified and no such system currently exists, the answers to several questions will indicate whether it is feasible to construct a system.

- *Is a human expert available and able to contribute?*

Clearly, no matter how valuable an ES might be, unless an expert can be found who combines an appropriate level of understanding with the time and motivation needed to create the system, no ES can be created. This is often the key challenge in preparing a new ES, since almost by definition there are few experts available in areas in which the ES is likely to be of the greatest value.

- *Is there a demand for expert advice that cannot be met?*

It is only worth spending time on the development of a silicon expert if real experts in the field are rare and there is some commercial or scientific value to be attached to the knowledge that they possess. Garage mechanics are thick on the ground in most countries, so although expert systems are used by the largest car manufacturers to help mechanics understand the details of their cars, an ES whose role was to diagnose the problem if a car simply would not start might be poorly used.

- *Is the human expert able to explain how to mange the tasks that the ES will be expert in?*

We often perform tasks without knowing quite how we manage to do them. Suppose that I learn a Chopin nocturne by heart. When I play it from memory, I do not read in my mind an image of every note in the score before I play it, nor do I know which notes to play because I have remembered how the piece should sound. I do not play by remembering the emotions that the piece evokes, or seeing or feeling where my hands should be on the keyboard at every moment. All of these factors, and probably more, combine to provide a mental package that enables me to play a sequence of thousands of notes in the right order and with (roughly) the right degree of expression over a period of several minutes, even though I cannot remember each note individually. How is it done? I don't know. Could I use my understanding of how to play Chopin to create an expert system? Undoubtedly not, since I use rules without even knowing what they are, and could certainly not formulate them in a way that an ES could use.

Therefore, it is not always possible to construct an ES, even if experts exist and are willing to cooperate. There are also difficulties that arise if the "correct" answer to a question as determined by an expert is colored by personal preference or cultural views. An ES that could provide advice about classical and modern art might have many uses, but if it started passing comment on whether a sheep pickled in formaldehyde is "good" or "bad" art, it might antagonize users. Judging whether a dead sheep is beautiful, or whether it is art at all, is very much a matter of personal preference.

- *Can the problem be solved by symbolic reasoning?*

Computers are well suited to the manipulation of numbers, but the ES relies on symbolic computation, in which symbols stand for properties, concepts, and relationships. The degree to which an ES can manage a task may depend on the complexity of the problem. For example, computer vision is an area of great interest within AI and many programs exist that can, without human assistance, use the output from a digital camera to extract information, such as the characters on a car number plate. However, automatic analysis of more complex images, such as a sample of soil viewed through a microscope, is far less simple even though the underlying problem is fundamentally the same. ES are useful in recognizing car number plates or widgets on a conveyor belt in an industrial production line; they are not (yet) of much use in analyzing complex images with unpredictable structure and definition.

- *Are there no "traditional" algorithmic methods that could provide an acceptable alternative to an ES?*

The ES is time-consuming to create and may require the services of a knowledge engineer in addition to the human expert. If a traditional approach can manage the task, or even a self-training AI program, such as an artificial neural network, there may be little to be gained by investing time and effort in an ES.

- *Are the data provided in a format that an ES can readily process?*

The electronic nose is an example of an area in which the complexity of the analysis may make it difficult to replace a human with an ES. Electronic noses combine a sensor array with a neural network to make judgments about the composition of complex mixtures, such as fuels, wines, and natural oils. In such tasks, they often beat human noses in accuracy (Figure 7.8).

If the aim is to recognize a particular sample of wine, or even a particular type of wine, an electronic nose may be suitable. However, when it comes to judging the quality of a wine, life is tough, even with the benefit of an electronic nose. A competent ES linked to an electronic nose would be able to distinguish a Shiraz from a Merlot and a poor South American wine from a good one, but finer distinctions are trickier. We would expect it to struggle to distinguish a "good" wine from an "exceptional" one, and to be clueless when asked to determine from the composition of a wine how many years it will need to be stored before it is at its best. It is not easy for a wine expert to explain how they are able to judge the quality of a wine or to determine for how many years it should be left to mature (cynics might suggest a personal interest in maintaining an aura of mystery about the subject). If the wine experts themselves cannot put into words their subjective views about wine, it is impossible to construct an expert system to do this.

- *Can the program be properly evaluated?*

The prototype ES should be evaluated by a group of potential users and, if possible, also by a panel of experts who have not been involved in its creation. The former can assess the look-and-feel of the software while the latter can more rigorously verify that the system is functioning correctly. This evaluation is best done without the expert panel having direct access to the ES. They may work by generating a query, which is then answered either by the ES or by a human expert, and the answer returned without indication of its origin. If the ES performs as well as, or better than the human expert, this

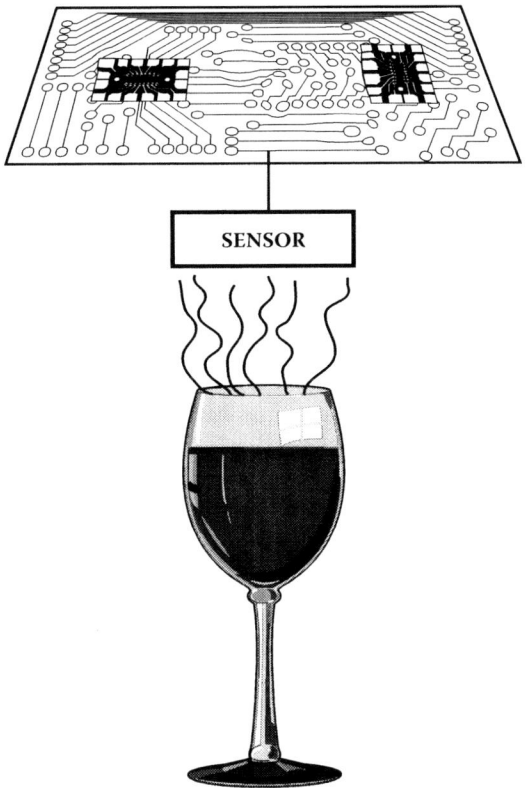

FIGURE 7.8
The electronic nose.

is strong evidence for its value.* The need for a panel of experts increases the number of people who will be involved in and, hence, the cost of the development of the system.

7.12 Applications

Experimental scientists, if they write computer programs, tend to be fond of FORTRAN, C, C++, Visual Basic, or Java. None of these is well suited to the construction of an ES, in which LISP and PROLOG are the favored languages. LISP (LISt Processing) is a simple and elegant language, which is still widely used in AI, while PROLOG (PROgramming in LOGic) uses easy-to-understand

---

* This is related to the Turing test. A program that converses with a user about a particular topic with such conviction and knowledge that the user is unable to determine whether he is dealing with computer software or another human, has passed the Turing Test.

assertions and rules, which makes programs written in the latter language straightforward to interpret. Fortunately for those who would like to build an ES but have no desire to learn further computer languages, a knowledge of neither LISP nor PROLOG is required to use most commercial ES shells.

The range of applications of expert systems in science is very wide, limited primarily by the considerable effort that is often needed to prepare one. Typical applications include the work of Pole, Ando, and Murphy on the prediction of the degradation of pharmaceuticals;[1] Nuzillard and Emerenciano's work on structure elucidation from two-dimensional NMR data;[2] the work of Patlewicz and co-authors on sensitization;[3] and of Prine in another area connected to safety, that of MSDS data.[4] The number of publications runs to several hundred per year, covering a great diversity of topics.

7.13 Where Do I Go Now?

Knowledge-based systems are such a fundamental part of the computer scientist's view of AI that it is no surprise that the large majority of textbooks on expert systems — and there are many — are intended for computer scientists rather than experimental scientists. However, a small number exist which do not presuppose a substantial knowledge of AI in the reader. Among them are the texts by Jackson,[5] Liebowitz,[6] Ignizio,[7] and Lucas and van der Gaag.[8] New books in the area appear regularly and, with an increasing recognition of the value of publications that are accessible to those outside mainstream computer science, the choice continues to expand.

7.14 Problems

1. Types of tasks suited to expert systems

   Consider the following tasks. Which, if any, of them would be suitable candidates for the construction of an expert system?

   a. Determining the chemical composition of different fuel oils through analysis of their infrared absorption spectra.

   b. Checking the shapes of cells seen through a microscope for possible abnormalities.

   c. Determining the Latin name for plants in a nursery.

   d. Finding a rich partner to marry.

   e. Preparing for an undergraduate examination in thermodynamics.

   f. Buying a sports car.

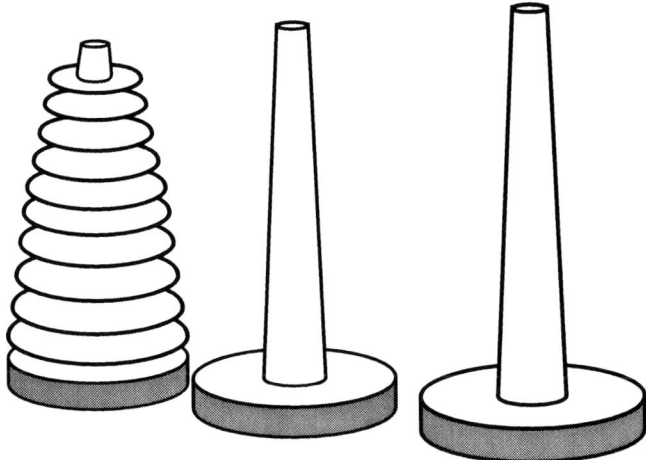

FIGURE 7.9
Tower of Hanoi.

2. Tower of Hanoi

   The Tower of Hanoi is a well-known problem in logic (Figure 7.9). A set of rings of different diameters, say, ten in number, is placed on one of three vertical rods, with the smallest ring on the top and rings of increasing diameter underneath it. The object is to move the complete set of rings from one rod to another, moving one ring at a time, without a ring of larger diameter ever being positioned above one of smaller diameter. The entire pile of rings must be transferred from one rod to a second, moving one ring at a time and using the third rod as necessary. Propose a set of expert system rules to accomplish this.

3. Funding research

   One of the most demanding exercises that scientists engage in is writing research proposals to obtain funding to support their work. Sketch out the structure of an expert system that would be able to help in this task.

References

1. Pole, D.L., Ando, H.Y., and Murphy, S.T., Prediction of drug degradants using DELPHI: An expert system for focusing knowledge, *Mol. Pharm.* 4, 539, 2007.
2. Nuzillard, J-M. and Emerenciano, V.D.P., Automatic structure elucidation through data base search and 2D NMR spectral analysis, *Nat. Prod. Comms.* 1, 57, 2006.

3. Patlewicz, G., et al., TIMES-SS — A promising tool for the assessment of skin sensitization hazard. A characterization with respect to the OECD validation principles for (Q)SARs and an external evaluation for predictivity, *Reg. Toxicity Pharmacol.*, 48, 225, 2007.

4. Prine, B., Expert system for fire and reactivity MSDS text, *Proc. Safety Prog.*, 26, 123, 2006.

5. Jackson, P., *Introduction to Expert Systems*, Addison-Wesley, Reading, MA, 1998.

6. Liebowitz, J., *Introduction to Expert Systems*, Mitchell Publishing, Santa Cruz, 1988.

7. Ignizio, J.P., *An Introduction to Expert Systems: The Development and Implementation of Rule-Based Expert Systems*, McGraw Hill, New York, 1991.

8. Lucas, P. and van der Gaag, L., *Principles of Expert Systems*, Addison-Wesley, Wokingham, U.K., 1991.

8

Fuzzy Logic

In the previous chapter, we met expert systems. Life is easy for an expert system (ES) when knowledge is perfect. In the rule:

```
R1: IF <the pH is less than 2> THEN <add 1 ml 0.1M NaOH
solution>
```

both the condition and the action are unambiguous. There is no uncertainty in application of the rule because, assuming that the pH can be measured with reasonable precision, it either is or is not less than 1, and if it is less than 1, the quantity and concentration of alkali to be added are precisely defined.

By contrast, suppose that the rule was instead:

```
R2: IF <the pH is low> THEN <add some alkali>
```

We now would need guesswork, prior knowledge, or just good luck to add the correct amount of alkali of the right concentration and to do so at the correct time. We can imagine an expert system struggling to deal with the imprecision of this rule and coming up with advice along the lines of:

```
"I'm not sure myself about the pH, but if you have the feeling
that it might be rather low, why not toss in a bit of alkali?
I'm fairly hopeful that this will do the job, but I don't really
know how much to suggest; a little bit sounds about right to
me, unless you have any better ideas."
```

Naturally, this advice is of no use at all, but fortunately there is help at hand in the form of fuzzy logic, which provides a way to reason from vague information.

8.1 Introduction

The vagueness in the second rule above and, consequently, the unhelpful imagined response of the ES, reflect the way that we use woolly language in everyday conversation:

```
Sally is quite thin.
Bill Clinton is very well known.
Overweight people should pay higher health insurance premiums.
```

These statements offer information of a sort, but they are imprecise. We know what the first two mean, but they are vague, while, in the third statement, the information provided is not factual, but is merely an opinion; thus all we know is that the speaker has a particular view about large people. None of the information that these statements provide could be incorporated into an ES using the kind of rules outlined in the last chapter because the system that we met there was not designed to deal with imprecision.

If we need to include this rather uncertain information in an ES, our first thought might be that vague statements could be "firmed up" by adding some quantitative information:

```
Sally has a waistline of 28 inches.
Ninety-one percent of people in America recognize Bill Clinton.
Anyone with a body mass index greater than thirty should pay
higher health insurance premiums.
```

The statements are now more precise, but in adding numerical data something has been lost. Neither of the first two statements has retained any judgmental quality. Does the speaker believe that Sally's waistline of 28 inches is thin, anorexic, or just slimmer than average? Does a 91 percent recognition rating qualify as very well known?

One might think that any loss of data that can be recognized to be judgmental should not concern us as scientists. After all, science is based overwhelmingly on measurement and quantitative information rather than on opinion, so perhaps scientists should be able to sidestep the problems caused by ill-defined statements. But science does not operate in vacuum-packed isolation, cut off from the rest of life. The points of contact between scientists and those working outside science are many and varied, hence, as normal conversation is vague, scientists must be able to handle that vagueness.

In fact, the problem of imprecision is far more widespread than might at first be apparent. It would be naive to imagine that vagueness arises only at the interface where the scientific and nonscientific worlds rub together. Vague language is common in science itself: reactions between ions in solution are recognized by chemists to be "very fast"; overtones in infrared absorption spectra are typically "much weaker" than the fundamental absorption; many colloids are stable "almost indefinitely"; lipid bilayers are an "important" factor in determining the functioning of cells.

To bridge the gap between the descriptive but imprecise language that we use in conversation and the more quantitative tools that are used in science, a computational method is required that can analyze and make deductions from imprecise statements and uncertain or *fuzzy* data. Fuzzy logic is this tool.

Fuzzy logic is often referred to as a way of "reasoning with uncertainty." It provides a well-defined mechanism to deal with uncertain and incompletely defined data, so that one can make precise deductions from imprecise data. The incorporation of fuzzy ideas into expert systems allows the development of software that can reason in roughly the same way that people think when confronted with information that is ragged around the edges. Fuzzy logic is also convenient in that it can operate on not just imprecise data, but inaccurate data, or data about which we have doubts. It does not require that some underlying mathematical model be constructed before we start to assess the data.

Despite the name, fuzzy logic is not in any way synonymous with muddled thinking. Instead, it is a technique that applies well-defined and logical mathematical procedures to fuzzy data, with the aim of offering the same level of advice that any expert system might provide. Systems that incorporate fuzzy logic are, as one might guess, known as *fuzzy systems*.*

8.2 Crisp Sets

Science is based on a bedrock of fact and definition:

- An amino acid contains both an amine group and a carboxyl group.
- Rotational energy levels in a diatomic molecule are described well by the expression:

$$E_J = BJ(J+1) - DJ^2(J+1)^2$$

- The energy of a proton in a magnetic field depends on the magnetogyric ratio for ^1H, which is 26.75×10^{-7} T^{-1} s^{-1}.

Statements of this sort represent *objective knowledge*, which comprises facts that are known to be true, together with generally accepted material, such as definitions and equations. Much scientific knowledge, including mathematical formulae and the values of fundamental constants, is clear-cut and unambiguous and, therefore, falls into this category.

Objective knowledge can be used to place objects into *classes* or *sets*. A class could be defined that consists of all the organic molecules that contain exactly three carbon atoms. If all objects in a group can be allocated to classes in a way that involves no uncertainty, so that the division is sharp and unambiguous, we call this a *crisp* division. Every organic molecule either contains

---

* Other fuzzy methods, such as Fuzzy Calculus, also exist, but are at present less widely used than Fuzzy logic.

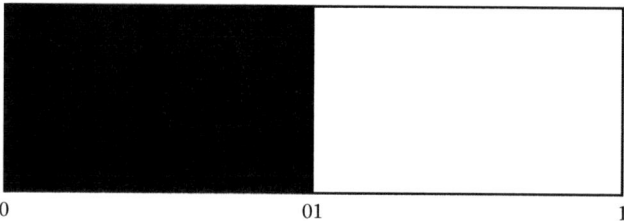

FIGURE 8.1
The range of logical values in Boolean logic.

three carbon atoms and so is a member of the set of three-carbon molecules, or it does not, thus this is a crisp set.

The *degree of membership* or the *membership value* of an object in a set measures the extent to which the object belongs in that set. For crisp sets, the only possible membership values are 0 and 1 (Figure 8.1). A membership of 1 tells us that the object is a member of that set alone, while a membership of 0 shows that it has no membership in the set.

<hr>

Box 1. Membership of crisp sets

$\mu_A(x) = 0$ if x is not within set A
$\mu_A(x) = 1$ if x is within set A

where $\mu_A(x)$ is the degree of membership of x in A.

<hr>

Thus, propanol, C_3H_7OH, has a membership of 1 in the three-carbon molecule class, while ethanol, C_2H_5OH, has a membership of 0 in the same class. As the membership in a crisp set must take one of only two possible values, Boolean (two-valued) logic can be used to manipulate crisp sets. If all the knowledge that we have can be described by placing objects in sets that are separated by crisp divisions, the sort of rule-based approach to the development of an expert system described in the previous chapter is appropriate.

8.3 Fuzzy Sets

In real life, information is often fuzzy. Consider the statement:

```
In a worst-case scenario, the run-off from agricultural land
may contain levels of fertilizers or animal waste that might
seriously contaminate the water supply.
```

This conveys information, but:

```
This water will kill you if you drink more than 100 ml of it.
```

gets the point across rather more directly.

A phrase such as "seriously contaminate" is known as a *linguistic variable;* it gives an indication of the magnitude of a parameter, but does not provide its exact value. *Subjective knowledge* is expressed by statements that contain vague terms, qualifications, probabilities, or judgmental data. Objects described by these vague statements are more difficult to fit into crisp sets.

The imprecision of a scientific statement may be due to *stochastic uncertainty,* with which every experimental scientist is familiar. This arises from the unpredictability of events or the presence of random error in measurement, and is pervasive in experimental science. Many types of instrument or measuring tools cannot be read perfectly. Temperature, height, mass, lifetime, intensity of absorption, the rate of radioactive decay — measurements of all of these are uncertain. "The car weighs 1800 kg" seems a precise enough statement until we ask whether the actual weight could be a kilogram or two under or over 1800 kg. If not, perhaps the weight is a few grams under or over the stated value. Or even a few milligrams under or over? It is clear that the weight must be uncertain to some degree, however small. Although stochastic uncertainty complicates most scientific experiments, it can be handled with the well-established methods of statistics. Stochastic uncertainty is unavoidable in many types of measurement, but not all, as precise measurements are possible when the parameter to be determined is quantized. The number of cars in a parking garage at any time, for example, can be measured with zero error.

Semantic uncertainty is the type of uncertainty for which we shall need fuzzy logic. Expressed by phrases such as "acidic" or "much weaker," this is imprecision in the *description* of an event, state, or object rather than its measurement. Fuzzy logic offers a way to make credible deductions from uncertain statements. We shall illustrate this with a simple example.

Example 1: Volatile Liquids

Suppose that an ES contains the rule:

```
IF <the sample is volatile> THEN <use GCMS for
analysis>
```

In an ES that cannot handle uncertainty, every liquid must be placed into either the set of volatile liquids or the set of all other liquids before the rule can be used, thus the term "volatile" must be defined. This can be done by specifying a cut-off temperature that separates the two sets:

```
Only liquids that boil below 40°C are volatile.
```

This unambiguous definition is represented by the crisp function in Figure 8.2.

The boiling points of all the liquids in which we are interested could also be shown on a diagram in which the volatile liquids lay within some closed boundary and all others lay outside of it (Figure 8.3).

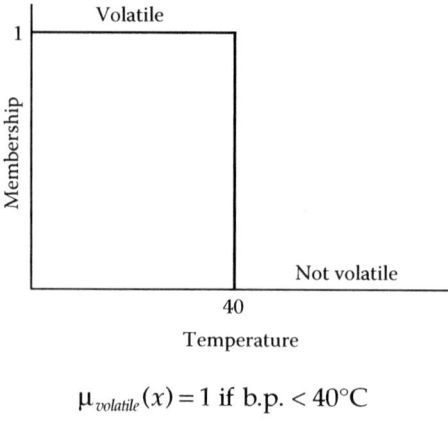

$$\mu_{volatile}(x) = 1 \text{ if b.p.} < 40°C$$

$$\mu_{volatile}(x) = 0 \text{ if b.p.} \geq 40°C$$

FIGURE 8.2
A crisp membership function for the determination of volatile liquids.

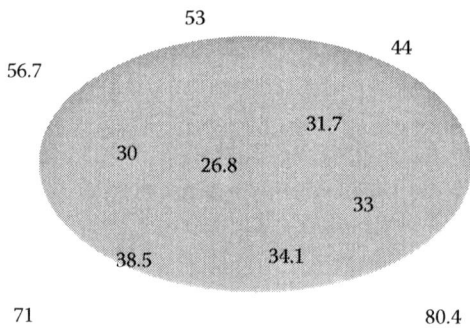

FIGURE 8.3
A representation of several liquids defined by their boiling points. Liquids within the shaded area are (defined to be) volatile; those outside it are not.

Now that the definition of a volatile liquid has been settled, the expert system could apply the rule. However, this approach is clearly unsatisfactory. The all-or-nothing crisp set that defines "volatile" does not allow for degrees of volatility. This conflicts with our common sense notion of volatility as a description, which changes smoothly from low-boiling liquids, like diethyl ether (boiling point = 34.6°C), which are widely accepted to be volatile, to materials like graphite or steel that are nonvolatile. If a human expert used the rule:

```
IF <the sample is volatile> THEN <use GCMS for analysis>
```

to conclude that a liquid with a boiling point of 38°C should be analyzed using gas chromatography-mass spectrometry (GC-MS), she would probably

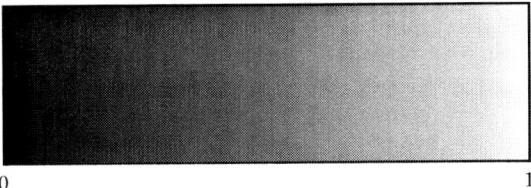

0   1

FIGURE 8.4
The range of logical values in fuzzy logic.

reach the same conclusion about a liquid whose boiling point is 41°C. The hard boundaries imposed by Boolean logic force an abrupt, and unreasonable, change in the recommended method of analysis when the 40°C boundary is crossed.

Fuzzy logic gets around this difficulty by replacing hard boundaries between sets with soft divisions. Objects are allocated to *fuzzy sets*, which are sets with fuzzy boundaries. The membership value of an object within a fuzzy set can lie anywhere within the real number range of 0.0 to 1.0 (Figure 8.4).

Box 2. Membership within a fuzzy set

$\mu_A(x) = 0$,   if x is not within set A
$0 < \mu_A(x) < 1$,   if x is partly within set A
$\mu_A(x) = 1$,   if x is completely within set A

An object may lie entirely within one fuzzy set and, thus, have a membership in that set of 1.0; entirely outside of it and, thus, have a membership of 0; or it may belong to two or more fuzzy sets simultaneously and have a membership between 0 and 1 in each. A liquid whose boiling point is 46°C may belong to the volatile class with a membership 0.45, and the not volatile class with a membership of 0.55.

The NOT function is defined by:

$$\mu_A(not\ X) = 1.0 - \mu_A(X)$$

In a scientific application, the sets to which an object might belong may describe physical or chemical observables that are in some sense continuously variable, such as "volatile," "acidic," or "green," but not all sets are so tangible, nor does the description of a set itself need to be inherently vague. We could create the sets "True" and "False" and, thus, define degrees of truth.

```
Being a scientist is a rewarding profession.
```

This statement might belong to the "True" set with a membership of, say, 0.9, and to the "False" set with a membership of 0.1. Statements that in Boolean logic are either true or false, but not both, can in fuzzy logic be simultaneously

TABLE 8.1

Fuzzy and Crisp Membership Values in the "Volatile" Set for Some Liquids

Chemical	Normal Boiling Point/°C	Volatile (Fuzzy Membership)	Volatile (Crisp Membership)
Acetaldehyde	21	1.0	1.0
Diethyl ether	34.6	0.88	1.0
Sulfur trioxide	44.7	0.51	0.0
Carbon disulfide	46	0.45	0.0
Water	100	0.0	0.0

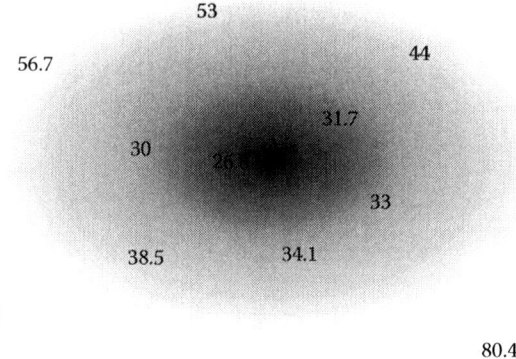

FIGURE 8.5
The boiling points of substances can be used to define the degree of membership (shown here as depth of shading) that a liquid has in a set.

true and false with a membership value in each set that measures the level of confidence that we have in the statement.

Table 8.1 compares the fuzzy membership and crisp membership values in the volatile set for a few liquids (Figure 8.5).

8.4 Calculating Membership Values

How is the membership of an object in a class determined? If it is obvious that an object lies completely in one class, or must have no membership in it, this is easy. We can confidently assign a membership of 0 in the volatile class to steel or diamond. How, though, do we know what membership in that class to give to a chemical, such as carbon disulfide, which has a boiling

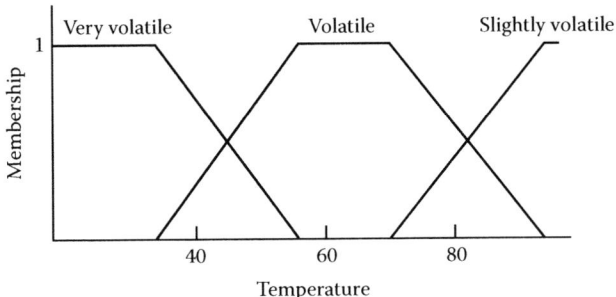

FIGURE 8.6
Membership functions for the sets "very volatile," "volatile," and "slightly volatile."

point of 46°C? A membership of about 0.5 in each of the volatile class and the not volatile class might seem about right, but the choice appears arbitrary.

Although it is the purpose of fuzzy systems to handle ill-defined information, this does not mean that we can get away with uncertainty in the allocation of membership values. If some of the membership values for liquids in a database were proposed by one person and the rest by a second person, the two groups of memberships could well be inconsistent unless both people used the same recipe for determining membership. Any deductions of the fuzzy system would then be open to doubt. In fact, even the membership values determined by just one person might be unreliable unless they had used a properly defined method to set membership values. The hold-a-wet-finger-in-the-air style of finding a membership value is not supportable.

To deal with this difficulty, we construct a membership function plot, from which memberships can be determined directly (Figure 8.6). The membership function defines an unambiguous relationship between boiling point and membership value, so the latter can then be determined consistently, given the boiling point.

The *x*-axis in a plot of a membership function represents the *universe of discourse*. This is the complete range of values that the independent variable can take; the *y*-axis is the membership value of the fuzzy set.

Membership can also be expressed in equation form by a statement, such as carbon disulfide is a member of the set of volatile compounds with a membership of 0.45:

$$mVOLATILE \text{ (carbon disulfide)} = 0.45 \tag{8.1}$$

8.5 Membership Functions

The membership functions in Figure 8.6 provide the basis for a consistent determination of membership values for liquids in three sets, "very volatile,"

"volatile," and "slightly volatile," but it might seem as though the functions themselves have been pulled out of thin air. In a sense they have. There is no recipe that must be followed to turn a concept that is as loosely defined as "very volatile" into a deterministic membership function. The shape and extent of these functions are both chosen in a way that is essentially arbitrary, but seems "reasonable" to the user.

Triangular (Figure 8.7), piecewise linear (Figure 8.8), or trapezoidal (Figure 8.9) functions are commonly used as membership functions because they are easily prepared and computationally fast.

However, there is no theoretical justification for using one of these shapes rather than another. One might suspect that there could be more of an argument for using a normal distribution, as that function arises naturally in the treatment of errors (Figure 8.10), but there is no theoretical justification for preferring this either over triangular or other shapes of function. Provided that the profile used correlates the membership value with the user's perception of what that ought to be to an acceptable degree, the shape of the function is not an important factor in the operation of a fuzzy logic system.

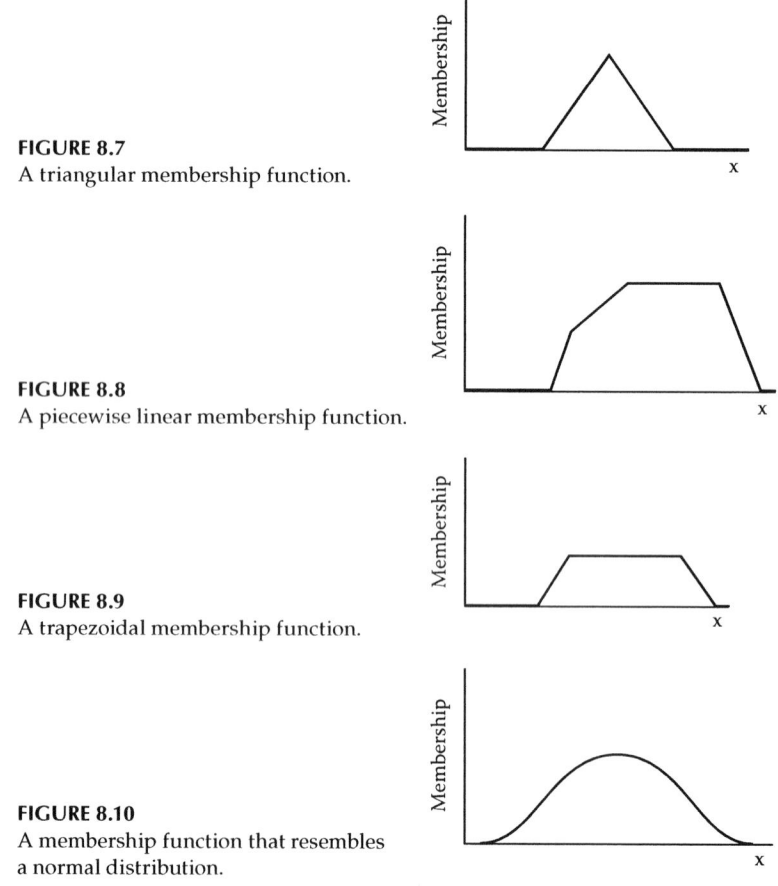

FIGURE 8.7
A triangular membership function.

FIGURE 8.8
A piecewise linear membership function.

FIGURE 8.9
A trapezoidal membership function.

FIGURE 8.10
A membership function that resembles a normal distribution.

The choice of membership function is largely arbitrary, but methods do exist that can help us choose a way to properly express the link between a subjective property and membership functions. Rather than the membership function being chosen by a single user, it can be designed by committee. We might ask several potential users of the system to sketch, or to describe, how they interpret a fuzzy concept. If a fuzzy system will be used to prevent the temperature in a furnace from rising "too rapidly," the scientists, who will be relying on the software to control the furnace, can be asked to provide their own interpretation of what "too fast" means in this context. This is more realistic than hoping that the builder of the fuzzy system will somehow divine their interpretation.

It also may be possible to design an experiment to provide data that can be used to determine the form of the membership function. Suppose that bacterial remediation is being used to reduce the level of pollution in the soil on an old industrial site. Remediation generally proceeds more slowly if the soil on the site is highly compacted. If measurements are made of the density of the soil in different parts of the site, a first fuzzy relationship can be established between an engineer's description of the degree of compaction of the soil and experimental density measurements. In a similar way, experimentally determined remediation rates in different parts of the site can be correlated with an environmental scientist's view of the progress of remediation as "satisfactory," "fast," etc., to give a second fuzzy relationship. Both of these relationships are established by users, but because there is a presumed link between rate of remediation and degree of compaction (albeit probably a rather noisy link), a fuzzy relationship can be derived from them between the engineer's view of the degree of compaction and the environmental scientist's judgment of how successful the remediation is.

Membership functions may be context dependent. We might describe the length of time required for the chemical analysis of a sample as short, medium, or long, but if many different types of samples must be processed, the meaning of these terms will be flexible, depending on what was known in advance about the identity of the sample. If the sample was known to be a pure element, a long analysis might require just twenty minutes. If it was a complex mixture, such as a sample of blood or a sample that required a complete DNA profile, a long analysis might require several weeks. The fuzzy piece of knowledge would then be written with a qualifier:

```
The analysis time is short for an unknown blood sample.
```

Through the definition of membership functions, ambiguity is removed. The vague, ill-defined notion of a concept such as volatility can be turned into something quantitative, which can then be manipulated mathematically. It may seem perverse to argue that the ambiguity has been removed when the form of a membership function has been chosen arbitrarily, especially if the single vague concept of "volatile" has been replaced by two equally vague concepts, "volatile" and "not volatile," or, even worse, by more than two vague concepts. It seems that uncertainty must have increased. However, whereas previously a chemical whose boiling point of 56°C fell in the same nonvolatile class as one whose boiling point was 100°C, by introducing the idea of a degree of membership, the level of volatility of these materials can now be distinguished. Provided that the same membership functions are used to treat all objects in an equivalent way, vagueness has been replaced by deterministic values that can be manipulated by the fuzzy system.

8.6 Is Membership the Same as Probability?

Membership of an object in a set tells us about the object's expected behavior; the total membership of an object in all sets is one. These properties may make us think of the probability that an object exhibits some particular behavior or characteristic because probabilities also sum to 1, so it is natural to wonder whether the degree of membership in a set might actually be the same as a probability.

Membership and probability have some points of contact, but they are not the same. In the probabilistic view, an object must be completely in one set or another and the probability defines the "chance" that the object will be a member of one particular class. If there is a 70 percent chance that a person is tall, he is more likely than not to be in the "tall" set, but if he is not in that set, he must be in a different one, such as "medium" or "very tall." In the fuzzy view, an object may span two or more sets, so the person would be "mainly" in the tall set and at the same time partly in the "medium" or the "very tall" set.

The most important difference between the rules of fuzzy logic and those of probability becomes apparent when we consider the membership of an object in two or more classes simultaneously. Let us suppose that the fuzzy statement:

$$\mu_{very\_toxic}(X) = 0.9 \tag{8.2}$$

tells us that some chemical X has a membership of 0.9 in the fuzzy set "very toxic." If X rapidly decomposes in basic solution, it might be a member of the "rapidly degrades" set with a membership of 0.8. Using probability theory, we would determine that the probability that X is a rapidly degraded, very toxic chemical is $0.9 \times 0.8 = 0.72$. By contrast, fuzzy logic would yield the conclusion that, if X is in the very toxic class with a membership of 0.9 and in the rapidly degraded class with a membership of 0.8, then X is a member of the set of rapidly degraded, very toxic materials with a membership value of the *lesser* of the two memberships in the individual classes to which X belongs, not their product, that is 0.8. We shall learn more of this in section 8.9.

8.7 Hedges

Suppose that we have defined a membership function for the "Low pH" set. Most acid solutions would be, to some degree, members of this fuzzy set. We may want to be able to qualify the description by adding modifiers, such as "very," "slightly," or "extremely" whose use allows us to retain close ties to natural language. The qualifiers that modify the shape of a fuzzy set are known as *hedges*. We can see the effect of the hedges "very" and "very very" in Figure 8.11.

The application of both of these hedges concentrates the region to which the hedge applies. Nonzero membership of the fuzzy set "very low pH" covers the same overall range of pH as membership of the group "low pH," even with the hedge in place, but the effect of the hedge is to reduce membership in the "low pH" set for those pHs that also have some membership in the "medium pH" set. The effect of "very very" is even more marked, as we would expect.

In order to ensure consistency among users of fuzzy logic, hedges have been defined mathematically. Thus, "very" is defined (arbitrarily but consistently within the field) as $\mu_A(x)^2$, "more or less" is defined as

$$\sqrt{\mu_A(x)}$$

and so on. Some examples of the mathematical form of hedges are given in Table 8.2.

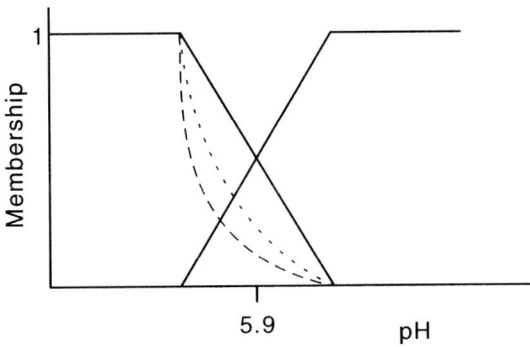

FIGURE 8.11
The effect of applying the hedges "very" (dotted line) and "very very" (dashed line) to the "low pH" set.

TABLE 8.2

Mathematical Equivalences of Some Hedges

Hedge	Mathematical Representation
A little	$\mu_A(x)^{1.3}$
Somewhat	$\mu_A(x)^{1.7}$
Very	$\mu_A(x)^2$
Extremely	$\mu_A(x)^3$
Very very	$\mu_A(x)^4$

8.8 How Does a Fuzzy Logic System Work?

Within a fuzzy system, an inference engine works with fuzzy rules; it takes input, part of which may be fuzzy, and generates output, some or all of which may be fuzzy. Although the role of a fuzzy system is to deal with uncertain data, the input is itself not necessarily fuzzy. For example, the data fed into the system might consist of the pH of a solution or the molecular weight of a compound, both of which can be specified with minimal uncertainty. In addition, the output that the system is required to produce is of more value if it is provided in a form that is crisp: "Set the thermostat to 78°C" is more helpful to a scientist than "raise the temperature of the oven." Consequently, the fuzzy core of the inference engine is bracketed by one step that can turn crisp data into fuzzy data, and another that does the reverse.

In summary the steps in a fuzzy system are

Box 3. Steps in a Fuzzy System

1. Input the data vector.
2. Fuzzify the input data.
3. Evaluate the fuzzy rules.
4. Aggregate the output of the fuzzy rules.
5. Defuzzify the output.

Example 2: Enzyme Kinetics

We will outline the way that a fuzzy system makes deductions using an example from enzyme kinetics.

The rate of a reaction catalyzed by an enzyme depends on several factors, in particular, the concentrations of enzyme and substrate, $[E]$ and $[S]$. As the concentration of enzyme increases, the rate rises, though as the Michaelis–Menten equation shows (equation (8.3)), the rate becomes independent of substrate concentration at high concentrations when the term $K_M/[S]_o$ becomes much less than 1.

$$v = \frac{k_b[E]_o}{1 + K_M/[S]_o} \quad \text{where} \quad K_M = \frac{k_a' + k_b}{k_a} \tag{8.3}$$

k_a and k_a' are the rate constants for the formation of the enzyme-substrate complex and k_b is the rate at which this complex collapses to give product.

The activity of many enzymes is pH-dependent because the enzyme may ionize in solution and the biological activity of unionized and ionized forms may be different. In this case, the rate of an enzyme-mediated reaction can be expected to depend on the acidity of the solution. If the enzyme can lose more than one proton as the pH increases (Figure 8.12), the rate of reaction as a function of pH may display a maximum if the forms of the enzyme in strongly acidic or strongly basic solution are inactive, but the intermediate, monoanion, is active. An example of this behavior is provided by fumarase (Figure 8.13).

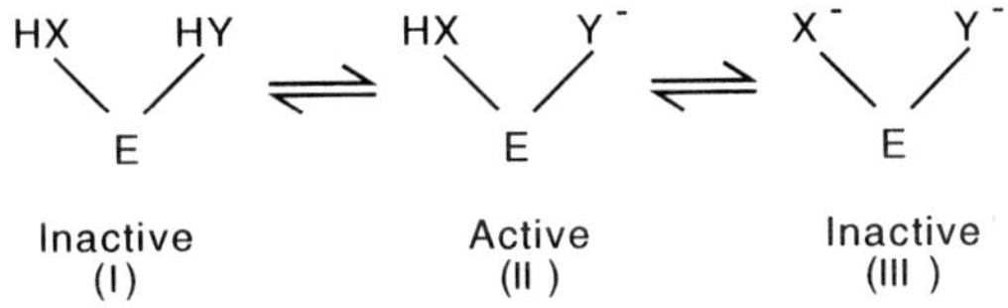

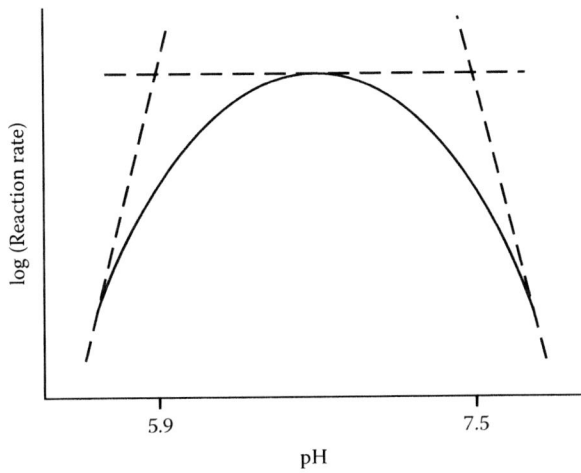

FIGURE 8.12
An enzyme in which two ionizing groups are present.

FIGURE 8.13
Activity of the enzyme fumerase as a function of solution pH.

We could express the activity of fumerase as a function of pH in the fuzzy rules:

R1: **IF** the pH is medium **AND** the enzyme concentration is high **THEN** the reaction rate is high.
R2: **IF** {the pH is low **OR** the pH is high} **AND** the enzyme concentration is high **THEN** the reaction rate is low.
R3: **IF** the enzyme concentration is low **THEN** the reaction rate is low.

8.8.1 Input Data

The inputs into the system are the pH and the concentration of enzyme, measured across all forms, both active and inactive. From these pieces of data, the system is required to provide an estimate of the reaction rate. Let us assume that the total concentration of enzyme is 3.5 mmol dm^{-3} and that the pH is 5.7 and use these values to estimate the rate of reaction.

8.8.2 Fuzzification of the Input

The rules that the fuzzy system uses are expressed in terms such as a "high" or a "medium" pH, while the experimental input data are numerical quantities. The first stage in applying these rules is to transform the input data into a degree of membership for each variable in each class through the use of membership functions.

The pK_a values for the two ionizations of fumarase can be determined experimentally to be approximately 5.9 and 7.5, thus these figures provide an appropriate way to define what, in the context of this problem, is meant by "medium" pH (Table 8.3).

TABLE 8.3

Relationship between pH and Membership of the Low, Medium, and High pH Sets

| | Membership | | |
pH	Low	Medium	High
5.0	1.0	0	0
5.4	1.0	0	0
5.8	0.63	0.37	0
6.2	0.11	0.89	0
6.6	0	1	0
7.0	0	1	0
7.4	0	0.64	0.36
7.8	0	0.12	0.88

If the pH of the solution is 5.7, from Figure 8.14 we can determine that the pH has a membership of 0.76 in the low pH set, 0.24 in the medium pH set, and 0.0 in the high pH set.

Similarly, we can use Figure 8.15 to determine that the enzyme concentration of 3.5 mmol dm^{-3} translates into a membership of 0.80 in the low concentration set and 0.20 in the high concentration set (Table 8.4).

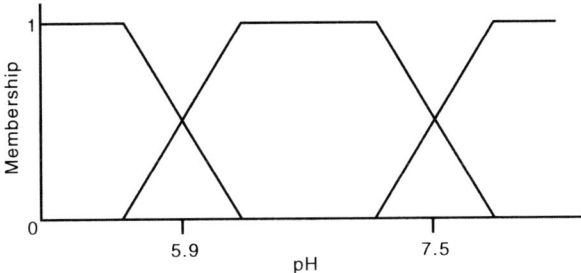

FIGURE 8.14
Membership of the pH in the sets low, medium, and high.

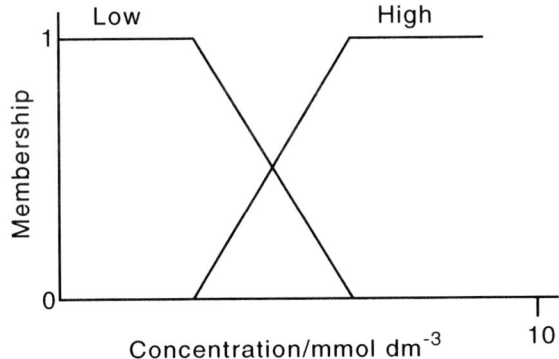

FIGURE 8.15
The relationship between concentration and membership in the low and high concentration sets.

TABLE 8.4

Relationship between Concentration of Enzyme and Set Membership

Concentration/mmol dm^{-3}	Membership	
	Low	High
2	1	0
3	0.94	0.06
4	0.64	0.36
5	0.33	0.67
6	0	1

8.9 Application of Fuzzy Rules

In a conventional expert system, the only rules to fire are those for which the condition is met. In a fuzzy system, all of the rules fire because all are expressed in terms of membership, not the Boolean values of true and false. Some rules may involve membership values only of zero, so have no effect, but they must still be inspected. Implicitly, we assume an OR between every pair of rules, so the whole rule base is

```
R1 OR R2 OR R3 OR ...
```

In order to reason using fuzzy data, a way must be found to express rules so that the degree of certainty in knowledge can be taken into account and a level of certainty can be ascribed to conclusions. This is done through fuzzy rules. A fuzzy rule has the form:

```
If A is x then B is y
```

where A and B are linguistic variables and x and y are linguistic values. For example:

```
IF the car is very cheap THEN the owner is probably a
University Professor.
```

Once the input data have been used to find fuzzy memberships, in the second step we compute the membership value for each part in the condition or *antecedent* of a rule. These are then combined to determine a value for the conclusion or *consequent* of that rule. If the antecedent is true to some degree, the consequent will be true to the same degree.

We have seen above that the fuzzified inputs are

$$\mu(\text{pH is Low}) = 0.76, \quad \mu(\text{pH is Medium}) = 0.24 \quad \mu(\text{pH is High}) = 0.0$$
$$\mu(\text{Concentration is Low}) = 0.80 \quad \mu(\text{Concentration is High}) = 0.20.$$

With these values, we can immediately evaluate rule R3.

```
R3: IF the enzyme concentration is low THEN the reaction rate
    is low.
```

Because the enzyme concentration has a membership of 0.8 in the low set, the antecedent has a membership value of 0.8, and we can conclude that the consequent of this rule has a membership of 0.8.

The remaining two rules have multiple parts, which must be combined to find a value that specifies the degree to which the entire condition holds.

Let us first tackle rule R1:

R1: **IF** the pH is medium **AND** the concentration is high **THEN** the
    rate is high.

If two antecedents are linked by **AND**, the degree to which the entire statement holds is found by forming their conjunction. This degree cannot exceed
the degree to which either part of the statement is true, thus the evaluation
required is the minimum of the two memberships.

$$\mu_A(X \text{ and } Y) = \text{minimum}(\mu_A(X), \mu_A(Y)) \tag{8.4}$$

The pH is medium to a degree 0.24 and the concentration is high to a degree
0.2. Thus, applying the **AND** operator, the rate is high to a degree 0.2.
    Rule R2 is

R2: **IF** {the pH is low **OR** the pH is high} **AND** the enzyme con-
    centration is high **THEN** the rate is low

Here we have both **OR** and **AND** operators. If two antecedents are linked
by **OR**, the degree to which the entire statement holds is determined by how
much the object belongs to *either* set, therefore is given by the maximum of
$\mu_A(X)$ and $\mu_B(Y)$.
    Thus,

$$\mu_A(X \text{ or } Y) = \text{maximum}(\mu_A(X), \mu_A(Y)) \tag{8.5}$$

The pH is low to a degree 0.76 and high to a degree 0, so the combined
membership is 0.76. This is combined with the membership of the enzyme
in the high class, which is 0.2, using **AND** to give a final value of 0.2 for the
antecedent. Thus, the consequent, i.e., that the rate is low, has a degree of 0.2.

8.9.1 Aggregation

It may happen that only a single rule provides information about a particular
output variable. When this is true, that rule can be used immediately as a
measure of the membership for the variable in the corresponding set. In the
enzyme problem, only one rule predicts that the rate is high, therefore, we
can provisionally assign a membership of 0.2 for the rate in this fuzzy class.
Often though, several rules provide fuzzy information about the same variable and these different outputs must be combined in some way. This is done
by *aggregating* the outputs of all rules for each output variable.
    When we have evaluated all the rules, an output variable might belong to
two or more fuzzy subsets to different degrees. For example, in the enzyme
problem one rule might conclude that the rate is low to a degree of 0.2 and
another that the rate is low to a degree of 0.8. In aggregation, all the fuzzy
values that have been calculated for each output variable are combined to
provide a consensus value for the membership of the output variable in each

fuzzy set to which it has access. This consensus is determined by both the degree of membership in each of the output sets and the width of that set in the universe of discourse, as we illustrate below.

The outputs of the three rules have been found to be

```
R1: Rate is high, degree of membership 0.2
R2: Rate is low, degree of membership 0.2
R3: Rate is low, degree of membership 0.8
```

Several recipes exist for aggregation. One of the simplest and most easily applied, due to Mamdani, is as follows. We note that, of the three rules that predict the rate, two of them give different degrees of membership in the "low rate" set. Should we use the higher of these two values or the lower to determine the overall prediction of the system? Let us initially skirt around the question by taking them both, using their average value of 0.5 to give a consensus membership. (You may spot the flaw in this procedure. We shall return to this point in a moment.)

This presumed membership of 0.5 in the "low" set must now be combined with the output from R1, which was that the rate is "high" to a degree of 0.2. To combine these, we require a membership function that relates the actual reaction rate to the fuzzy descriptors "low" and "high" (Figure 8.16).

The memberships in each category are combined by determining the area within each membership function that lies below the line defined by the membership value (Figure 8.17). Thus, the degree to which the rate is low, 0.5, defines a region in Figure 8.17 of a size determined by the membership of the rate in the "low" set. Similarly, the membership in the "high" set determines a second region in the figure. The figure can now be used to defuzzify the conclusions of the fuzzy system and turn them back into crisp numbers.

A defuzzifier is the opposite of a fuzzifier; it maps the output sets into crisp numbers, which is essential if a fuzzy logic system is to be used to control an

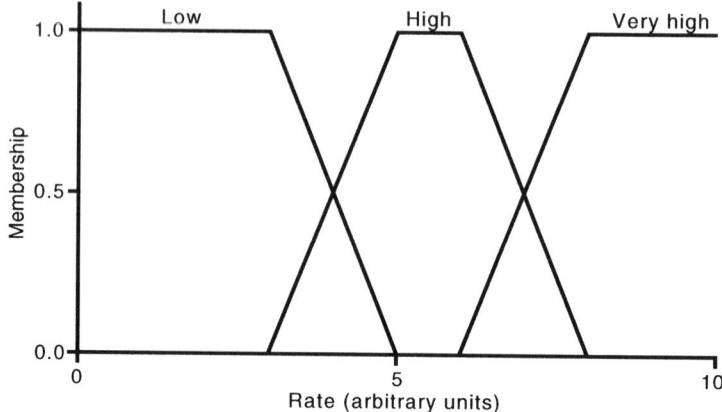

FIGURE 8.16
Membership functions to convert between experimental reaction rates and fuzzy sets.

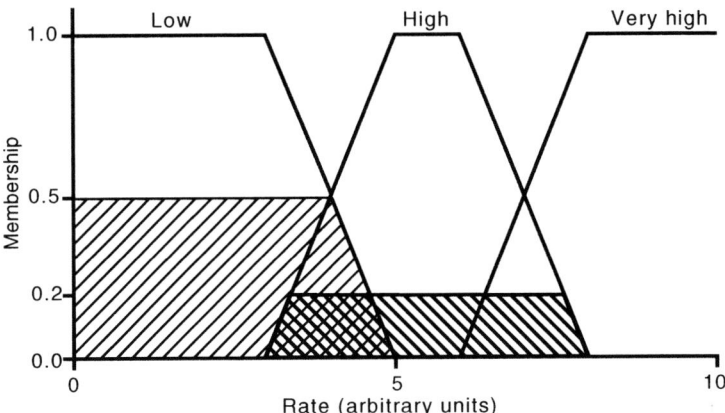

FIGURE 8.17
Defuzzification of set memberships to give a predicted reaction rate.

instrument. It selects a representative element from the fuzzy output created in the aggregation step.

The most popular recipe for defuzzification, although not the fastest, is the *centroid* method. This finds the vertical line, which would divide the aggregated set determined in the previous step, into two equal portions. The center of gravity (cog) is defined by:

$$cog = \frac{\int_a^b \mu_A(x)x\,dx}{\int_a^b \mu_A(x)\,dx} \tag{8.6}$$

The expression in equation (8.6) is applied to all the shaded areas in Figure 8.17. We can break these areas into rectangles and triangles. Starting from the left, we can identify a rectangle that starts at rate = 0 and finishes at rate = 4; this gives the first term in equation (8.7). The shaded area in the "low" membership set is completed by a triangle from rate = 4 to rate = 5. Moving to the "high" set, we must first find the area of a triangle that runs from rate = 3 to rate =3.4, and so on. The numerator in equation (8.6) thus is given by:

$$\int_0^4 0.5x\,dx + \int_4^5 (-0.5x + 2.5)x\,dx + \int_3^{3.4} (0.5x - 1.5)x\,dx + \int_{3.4}^{7.6} 0.2x\,dx +$$

$$\int_{7.6}^8 (-0.5x + 4)x\,dx$$

$$= 4.00 + 1.09 + 0.13 + 4.61 + 0.31 = 10.14 \tag{8.7}$$

while the denominator is

$$\int_0^4 0.5dx + \int_4^5 (-0.5x+2.5)dx + \int_3^{3.4} (0.5x-1.5)dx + \int_{3.4}^{7.6} 0.2dx +$$

$$\int_{7.6}^8 (-0.5x+4)dx$$

$$= 2 + 0.25 + 0.04 + 0.84 + 0.04 = 3.17 \qquad (8.8)$$

The center of gravity, therefore, can be calculated to be

$$10.14/3.17 = 3.2$$

and the predicted rate of reaction is 3.2. The value generated by this procedure seems to be reasonable, but our initial averaging of the two membership values for the membership in the "low" set was a dubious way to proceed. This is because the result of that averaging, a membership of 0.5 in the "low" set, was given exactly the same weighting as that allocated to the membership of 0.2 in the "high" set generated by R1, even though two rules contributed to the low set membership and only one rule to the high. This is hard to justify. If a dozen rules predicted a low rate, with varying degrees of confidence, and just one rule predicted a high rate, we could be fairly sure that the rate really is low because so many rules were voting for this. We would be much less sure if only one rule predicted a low rate. Therefore, somehow we must take into account each rule individually.

This is easily done. Rather than averaging the values of all consequents that predict membership of the same fuzzy set, we include separately the membership values for every rule in the calculation of equation (8.6). As Figure 8.18 shows, there are now three areas to consider, so the numerator in equation (8.6) becomes:

$$\int_0^{3.4} 0.8xdx + \int_{3.4}^5 (-0.5x+2.5)xdx + \int_0^{4.6} 0.2xdx + \int_{4.6}^5 (-0.5x+2.5)xdx +$$

$$\int_3^{3.4} (0.5x-1.5)xdx + \int_{3.4}^{7.6} 0.2xdx + \int_{7.6}^8 (-0.5x+4)xdx$$

$$= 4.62 + 2.52 + 2.12 + 0.19 + 0.13 + 4.61 + 0.31 = 14.5 \qquad (8.9)$$

and the denominator is now:

$$\int_0^{3.4} 0.8dx + \int_{3.4}^5 (-0.5x+2.5)dx + \int_0^{4.6} 0.2dx + \int_{4.6}^5 (-0.5x+2.5)dx +$$

$$\int_3^{3.4} (0.5x-1.5)dx + \int_{3.4}^{7.6} 0.2dx + \int_{7.6}^8 (-0.5x+4)dx$$

$$= 2.72 + 0.64 + 0.46 + 0.04 + 0.04 + 0.84 + 0.04 = 4.78 \qquad (8.10)$$

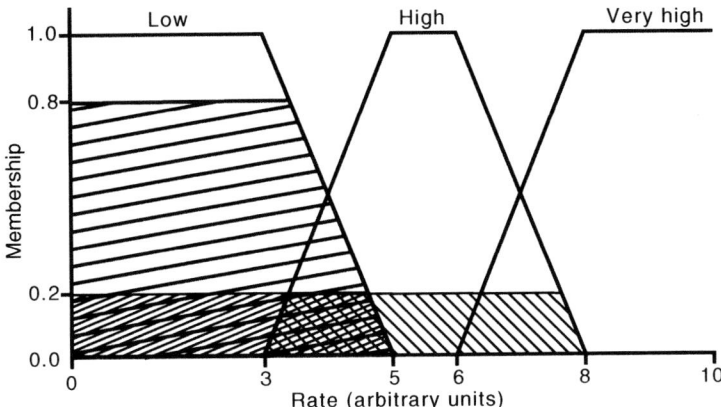

FIGURE 8.18
A better defuzzification method to give a predicted reaction rate.

The predicted rate is, consequently, $14.5/4.78 = 3.03$ and we note that in the revised procedure the overall rate is found to be lower, as we might have expected because two rules are voting for a low rate.

The center of gravity is a reliable and well-used method. Its principal disadvantage is that it is computationally slow if the membership functions are complex (though it will be clear that if the membership functions are as simple as the ones we have used here, the computation is quite straightforward). Other methods for determining the nonfuzzy output include center-of-largest-area and first-of-maxima.

8.10 Applications

An early application of fuzzy logic was in the control of a cement kiln and applications to similar or identical systems have continued to appear (see, for example, Jarvensivu et al.[1]). This is an example of a practical application in which generation of high quality product may depend on the skill and experience of a small number of workers. Even when computer control is available, workers may describe the way that they adjust conditions to provide a high-quality product by statements such as, "The kiln rotation rate should be lowered slightly," and that, in order to compensate, "The temperature should be diminished just a bit." Fuzzy logic now has a lengthening track record of use in such situations.

Control problems represent a major area of application for fuzzy logic since reliable process control may rely on the long-term expertise of one or a few people, and those people may be able to frame their knowledge of the system only in imprecise terms. A typical example is the control of pH in a crystallization reactor.[2] A similar application was described by Puig and co-workers, who used this method to control the amount of dissolved oxygen in a wastewater treatment plant.[3] In other control problems that are rarely under the control of people but for which the control problem may be difficult to program in exact terms, fuzzy logic can again be of value, e.g., to control autofocus in cameras[4] or to reduce the effect of camcorder vibration on picture quality.

A recent review presents a broad overview of the use of fuzzy logic in biological sciences. Biological applications represent a rich field for fuzzy logic, as a recent review suggests.[5] Du and co-workers have used fuzzy systems to model gene regulation,[6] and Marengo's group has considered the use of fuzzy logic in analysis of electrophoresis maps.[7] Medical applications are also increasingly common, as exemplified by the work of Chomej and co-authors.[8]

8.11 Where Do I Go Now?

Fuzzy logic is often presented as an extension in books that cover expert systems. Few texts exist in which the applications of fuzzy logic to scientific problems are described, but several texts include more general discussions of the principles and practical implementation of this method. Among the best is Negnevitsky's text on intelligent systems.[9]

8.12 Problems

1. The rate of enzyme-mediated reactions, like most other types of reaction, depends on temperature. Over a limited temperature range, the reaction may follow the Arrhenius equation:

$$k = Ae^{-\frac{E_a}{RT}}$$

in which k is the rate constant for the decomposition of the enzyme-substrate complex, E_a is the activation energy for that step, and A is the Arrhenius constant. Reactions involving biochemicals may be complicated by the sensitivity of these molecules to heat. Proteins are readily denatured by a rise in temperature as the protein unfolds and the geometry of the active site is disturbed.

Expand the list of rules given for the reaction mediated by fumerase that was discussed above so as to include both the rate increase with temperature as described by the Arrhenius equation and also the possibility of deactivation of the enzyme at higher temperatures. Consider whether it is necessary to incorporate the Arrhenius equation directly into a fuzzy rule or whether it is better to use this outside the fuzzy system.

References

1. Jarvensivu, M., Saari, K., and Jamsa-Jounela, S.L., Intelligent control of an industrial lime kiln process, *Contr. Eng. Prac.*, 9, 589, 2001.
2. Chanona, J., et al., Application of a fuzzy algorithm for pH control in a struvite crystallisation reactor, *Water Sci. Tech.*, 53, 161, 2006.
3. Puig, S., et al., An on-line optimisation of a SBR cycle for carbon and nitrogen removal based on on-line pH and OUR: The role of dissolved oxygen control, *Water Sci. Tech.*, 53, 171, 2006.
4. Malik, A.S., and Choi, T.S., Application of passive techniques for three-dimensional cameras, *IEEE Trans. Cons. Elec.* 53, 258, 2007.
5. Torres, A. and Nieto, J.J., Fuzzy logic in medicine and bioinformatics, *J. Biomed. Biotech.*, 91908, 2006.
6. Du, P., et al., Modeling gene expression networks using fuzzy logic, *IEEE Trans. Sys. Man Cybernet. Part B – Cybernet.*, 35, 1351, 2005.
7. Marengo, E., et al., A new integrated statistical approach to the diagnostic use of two-dimensional maps, *Electrophoresis*, 24, 225, 2003.
8. Chomej, P., et al., Differential analysis of pleural effusions by fuzzy logic based analysis of cytokines, *Respir. Med.*, 98, 308, 2004.
9. Negnevitsky, M., *Artificial Intelligence: A Guide to Intelligent Systems*, 2nd ed., Addison-Wesley, Reading, MA, 2005.

9

Learning Classifier Systems

Classifier systems are software tools that can learn to control or interpret complex environments without help from the user. This is the sort of task to which artificial neural networks are often applied, but both the internal structure of a classifier system and the way that it learns are very different from those of a neural network. The "environment" that the classifier system attempts to learn about might be a physical entity, such as a biochemical fermentor, or it might be something less palpable, such as a scientific database or a library of scientific papers.

Classifier systems are at their most valuable when the rules that describe the behavior of the environment or its structure are opaque and there is no effective means to determine them from theory. An understanding of the environment is then difficult or impossible without the help of some learning tool. Classifier systems behave as a kind of self-tutoring expert system, generating and testing rules that might be of value in controlling or interpreting the environment. Rules within the system that testing reveals are of value are retained, while ineffective rules are discarded. In a classifier system, the rules evolve whereas, in an expert system the rules would be integrated into the software.

Classifier systems are in some respects a scientific solution looking for a problem because they are currently the least used of the methods discussed in this book. However, their potential as a disguised expert system is substantial and they are starting to make inroads into the fields of chemical and biochemical control and analysis.

9.1 Introduction

Scientific control problems are widespread in both academia and industry. Large segments of the chemical industry are devoted to the synthesis of bulk chemicals and the plants in which synthesis is performed, run with a high degree of automation. For most practical purposes, any chemical plant that operates on an industrial scale runs completely under automatic control.

A similar level of automation is found in the biochemical industry. Although the volumes of production of biochemicals are smaller by several orders of magnitude than those of bulk chemicals, companies that operate fermentors and other types of biochemical reactors must still work within the constraints imposed by multiple goals: the products must have a minimum level of purity, their composition must be consistent from batch to batch, and they must be produced with the maximum possible yield. At the same time the operation must be safe, have the minimum possible need for human intervention, and have a low demand on power and other services. The key to meeting these requirements is reliable process control.

Not long ago, evidence of sound process control would be the sight of a few computers running part of the operation and some engineers standing around with clipboards who would oversee what was going on. The extent to which process control has gained in sophistication since then is exemplified by the "lights out" laboratories that large pharmaceutical companies now operate. In these laboratories everything one would expect to see in a modern science laboratory is present, except that there are no people. The laboratories are packed with scientific instruments, which run automatically and continuously, robotic samplers that feed the instruments, and computers to ensure that everything runs smoothly; yet no technicians are needed to feed in samples or control instruments. Humans do enter the laboratory, but only to complete tasks like refilling reagent containers or dealing with faulty equipment. Every other operation, from sample preparation or choice of reaction conditions to the interpretation of spectra, is performed under the direction of computers.

The development of the lights out laboratory has been driven by economics: computers are cheaper than a technician, will work long hours without complaining or asking for a comfort break, do not need to spend time away at Disney World® every August, and may be more consistent in their performance. The laboratories are a notable modern example of the way in which process control is being taken over by computers, but automatic control has been long recognized as one of the best ways to increase efficiency and ensure safety.

The principal reason why automatic control has these advantages over human control is that the synthesis or isolation of even a simple chemical, such as ethylene oxide, on an industrial scale is a complex technical problem. Figure 9.1 shows a schematic of a section of an industrial unit for the production of this compound. Whenever the plant is operating, a stream of messages floods into the control center, reporting the temperature in the reactors, the pressure in the flowlines, the purity of the product, and other data. The rate at which messages appear may be so great that no single human could even read them all, let alone process them efficiently and effectively; thus messages from sensors are collected and analyzed by computers in real time in order to assess the current state of the plant. Necessary action, such as turning on a heater or reducing the pressure in a flowline, can then be taken by the controlling computers so that the process proceeds smoothly, safely, and efficiently.

Computers can provide a near instantaneous response to any change in operational conditions, thus enhancing operating efficiency. However, while a quick response is desirable, making the wrong decision is not. The

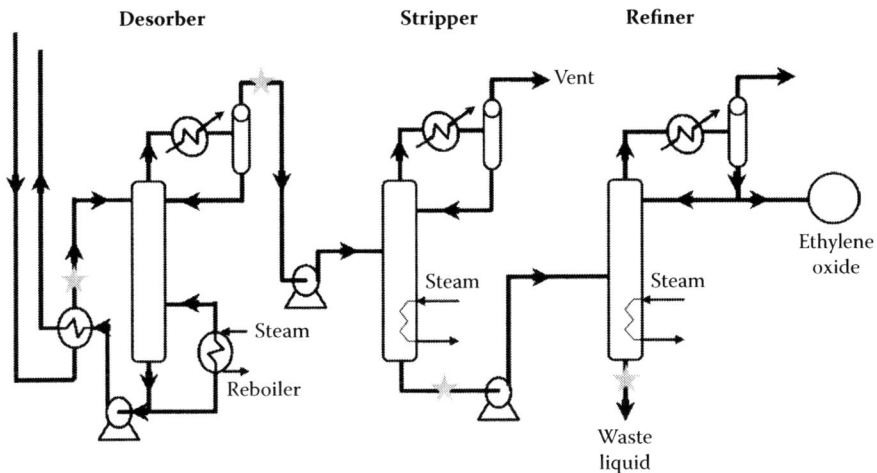

FIGURE 9.1
A portion of an industrial plant that produces ethylene oxide.

response must be appropriate so the computers that control the plant need to know what they are doing. Some algorithmic understanding of the processes that are occurring within the plant in the form of a model of those processes is required, in order that any adjustments to operating conditions made in response to messages coming from the sensors are suitable.

The understanding that the computers controlling the equipment might possess could be contained within a kinetic and thermodynamic model that encapsulates the detailed chemistry of the process. This would include a description of all the reactions that might be expected to occur under the conditions achievable in the plant, together with a list of the relevant rate constants and activation energies for each reaction. In addition, process variables, such as maximum flow rates or pump pressures that are needed for a full description of the behavior of the system under all feasible conditions, would be provided.

Even for a quite simple process, this might be a fearsome model, but in favorable circumstances, when changes in operating conditions within the plant are at least an order of magnitude slower than the rate at which the kinetic and thermodynamic computer model runs, it should be possible to use the model to predict the behavior of the reactor. Assuming that it is possible to build such a chemical model, the controlling computers might rely upon an expert system to monitor and direct the process; such a system would incorporate a set of rules derived from an inspection of, and interaction with, the model so that the appropriate response is provided to any possible combination of system conditions. The reactor in this scenario is controlled by explicit programming for every possible combination of conditions that might arise in the reactor. If the model on which the expert system relies is sufficiently detailed and robust, software built on that foundation should be adequate to control the plant.

What if the behavior of the system cannot be reliably predicted? In the combustion of air–fuel mixtures in a car engine or a power station, the number of distinct chemical reactions taking place may exceed one thousand; real-time modeling of such a system is not feasible.

If the available computational models are insufficiently detailed so that behavior is too uncertain to predict, or if the only model that can be constructed is so detailed that its execution is unacceptably slow, we cannot expect to be able to use an expert system to control the process under all circumstances. Fortunately, there is an alternative — a software model that can *learn* how to control a system rather than needing to be told how to do so.

This software model is a *learning classifier system*. Because classifier systems learn, they can be applied to the control of a dynamic system, such as a reactor or an instrument, which must process various types of samples under unpredictable conditions, even when the rules required for successful control are unknown.

As we shall see in this chapter, the classifier system is given no advance knowledge to help it plan what to do. Data on the chemical reactions that may be occurring or the values of rate constants are not provided; therefore, it is wholly dependent for its success upon learning, which it manages by investigating the environment. This is a very different approach from one that relies on a comprehensive theoretical model of the reactions in the system.

This seems like a promising method of attack, but the news is not all good. The learning system may generate a model that is difficult to interpret, just as was the case with artificial neural networks, and it is then hard for outsiders to check the system for completeness. If a reputedly "fully trained" learning system is put in control of a chemical plant that then moves into a state that the learning system has never before encountered, the reaction of the system will be unpredictable and may be dangerous. In addition, a model based on an expert system that uses chemical properties and rules that can actually be inspected may garner more trust from the technicians who must rely on it, while they may be suspicious of an evolving system that can only learn from the conditions of which it has had experience and is unable to explain what it is doing.

Despite these reservations, classifier systems have the potential to outperform model-based systems, particularly when theoretical models are weak.

9.2 A Basic Classifier System

A classifier system (CS) is a type of *agent*. An agent is a software object that accepts messages from the outside world, analyzes the content of those messages, and reacts to them, either by generating a response that is designed to bring about some particular result in the environment in which the agent finds itself, or by changing its own behavior in a way that the agent believes will be beneficial.

The CS is aware of what is happening in its surroundings through messages that are sent to it from *detectors* (pressure, temperature, or pH sensors in a reaction vessel, perhaps). It responds by sending requests into the environment, asking that changes, such as turning on a heater or adding some acid, be made to meet a defined goal. The changes are put into effect in the environment, which then returns a *reward* whose size is proportional to the degree to which the changes were successful in meeting the goal. The software operates in such a way that it tries to maximize the reward it receives, and because a high reward is provided when the environment is properly controlled, the interests of the software and of the environment are aligned.

This trial-and-error approach to training software is known as *reinforcement learning*, a fundamental paradigm in machine learning. In reinforcement learning, a system does exactly what we have just described, interacting with its environment in such a way that it tries to maximize any payoff that the environment can provide. The system knows nothing about the details of the task that it is carrying out, but concentrates entirely on maximizing its reward; thus the same basic software could be used to run a chemical plant through control of valves and heaters, or stabilize the temperature in a greenhouse by opening and closing windows.*

Computer scientists use reinforcement learning to train robots equipped with a simple vision system that are seeking a source of electric power to recharge themselves. The inputs to the software that resides in the robot include signals from position and light sensors on the robot and these inputs prompt the software to generate instructions to actuators that move the robot or change the direction in which it faces. If the robot moves toward, and eventually reaches, a power outlet, it receives a reward. The expectation is that in due course the robot will learn to associate an image of the power outlet that its sensors provide and movement toward that outlet with a reward.

A chemist or a biochemist might want this sort of software to do something a little more practical, like controlling a biochemical reactor or analyzing a medical database. In the former case, the inputs would provide details of the progress of the reaction and conditions within the reactor, while the outputs would be instructions to bring about physical changes, such as opening a valve to add reagent or increasing the flow of coolant through a reaction vessel jacket.

---

* We have been here before. The genetic algorithm attempts to find the string of highest quality without knowing anything about what the string actually "means" in the outside world. An artificial neural network attempts to find the set of connection weights that minimize the difference between the network output for a test set and the target responses. And a self-organizing map modifies its weights vectors so that every sample pattern can find a node whose weights resemble those of the pattern, though the algorithm knows nothing of the significance of its weights. In each case, the scientific interpretation of the sample data is of no consequence to the AI algorithm. It is evident that this underlying indifference to the meaning of the sample data is one of the reasons that AI systems are so flexible.

The most useful classifier systems learn without any direct help or instruction from the user. We shall meet these in the second part of this chapter, but first we consider how a CS programmed in advance could be used to control the temperature of a reacting mixture in a vessel.

9.2.1 Components of a Classifier System

A CS is no more than a chunk of computer code, but the software is not isolated from the environment in which it operates. Indeed, were there no environment in which to work, the system would have nothing to do. We therefore include both the software and its surroundings among the components of the CS (Figure 9.2).

The components in a CS include:

1. The *environment*, which consists of everything outside the software itself that the system is trying to control or analyze.*

2. *Receptors* that monitor the environment and generate *input messages*; these inform the CS about conditions in the environment. The input messages and *post-it notes* from the classifier itself are gathered on an *input list*.

3. The *classifier system*, which analyzes the input messages, creates messages for output, and places them on an *output list*.

4. *Effectors*, which are most commonly pieces of physical equipment through which the CS can interact with the environment.

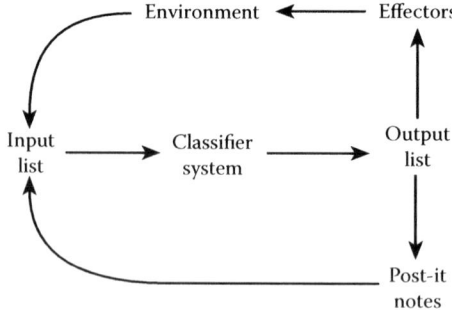

FIGURE 9.2
The components that comprise, and the flow of messages within, a classifier system.

---

* In this discussion, we assume that the environment is a physical entity, but other environments may be used. It could be a database, a stream of messages, or any object that can provide input to the classifier system, accept output from it, and pass judgment on the quality or value of that output.

The components engage in the following steps in what is known as the *major cycle* of the system:

Major cycle steps

---

1. Collect messages from the environment on the input message list.
2. Compare every input message against every classifier to identify matches.
3. For every match, generate some output from the classifier and attempt to place the output on the output list.
4. Read the output messages and use them to bring about change in the environment.
5. Receive a reward from the environment and divide it up among the classifiers.
6. Go to 1.

Steps 2 and 3, in which the environment plays no direct part, can be run independently for each prototype rule in the CS; thus the system is well suited to implementation on parallel processor machines. In the next section, we consider the components that form the system in more detail.

9.2.1.1 The Environment and Messages from It

Imagine that we wish to control automatically a biochemical reactor in which fermentation is taking place (Figure 9.3).

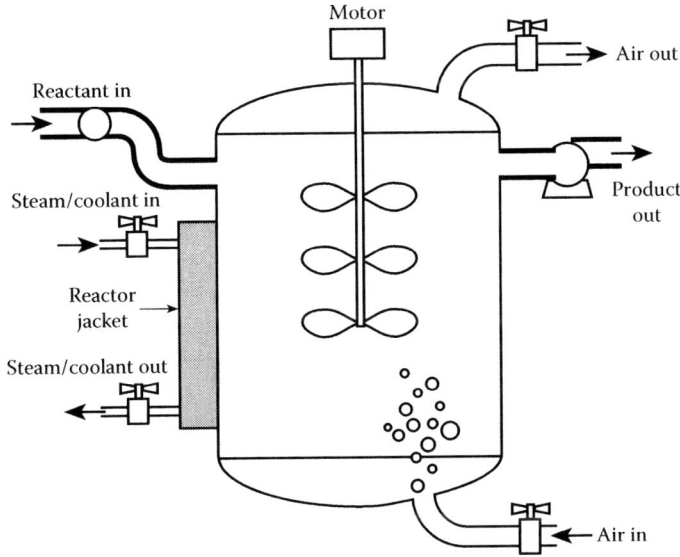

FIGURE 9.3
Schematic of a biochemical reactor.

For healthy fermentation, several properties of the fermenting medium must be controlled. We shall focus on control of the temperature. Minor departures from the optimum temperature will reduce the efficiency of operation or the rate of fermentation. Larger fluctuations might cause more serious disruption to the process by killing the active organisms so that the reactor needs to be drained and the fermentation restarted from scratch. We shall define the optimum temperature range for fermentation as being from T_{low}^{opt} to T_{high}^{opt} and the extreme limits, beyond which the viability of the organisms is threatened, as T_{too_hot} and T_{too_cold}. A CS could be used to control conditions in the reactor by directing the operation of heaters and coolers to maintain a suitable operating temperature.

To do this, the software must be aware of conditions within the reactor; therefore, we shall arrange that it is fed with regular messages from a digital thermometer in contact with the fermenting medium. The thermometer, and any other probes in the vessel, send messages to the software in the form of binary strings, such as:

```
0010100   11101111   010100010
```

in which each group of digits comprises a single message.

These messages that report on the conditions in the reactor have two parts (Figure 9.4). The first is an identifier that specifies which probe has sent the message. In the reactor, several properties of the system may be monitored continuously by different types of probes. Along with the temperature of the reaction mixture, data might be provided on the pH of the mixture, the pressure within a pump, or the rate at which material is leaving the reactor through a draw-off pipe. An identity tag, therefore, is attached to every message that arrives at the CS so that the software can determine which probe sent it; the rest of the message is quantitative information from the detector.

The total length of the message is the same irrespective of which probe was responsible for it. The length is determined by the number of different detectors that are available, which determines the length of the identifier tag and by the precision to which the data are specified.

Let us assume that messages from the digital thermometer are tagged with the identifier "00." The rest of the message conveys information about the current temperature, T. Table 9.1 shows how the message sent from the digital probe provides information about reactor conditions.

It might appear that to use a short binary string to represent a message from a sensor such as a temperature probe, which monitors a property that can vary continuously over a real-number range, is to throw information

$$\lfloor 010 \rfloor \quad \lfloor 100010 \rfloor$$
Identifier   Data

FIGURE 9. 4
The format of an input message.

TABLE 9.1

The Meaning of Messages Sent by the Digital Thermometer

Message	Condition
00111	$T \leq T_{too\_cold}$
00100	$T_{too\_cold} < T < T_{low}^{opt}$
00011	$T_{low}^{opt} \leq T \leq T_{high}^{opt}$
00001	$T_{high}^{opt} < T < T_{too\_hot}$
00000	$T \geq T_{too\_hot}$

```
Input Message List
1000101
0110101
1101010
1101010
1001010
0111011
. . . . .
```

FIGURE 9.5
The input message list.

away. We shall soon need to compare the binary input messages with other binary strings, however, and use of this format makes this comparison simple, fast, and unambiguous.

9.2.1.2 Input Message List

As messages from the sensors arrive, they are collected onto an *input message list*. Messages are not dealt with by the software immediately when they arrive, but instead accumulate on this list until the CS is ready to process them. By assessing a group of messages as a whole (Figure 9.5) rather than dealing with each message as soon as it appears, the CS can get a more complete picture of the current state of the environment. This allows for more precise control because the best course of action at any time will depend on an assessment of all factors, such as the temperature, pH, and oxygen concentration that characterize conditions in the reactor.

9.2.1.3 The Classifier List

When the CS is ready to start processing messages, it freezes the contents of the input message list, then reads and assesses its contents. The messages are fed, one at a time, into the *classifier list*, which is a large collection of rules.* The classifier list is the only part of the system that can be modified, thus this is where the memory of a CS resides. Rules in a CS have the same format as those used by expert systems:

```
IF <condition> THEN <action>
```

The <action> in every case is the generation of a message, so the processing of the input messages leads to zero or more output messages being created. These are placed on an *output message list.*

Even for a small-scale problem, the classifier list may be long. This has the advantage that, since many classifiers may be able to generate output messages at one time, the system implicitly includes several different ways of responding to messages from the environment (in other words, it can incorporate several different hypotheses about how the environment behaves). This helps the CS to cope with uncertainty when there may be insufficient information on the input message list to determine precisely what state the environment is in.

9.2.1.4 The Output Interface

Every classifier reads, and may respond to, every input message; hence the number of messages that may be placed on the output list could be large. In order for the CS to make sense of all these messages and bring about change in the environment, the system has an *output interface.* Messages on the output list are delivered to this interface, which sorts them, resolves any conflicts within them, and sends the messages to the appropriate effectors. The effectors, which may be any kind of object that can bring about change in the environment, act upon these messages, implementing the changes requested by the CS, such as increasing the stirring rate or changing the rate of flow of oxygen into the reactor.

9.3 How a Classifier System Works

9.3.1 Gather the Input

The first step in running a classifier is to gather messages on the input list. Probes and sensors in the environment do not synchronize their dispatch of messages, but just send them whenever they feel the urge; thus messages from the environment begin to pile up in the input list until the CS is prepared to deal with them.

---

* The set of rules that comprise a CS is sometimes referred to as a *production system,* as they have the form of production rules.

9.3.2 Compare Input Messages with the Classifiers

The classifier list is at the heart of a CS. Each classifier is a string of characters that can be interpreted as a rule, cast in the form IF <condition> THEN <action>. Following the style of Riolo's classifiers, we shall write a classifier as:

condition/action

The condition part of the classifier is constructed from the ternary alphabet [0, 1, #], so each of the following is a valid condition string:[*]

```
11001   10##0   11000   #####
```

A classifier (often) contains more than one condition and (rarely) contains more than one action, thus the most general form of a classifier is

condition₁, condition₂, ... conditionₙ/action₁, action₂, ... actionₘ

When the classifier system is ready to tackle messages on the input list, the first message is extracted and processed by the first classifier. Each digit of the entire input message, including the identifier tag, is compared with the corresponding digit in the condition of the classifier to determine whether the digits in the message and the condition match in every position. Because all digits are compared, position-by-position, the length of the condition string in the classifier must equal that of the input message. If the # sign appears in the condition, this is interpreted as a "don't care" symbol, so a # in the classifier matches either a 1 or a 0 in the message.

Thus, the input message 01101 matches each of the conditions 01101 01##1 and ##### but does not match #1110. Note that the presence of the identifier tag at the start of a message and the requirement that this must match the corresponding digits in the classifier conditions means that some classifiers will match and, therefore, deal with only messages from a particular probe.

Multiple condition classifiers are satisfied only when all of the conditions are met by messages on the current input list. It is not necessary that the different conditions are satisfied by different messages, although they may be. Thus, the message 11001 will satisfy both conditions in the classifier 110##, 1##01/action.

One or more conditions in a classifier may be preceded by a "~" character; this is interpreted as negation of the condition, thus the condition is satisfied only if there is no matching message on the input list. Hence, the classifier:

```
11001, ~01### / action
```

---

[*] This ternary coding has some limitations; in particular, it may affect the ability of the system to derive general rules. The coding described in this chapter, which relates to work by John Holland, has been successfully used in a variety of applications and has the advantage of simplicity. Readers who wish to explore alternative ways of representing classifiers will find them described in recent papers on classifier systems.

will be satisfied only if (1) the message `11001` does appear on the input list and (2) there are no messages on the input list that start with 01.

9.3.3 Generate Output Messages

If the classifier condition and the input message match at every position, or if all conditions in a multiple condition classifier are matched by some combination of input messages, the classifier is said to be *satisfied* by the message(s), the action part of the classifier is triggered and it generates an output message.

Just like the condition, the classifier's action is formed from the ternary alphabet [0, 1, #]. If it appears in the action, the symbol # acts as a "pass through" operator; wherever # is present, it is replaced by the digit that occurs in the corresponding position in the input message.

Example 1

If the input message list consists of the messages:

`01010  10010  00111  00010`

and the classifiers are

`0010#/#0001, #001#/11111, 00101/10##1, 0##10/1##10,`
`01##0/1#001`

the output messages shown in Table 9.2 will be generated.

The messages `11010`, `11001`, `11111` (twice) and `10010` will then be posted onto the output list. A single input message may be, and often will be, matched by more than one classifier. Equally, a single classifier may be satisfied by several input messages. If the classifier has more than one condition, each input message may give rise to a separate output message. If the input messages were

TABLE 9.2

Output Messages Generated by a Small Group of Classifiers (see text)

Input message...	...is matched by the classifier...	...and generates the output
01010	0##10/1##10 and 01##0/1#001	11010 and 11001
10010	#001#/11111	11111
00111	None	None
00010	#001#/11111	11111
00010	0##10/1##10	10010

`11001` and `10011`

The classifier

`11#01, #0010/01###`

would generate the two output messages

`01001` and `01011`

It is easy to see that multiaction classifiers could give rise to large numbers of output messages, so these are used sparingly.

The order in which output messages are generated and placed on the output list is of no consequence, so the system can process information in parallel without worrying about whether one classifier should inspect the input list before another gets the chance. The number of messages posted to the output list is, in Example 1, equal to the number of classifiers, but in general the input and the output lists need not be equal in length either to each other or to the classifier list.

Output messages have a format that is similar to input messages; they are binary strings and are divided into an identifier and a data portion. We shall assume that any output message that starts with `01` is an instruction to a heater in the reactor. The remaining digits in the message indicate the power to be supplied to the heater, as specified in Table 9.3.

Thus, input message `00111` (interpreted as "the temperature is below T_{too_cold}") would satisfy the classifier `00111/01111`. The classifier would output the message `01111`, which would be interpreted as "turn on the heater" (the `01` part of the action) "to maximum power" (the `111` part). In this way, an input message that indicates a need for action has generated a suitable

TABLE 9.3

Relationship between an Output Message and the Action Taken in the Environment

Output Message	Action	Heater Power
01000	Turn on heater	0 (minimum)
01001	"	1
01010	"	2
01011	"	3
01100	"	4
01101	"	5
01110	"	6
01111	"	7 (maximum)

response. In effect, the identifier tag ensures that the message is delivered to the classifier(s) designed to cope with it; therefore, the classifier 01##0/101#1 will only read and respond to messages sent from the temperature probe whose tag is 01.

9.3.4 Update the Environment

Each input message is compared with each classifier in turn. Once any triggered classifiers have created their output, the output message list is complete.

Before the contents of this list are sent on to the effectors, all messages are checked to determine whether there are any conflicts. These can readily arise: one classifier might request that a heater be turned on, while a second asks that it be turned off. (A simple conflict-resolution procedure will be discussed in section 9.5.1.) Once conflicts among the output messages have been resolved, the output messages that are instructions to the environment (some messages on the output list may have other meanings, as we shall learn shortly) are sent to the appropriate effectors, which respond by turning on or off heaters or chillers, adding reagent, and so on.

9.4 Properties of Classifiers

9.4.1 Generalist and Specialist Classifiers

In a system that shows complex behavior, the CS must be able to generate many different output messages so that it can control the system under all circumstances, or if it cannot generate a large number of distinct messages, it must at least be able to form a wide variety of *combinations* of output messages. The more varied the output messages that are created, the richer the range of behavior that the CS can handle.

This might suggest that a classifier that generates thousands of output messages is to be preferred over one that generates a more modest number, but we can see why this is unlikely to be true by considering the action of classifiers such as:

#####/10010   ##1#1/00###

The condition part of these classifiers contains several # symbols, so the classifiers will be satisfied by many input messages and, therefore, generate numerous output messages.

Classifiers that contain many # symbols in their condition part are known as *generalists*, as they will be satisfied by a wide variety of different input messages. The first classifier given above is the supreme generalist, since any message turns it on. By contrast, classifiers whose condition part contains few # symbols compared to the number of 0s and 1s, or contains none at all, are *specialists*, triggered only by one or a few messages. Although these classifiers will rarely be called into action, they may nevertheless be of particular value, providing the right response to deal with specific environmental conditions. We can expect them to be a valuable component in the classifier set.

Example 2

Table 9.1 shows that both the input messages 00111 and 00100 indicate that the temperature is too low. The classifier 001##/01### would be satisfied by either of these messages. In response to the message 00111, which indicates that the temperature is below the safe limit so must immediately be raised, the classifier would output 00111 (turn on the heater to full power). The message 00100 indicates that the temperature is above the lowest limit, but below the optimum range, and the classifier would generate the message 01100 to turn on the heater to a medium power. The # symbol, therefore, allows for the creation of multipurpose rules, whose action is able to adapt to different input messages. The ability to match a number of different input messages has the advantage that several classifiers can be replaced by just one, thus increasing efficiency.

As we have seen in Example 2, a single generalist may sometimes do the job of several specialists. This can reduce the size of the classifier list, but it may have the opposite effect on the output list. The classifier #####/10010 matches every input message, so will on its own post as many output messages as there are messages in the input list, thus flooding the output list with the message 10010. When the combined action on the environment of all messages in the output list is favorable, we might anticipate that this classifier will share in any reward that the environment provides because the classifier always generates an output, even though, because the messages that it produces are entirely unrelated to conditions in the environment, they cannot be of much help in controlling the reactor.

9.4.2 Classifier Specificity

The distinction between specialist and generalist classifiers is quantified by the classifier *specificity*. The specificity of a classifier is determined by the number of # symbols that it contains; thus, it is a function of the form of the classifier. The specificity is larger the greater the number of 0s and 1s in the string, and is equal to the ratio of the total number of 0s and 1s, b_i, to the total length of the string, l_i:

$$k_i = \frac{b_i}{l_i} \tag{9.1}$$

In a complex environment, an effective set of classifiers will contain both specialists and generalists. Figure 9.6 illustrates how these rules can work together to control the environment.

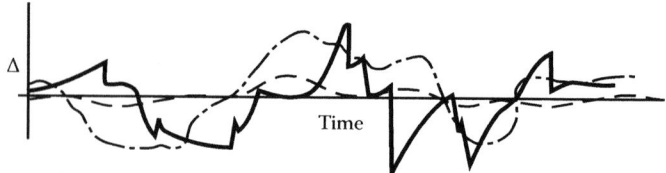

FIGURE 9.6

The variation with time of an environment parameter such as pH. Horizontal line: represents perfect control; bold line: environment under the control of specialist classifiers only; dot–dashed line: environment under the control of generalist classifiers only; dashed line: environment under the control of both generalist and specialist classifiers.

Generalists are frequently triggered, so in every cycle there may be many output messages from the generalists trying to nudge the environment in some new direction. However, because the generalists may deal poorly with less common situations, the environment may gradually move away from its optimum point before slowly returning. By contrast, if the classifier system contained only specialists, the output list would often be empty as few classifiers would be triggered and the environment would drift unchecked until a specialist rule was eventually satisfied. This might produce a sharp correction in the state of the system and move it significantly back toward the optimum point. The combination of both types of classifiers gives the system the best chance of being well controlled at all times, as shown by the dashed line in Figure 9.6.

9.4.3 Classifier Strength

The strength of a classifier measures how useful it is within the CS. When the classifier is triggered and generates an output message, does that message produce a response that is useful and appropriate in the environment? While the specificity of a particular classifier is determined by the form of the classifier and, therefore, is fixed, its strength changes in response to feedback from the system that indicates how valuable the classifier is, within each pool of output messages, in producing the desired behavior in the environment.

Recent work has seen a distinction emerge between the strength of a classifier and its fitness. Over a long period, the strength, which measures the value of the classifier to the system, becomes proportional to the average amount of reward that the classifier can expect to receive from the environment. The fitness provides a handle that the algorithm which manipulates and modifies classifiers (see section 9.5) can use to guide its operation.* A high fitness indicates that the classifier contains unusual and potentially valuable information and should be manipulated by the discovery algorithm at a high rate. Conversely, a low fitness indicates that the classifier is "nothing special," thus the discovery algorithm need not process it frequently.

---

* The term "fitness" is used because, as you might have guessed, the genetic algorithm is the favorite tool for this manipulation.

9.5 Learning Classifier System

The decision-making engine in the CS is the set of classifier condition–action rules; therefore, the key to a successful application is a well-constructed set of rules. If the control problem is straightforward, the necessary classifiers could, in principle, be created by hand, but there is rarely much point in doing this. A single classifier is equivalent to a production rule, the same structures that form the basis of most expert systems; if a set of classifiers that could adequately control the environment could be created by hand, it would probably be as easy to create an equivalent expert system (ES). As an ES is able to explain its actions but a CS is not, in these circumstances, an ES would be preferable.

The CS comes into its own when the rules that are needed to control the environment are only partly known, or are completely unknown, so that a comprehensive set of ES rules cannot be constructed by hand. If we cannot create the ES by hand, it must also be impossible to create the CS by hand; some other method of creating the classifiers must be found. This is the realm of the learning classifier system (LCS) in which all classifier systems of value lie.

Raw AI algorithms are empty shells that require training or instruction before they can be used. When an LCS is applied to a control problem for which the classifier rules are unknown, the system starts with an initial pool of classifiers that are random binary strings; from this pool useful rules must be derived. The application to a control problem of a system that is in this state, and, therefore, that contains no knowledge, is likely to be highly chaotic. Numerous output messages will be generated, many of which will conflict; the environment has no way of knowing which, among all the noise, are messages that are potentially useful and which should be ignored, hence, "control" is anything but.

To make sense out of this chaos, a method is needed to pick out the rules that will be useful in controlling the system from among the large number that are worthless. The process of identifying productive classifiers relies on a mechanism that provides rewards to those rules that are helpful with a penalty for those that are not. The better rules then gradually emerge from the background noise.

9.5.1 Bids

The crucial first step in identifying useful classifiers is to require classifiers to take a risk when they post a message to the output list. When the input message list is checked against the classifiers, classifiers that are satisfied will attempt to generate at least one output message.* So far we have assumed that, if a classifier is satisfied by an input message, it always posts an output message. However,

---

* Any classifier with more than one action part could post more than one output message per input message.

a free-for-all in which every satisfied classifier posts a message is not desirable because this will tend to favor classifiers that post output messages whatever the input. If there is no cost associated with posting a message, classifiers that are frequently triggered by input messages will benefit because the more messages they post, the greater the chance that they will receive a reward.

Classifiers that are satisfied by an input message, therefore, are required to make a bid for the privilege of posting an output message. The bid is

$$Bid = C \times S_i(t) \times k_i \tag{9.2}$$

where C is a constant, $S_i(t)$ is the strength of classifier i at time t and k_i is its specificity. A classifier will be allowed to post a message if its bid is above a certain level. The probability that this level will be exceeded is proportional to the ratio of the bid of one classifier to the total of all bids:

$$p_i(t) = \frac{bid_i(t)}{\sum_j bid_j(t)} \tag{9.3}$$

Since generalists have low specificity, their bids for a given classifier strength will be lower than bids from specialists and they may fail in their bids unless they are of very general utility, so often receive a reward and, thus, have high strength. Notice that the generalist `#####/action` has a specificity of zero, so if this classifier is a member of the classifier list, any bid it makes cannot succeed.

The classifier requires some resources that can be used in the bidding and for this purpose it uses a portion of its strength when it bids. If the bid is large enough to succeed, the message appears on the output list and the bid that has been made is deducted from the classifier strength. If the bid fails, no message is posted and the strength remains unaltered. In due course, if the combined effect of all output messages on the environment is favorable, the reward that the environment returns will enhance the strength of the successful bidders. The next time the classifier bids, it will be stronger, its bid will be more likely to succeed and the environment will benefit again. On the other hand, if the message that a classifier posts is not helpful, the classifier may receive no reward, its strength will fall because part of it has been used to make the bid and eventually its bids to post its unproductive messages will fail. In this way the output of the system is gradually filtered so that useful messages predominate while unwanted messages are eliminated.

If conflicts arise when the messages are sent to the effectors, recourse is made to the bids that each classifier made to post the message. The probability that one message is accepted rather than another is proportional to the ratio of the bids made by the classifiers when the messages were posted:

$$B_i(t) / \left\{ B_i(t) + B_j(t) \right\}$$

9.5.2 Rewards

Once the output list is complete, messages that are intended for effectors are sent into the environment where they are acted on and, as a consequence, the environment changes. If the combined actions of the effectors produce a desirable change, for example, by lowering the pressure in a pipeline that had been identified by pressure sensors as operating above the prescribed pressure limits, the environment provides a positive reward to the classifier system, which is shared among all the classifiers that succeeded in posting a message.

In the simplest of classifier systems, the reward from the environment is divided up equally among all those classifiers that were successful in posting output messages (recognizing that classifiers that post several messages will get a proportionately larger share of the pool). It may seem unfair that all classifiers that made successful bids get the same share of the payoff when some of the messages will be more helpful in bringing about the necessary change in the environment than others, which will be of lesser value or be counterproductive. However, the CS has no way of knowing which messages were most beneficial as the environment returns a single, global reward, thus the system cannot selectively reward some classifiers and ignore others.

If classifiers are coupled (see section 9.5.5), one of the classifiers is not satisfied by a message from the environment, but by a message from a second classifier generated during the previous major cycle. When this happens, after receiving its share of the reward from the environment, the first classifier then passes an amount equal to its bid to the second classifier, whose message satisfied it.

After many cycles, productive classifiers will have gained significant strength, but those classifiers that are not of value in controlling the environment will have accumulated little. The reward scheme thus allows us to identify the more useful classifiers, but how can we persuade the system as a whole to learn?

To bring about learning, there must be a mechanism that can modify the contents of the classifier list so that classifiers that serve no useful purpose can be removed and possibly replaced by others of greater value. This can be accomplished in two ways: *rule culling* and *rule evolution*.

9.5.3 Learning through Rule Culling

The simplest way to create a set of productive classifiers is through rule culling. A large pool of classifiers is created, either by generating the condition and action parts for numerous classifiers at random or by creating every possible classifier of the required length that could be constructed from the ternary alphabet.

Once the classifiers have been created, the system is started up and its behavior is monitored. Classifiers that produce useful behavior, such as turning on a chiller if the temperature is too high, receive rewards from the environment.

Those that give unhelpful behavior, or simply do nothing, receive few or no rewards. Periodically, the strength of all classifiers is inspected and those of the lowest strength are discarded. In this way the classifier list gradually learns helpful behavior and shrinks toward a manageable size.

Because all we need to do in this process is to create a large pool of classifiers, rank them by their performance, and then throw away the worst, this process is easy to perform. However, though culling is conceptually simple, it is limited in applicability. The most serious limitation is that no new classifiers are created. If a crucial rule is missing from the initial list of classifiers, it will never appear. The CS will then only be able to struggle on, doing as well as it can with the classifiers that are available to it.

If the control problem is of modest scale, the number of distinct classifiers may be sufficiently small that every possible variant can be included in the initial population; however, for a system of even moderate size, with just a few inputs and effectors, this is a forlorn hope.

For classifiers described using a ternary alphabet, the number of different classifiers increases as 3^n, where n is the combined length of the condition and actions parts (Figure 9.7), so a classifier list that contains all permutations of the ternary alphabet quickly becomes too large to manipulate. If the number of probes in the reactor is 6 and the messages that they send have a resolution of 1 part in 256, if the number of effectors is 4 and their resolution is 1 part in 64, the number of different classifiers is

$$3^3 \times 3^8 \times 3^2 \times 3^6 = 3^{19} = 1.2 \times 10^9$$

which, by several orders of magnitude, is too large a pool with which to deal. As a consequence, the creation of an exhaustive list of classifiers is rarely a realistic option.

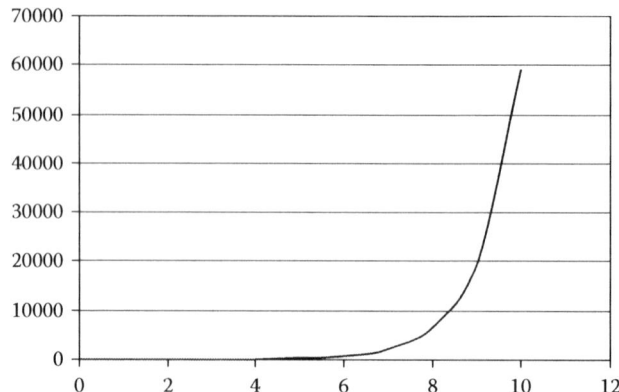

FIGURE 9.7
The number of distinct classifiers as a function of classifier length.

9.5.4 Learning through Evolution

In almost every problem of meaningful scale, the number of possible classifiers is far too large for rule-culling to be feasible; instead, a dynamic algorithm is required that can run the CS, assess the quality of the classifiers, remove the poor ones, and create new, potentially valuable replacements. This is accomplished by using a genetic algorithm (Chapter 5) to evolve the classifier list.

There are three steps in the evolution of the classifier list: (1) the identification of useful classifiers, (2) the creation of new classifiers that may be of value, and (3) the removal of classifiers that serve no useful purpose. This is just the sort of task for which the genetic algorithm is designed. It assesses individuals on the basis of their quality, selects the better individuals, and from them creates new, potentially better, individuals.

The steps in the combined genetic algorithm-classifier system are given below.

Combined Genetic Algorithm–Classifier System

1.
   a. Randomize all classifiers.
   b. Place messages from the environment on the message list.
   c. Compare each message with the condition part of each classifier. If a classifier is satisfied, place the output on the output message list, but only if the bid it makes is sufficient.
   d. Tax all bidding classifiers.
   e. Once all messages have been read, delete the input message list.
   f. Sweep through the output message list checking for any messages to effectors.
   g. Request the effectors to carry out the appropriate action.
   h. Receive a payoff from the environment and distribute this to classifiers that have posted to the message list.
   i. Occasionally run an evolutionary algorithm (step 2) to generate new classifiers from the old based on classifier fitness.
   j. Return to step 1b.
2. Operate the GA by:
   a. Picking classifiers semistochastically from the classifier list to act as parents, using an algorithm in which the fitness of a classifier is based on its accumulated reward.
   b. Copying the chosen classifiers into a parent population, then using the genetic operators of crossover and mutation to prepare new classifiers.
   c. Replacing some of the old classifiers by new classifiers and continuing.

In a standard genetic algorithm, the genetic operators are invoked every cycle, but this would be inappropriate within a classifier system. The value of an individual classifier cannot be reliably assessed in a single cycle because classifier strength adjusts slowly. Only generalist classifiers are likely to post messages every cycle; specialist classifiers will rarely post messages. Because a classifier can gain a reward only when it posts a message, there may be long periods when a classifier is not triggered because the conditions that it addresses have not arisen. It is then unable to receive any reward. Therefore, the LCS is run for many cycles to give all classifiers their chance to play their part in control of the system and gain a reward.

After the chosen number of cycles has passed, the genetic algorithm is applied to the set of classifiers. The fitness of each classifier may be related directly to its strength, or the fitness may be determined by combining classifier strength with other factors, such as the specificity. The usual GA operators are applied to create a new population of classifiers, which is then given the opportunity to control the environment for many cycles. The process continues until overall control is judged to be adequate under all circumstances.

9.5.5 Coupled Classifiers

When the genetic algorithm creates new classifiers, it may mutate randomly any part of the condition and the action segments of the classifier. New classifiers created by the genetic algorithm, when satisfied by an input message, may generate an output message whose identifier tag does not match that of any effector; thus there is no part of the environment to which the message could be sent. Such messages are not just meaningless chatter, even though the message cannot influence the environment. Output messages that are not sent to effectors are merely recycled into the input list, where they will be read by the classifiers in the next cycle. These *post-it notes* provide a mechanism by which classifiers can be *coupled*.

As an illustration of how this can happen, suppose that the classifier list contains the following entries:

01001/00001   00001/00111   00#11/0101#

and the input messages are

00110   11110   10100   01001

The first classifier is satisfied by the message 01001 and posts the message 00001 to the output list. If this is not a message to any effector, it will be transferred to the input list on the next major cycle and read by all the classifiers in that cycle. The message 00001 satisfies the second classifier, which would post the message 00111; in turn, this message activates the third classifier in the next cycle.

Coupled classifiers can perform complex calculations, such as handling temporal information. Recall that the input message list delivers a snapshot of current conditions in the reactor. If the reactor contains a pH probe, the message that the probe sends will define the latest pH of the reaction mixture.

Suppose that, if the pH in the reactor is varying randomly near pH 7, we wish to take one action in the environment, but, if it is drifting continuously down or up, some other action is required. A single measurement of the pH gives no indication of how the pH is changing. However, if one classifier reads the current pH, then leaves on the output list a post-it note indicating what the pH is, that message can be picked up by a second, multicondition classifier in the next major cycle. This information can be combined with data about the new pH to determine in which direction the pH is moving and a further post-it note deposited that provides the information to another classifier. If the process is repeated, classifiers can in this way not only get a picture of the current state of the reactor, but also how it changes with time and, therefore, take the appropriate action.

There is, of course, no more of a guarantee that coupled classifiers will produce messages of value than any other classifier. It is also possible that a classifier will produce an output message which will trigger the same classifier in the next cycle. Such self-exciting messages are sometimes known as hallucinatory messages. However, the addition of coupled classifiers gives the overall system much greater capabilities and can lead to very complex behavior.

9.5.6 Taxation

One final element is needed to ensure smooth evolution of a CS. In a large set of classifiers, there may be a number that are never satisfied by an input message and so never bid; others may be so indiscriminate that they put in a bid every cycle, but have so little strength that they never succeed in posting any message. Neither type of classifier will contribute much to the smooth working of the overall system, except possibly by providing a pool of genetic material that may be accessed by the GA. Their presence will only expand the classifier list and slow operation of the algorithm.

It is, therefore, common practice to impose a tax on the strength of the classifiers each cycle. Taxes may include a head tax, which is applied to every classifier, and a bid tax, which applies to all those classifiers that make a bid during the current cycle, irrespective of whether or not the bid succeeds. The bid tax is designed to slowly weaken those classifiers that bid repeatedly, but never do so with enough conviction to succeed. The head tax applies to every classifier so that those that are so hopeless that they never even bid also lose strength and will eventually be removed.

The taxes are low, for if they were not, the loss of classifiers would be high. Especially vulnerable would be specialist classifiers which may need to wait many cycles before being triggered by any input message and, therefore, having the chance to gain a reward. By the imposition of taxes, the number of classifiers is kept to a workable level, which both increases execution speed and makes the manipulations of the genetic algorithm more efficient.

9.6 Applications

At one stage in the early 1990s it appeared that, after a promising start, classifier systems would quietly fade away from the AI landscape. Although their principles are relatively straightforward, their behavior is complex and does not lend itself easily to a theoretical analysis. In addition, like neural networks, they are computationally greedy during training and long periods may pass in training during which they seem to make only slow progress.

Furthermore, although this chapter has concentrated on the use of a CS in the context of control problems, no chemical company synthesizing ethylene oxide in bulk, or any other chemical for that matter, would allow a CS to play around with its chemical plant for the days or weeks that might be necessary for the system to learn how to control the plant safely (bearing in mind the need to clear up the occasional mess after the CS makes a mistake in its learning and the plant explodes or burns down).

So what are the prospects for the CS, if indeed it has any?

Recently there has been a resurgence of interest as their operation is better understood and as the range of areas to which they could be applied starts to grow. Increases in computer power have also helped, since while a three-day wait to see if a CS is making progress may be excessive, a wait over the lunch hour to see what is happening is quite reasonable. The area in which they have the greatest potential, process control, is just the area in which training is most problematic. However, alternatives exist to giving a CS a complete, functional process line to investigate. The classifier list could be built incrementally, by running it successively on subsections of the entire system. It may be able to learn by interacting with a simulation, so it can get a satisfactory understanding of the process without destroying what it should be controlling. It may be possible for portions of a CS found useful in one application to be used as the core for a second.

Despite the revival in interest and what appears to be considerable potential, the number of applications remains low; there is considerable scope for development in this area. Attempts to use CS in the control of traffic lights, sequence prediction in proteins, and the analysis of medical data all suggest that the range of possible applications is wide — the field is just opening up.

9.7 Where Do I Go Now?

A helpful starting point for further investigation is *Learning Classifier Systems: From Foundations to Applications*.[1] The literature in classifier systems is far thinner than that in genetic algorithms, artificial neural networks, and other methods discussed in this book. A productive way to uncover more recent developments in this field is to search for papers by, and references to, Rick Riolo, Lashon Booker, Richard Belew, and Stephanie Forrest, who have been active in the field for a number of years.

9.8 Problems

1. The rates of many enzyme reactions are strongly dependent on both pH and temperature. Construct a CS that learns to keep conditions within a simple reactor within the limits $6 < pH < 9$ and $27 < T < 41$. You will need both to write the classifier system itself and a small routine that represents the environment. Test the operation of your system by including a method that periodically adds a random amount of acid or base, or turns on a heater or chiller for a short period.

References

1. Lanzi, P.L., Stolzmann, W., and Wilson, S.W., (Eds.) Learning classifier systems: From foundations to applications, *Lecture Notes in Artificial Intelligence 1813*, Springer, Berlin, 2000.

10

Evolvable Developmental Systems

Nawwaf Kharma

Associate Professor, Department of Electrical and Computer Engineering, Concordia University, Montreal, Canada

10.1 Introduction and Motivation

Evolutionary computation, in all of its flavors (genetic algorithms [GA], genetic programming [GP], evolutionary strategies [ES], and evolutionary programming [EP]), offers an attractive paradigm for the optimization of and, to a lesser degree, design of scientific and engineering artifacts. They offer the practitioner the opportunity to utilize artificial evolution rather than human intuition as a tool for searching unknown or partially characterized spaces of high dimensionality for the purpose of identifying optimal or acceptable solutions to well-defined problems.

Despite a mass of research activity in evolutionary computation (EC), activity that has led to solid theoretical results and realistic applications, there are still a number of perennial irritations that almost all EC techniques suffer from. First, the serious computational cost of evaluating numerous potential solutions (or individuals), over hundreds of iterations (or generations), places pragmatic and sometimes formal limitations on the use of EC in real-world applications with time-sensitive outputs, such as online multiprocessor scheduling. This real limitation deters many potential users from using or even considering the use of EC in heavy-duty engineering and scientific applications.

Second, EC techniques are, on the whole, heuristics that do not guarantee the discovery of optimal answers within certain time spans. Engineers like certainty: reliability and predictability. Hence, unless and until EC researchers are able to provide techniques with proven and predictable performance, it is unlikely that responsible engineers will use EC for optimization or design purposes, except for off-line or exploratory applications. This has resulted in a large and widening gap between the latest developments in EC (e.g., evolvable developmental systems [EDS]) and the actual software tools being offered as part of popular software packages (e.g., Matlab®).

Finally, we come to the issue that is of most relevance to this chapter — representation, or the mapping between the genomic encoding of a possible solution and the actual phenomic solution. Each major variant of EC, except possibly for EP, has a standard form of genomic representation. The most popular EC techniques are GA and GP: In essence, the GA uses a linear concatenation of the dimensions of the search space, while GP uses a grammar-tree representation, which represents expressions of predefined operators and operands. This standardization is a necessity because it allows theoreticians to prove certain general properties of a particular variant of EC. It also permits practitioners to focus their attention on the adaptation and tuning of a given EC technique for a specific problem.

However, both linear and tree-based representations are not always ideally suited to the solution (e.g., optimization or design) of all or even most scientific or engineering problems. For example, search spaces with hundreds of *potentially* relevant dimensions as well as scalable problems with *growing* dimensions present the usual EC representational formats with unusual challenges. Even relatively straightforward scheduling problems with a finite and small number of dimensions have required many designers to invent convoluted mappings between a standard genomic representation (e.g., a string of some form) and the necessary solution format (e.g., process schedule). The point is that users of EC techniques are often forced to expend considerable amounts of creative energy on molding an intuitive solution format into a standard EC representational format. This is usually followed by some amendment to the standard genetic operators in order to make them more effective.

10.2 Relationship between Evolution and Development

To deal with the problem of limiting representations and also out of inspiration by biological development, EC researchers have been breaking out of the representational mold in several high-profile studies and have sought fundamentally novel representational schemes, which introduce elaborate mappings between genome and phenome. The focus of this chapter is the body of EC work that utilizes well-defined indirect mappings between genome and phenome. These mappings come in different guises, including (1) production rules, (2) cellular growth mechanisms, (3) grammar trees, (4) random Boolean networks and cellular automata, and (5) other methods. All of these methods shift complexity from space or size of the genome, to time or a process that generates the phenome, iteratively, from a smaller genome.

The key processes linking genotype to phenotype are (1) genetic decoding from genome to phenome and, let us not forget, (2) the assignment of a fitness value to the genome following the evaluation of the corresponding phenome. This entails two things: (1) the only feedback going back from the

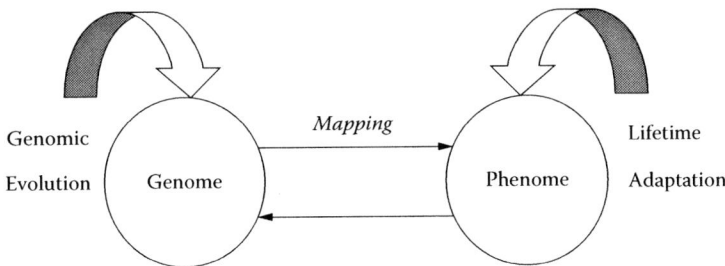

FIGURE 10.1
Relationship between genome and phenome and how each is adapted.

phenome to the genome comes in the form of a single quality associated with the behavior of the completely developed phenome in a given environment, and (2) any adaptation that the phenome (or individual) experiences during its *lifetime* is not passed forward to any future generations (via the genome) (Figure 10.1). This is in keeping with current theories of Darwinian evolution, as opposed to Lamarckian inheritance of acquired characteristics.

A made-up example of a developmental mapping may be presented using production (or rewrite) rules.

Axiom: X (may be viewed as the genome).

Rewrite rules and interpretation (together acting as the developmental mapping):

$$X \rightarrow LBR$$

$$L \rightarrow A2$$

$$R \rightarrow 1C$$

Interpretation: A letter represents a node (e.g., a neuron) and the integer to the right of a node determines connectivity. Say that a positive integer to the right of a node, such as 2, means that that node feeds into the second node on its right; a negative integer to the right of a node, such as –1, means that node connects to the first node on its left; a zero to the right of a node means that node connects to itself.

Hence, a genome containing X (and nothing else) would be expanded into A2B1C, which is interpreted as a phenome of 3 nodes, A, B, and C, with the outputs of A and B providing the inputs for C.

This simplistic example illustrates almost all core characteristics of developmental mappings (though of only the *rewrite rules* type):

1. A small genome and elaborative mapping procedure.
2. A multistep procedure that defines how a phenome is constructed, not what a phenome is to look like.

3. The lack of a systematic method of designing a developmental procedure and then embedding it within the genome.

Indeed, the last point highlights the fact that there exists no method that would allow one to go back from (1) a well-defined phenomic form and (2) a particular developmental approach (e.g., cellular automata) to a developmental program, which is embeddable within the genome. The process of devising a developmental mapping, for a given problem, has thus far been an art practiced by the few, rather than a technique taught to the many. Nevertheless, having a developmental mapping almost always involves:

- A smaller genome that generates a potentially larger genome.

- A developmental process that can, during growth, take into consideration feedback from the environment in order to determine the best final properties (e.g., size) of the *adult* phenome.

- A generative encoding that is intuitively more appropriate for phenomes, which humans are used to, to *generate* incrementally rather than *describe* fully prior to realization. For example, it is more natural for a human to describe the making of a cake in terms of a generative procedure or recipe rather than describe the chemical properties of the resulting mixture.

- The amplifying effect of the developmental procedure, which associates possible small changes in the genome with much larger changes in the phenomes.

- The potential for self-regeneration. If the developmental procedure is not (completely) switched off after the phenome reaches maturity, then it is possible to include a mechanism in the phenome that would detect and correct some perturbations to the adult phenome.

In short, any evolutionary computation technique that includes a developmental mapping of any form is considered to be a form of EDS, which comes in various forms; some are still to be invented. Devising a developmental mapping is no child's play, and there is not yet a systematic method for achieving that. However, EDSs have many potential advantages, which include compact genomes, scaling and self-regeneration. For this, and other reasons that we are still discovering, devising a developmental mapping may be a rewarding endeavor for EC researchers, at least for some applications.

An EDS is any EC technique with a non-1:1 mapping from genome to phenome. These types of techniques, by virtue of various types of mapping (e.g., rewrite rules, tree grammars, or cellular growth mechanisms) appear to offer their users certain advantages over conventional 1:1 mappings, including scalability and evolvability. Due to these, and other interesting properties of some (but not all) realizations of EDS, there is growing interest in the field.

There are two reasonably well-known papers that propose means of partitioning the space of work in the area. From Kumar and Bentley's point of view,[1] there are three types of mappings, which they call *embryogenies* (or mappings): external, explicit, and implicit. Though this taxonomy focuses on the essential distinguishing feature of an EDS (i.e., its mapping), it fails to provide a clear division between explicit and implicit mappings. The other taxonomy is that of Stanley and Miikkulainen.[2] It takes as its basis the underlying biological features of the developmental process. It is a valiant attempt at proposing something truly novel and inspiring. However, we feel that the taxonomy is somewhat arbitrary, complex, and too "biological" for it to be of immediate use to all parties interested in EDS.

The rest of the chapter is mainly composed of five diverse examples of application of evolutionary developmental methods to problems in engineering and computer science. The chapter concludes with an attempt to foresee some future trends in EDS research and application, followed by a very short story that we hope will entertain and perhaps inspire.

10.3 Production Rules and the Evolution of Digital Circuits

10.3.1 Description of Sample Problem

The purpose of the work by Haddow, Tufte, and Van Remortel[3] is to artificially evolve a design satisfying a predetermined functionality for implementation in a configurable electronic device (or chip), such as a field programmable gate array (or FPGA).

Artificial evolution, such as genetic algorithms, evolves a population of potential solutions (or individuals) to an optimization or design problem, over many iterations (or generations). For every generation, each individual is evaluated for quality (or fitness). For a population of one thousand individuals, evolving over one thousand generations, this entails a million (or fewer) fitness evaluations. In addition, a single fitness evaluation may involve the execution of a large module or separate program. This makes fitness evaluation a computationally costly part of a GA run. Many GA practitioners, hence, divide a GA iteration into (1) fitness evaluation and (2) the rest of the evolutionary loop, or simply the evolutionary loop.

In line with that division, one can also divide the various approaches to circuit evolution into: (1) on-chip evolution, (2) on-computer evolution, and (3) hybrid methods. In on-chip evolution, the whole GA run, including fitness evolutions, is carried out onboard the configurable logic device. In on-computer evolution, the GA is run on a computer and only when the design is optimized is it downloaded onto the configurable chip. In hybrid approaches, part or most of the evolutionary loop is run on the computer, while the rest, particularly fitness evaluation of individuals (or potential designs), is carried out onboard the programmable device.

10.3.2 Representation of Potential Solution

In 2001, Haddow and Tufte introduced the idea of an "Sblock."[4] As shown in Figure 10.2, an Sblock is made of two units. There is the combinatorial logic unit, which has five binary inputs (I_0-I_4) and one binary output (O). Four of the inputs to the logic unit are external to the Sblock and are used to connect it to its neighbors, as we shall see. The fifth input comes from the output of the other unit (explained below). The relationship between the output and the inputs is defined in a truth table, which provides the value of the output for every possible combination of input values. The other unit in an Sblock is a memory unit (called a D-flip-flop), which has one input and one output. The input to the memory unit comes from the output of the logic unit and is kept in "memory" for one clock cycle. As such, the Sblock has both logic and memory capabilities, and a properly connected network (such as a grid) of such Sblocks is sufficient to realize any combinatorial or sequential binary functionality, and, hence, any digital circuit.

A two-dimensional grid of programmable Sblocks may be realized in a programmable logic device or a programmable logic device may be fabricated specifically for that purpose. In either case, an Sblock grid becomes a programmable device, which can either be configured manually or by using some intelligent algorithm, such as a GA. What needs programming are the functionality of the Sblock and the connectivity of the Sblock. As stated above, the functionality of an Sblock is specified using a truth table. However, since the four external inputs and the single output of an Sblock come from and go to its four neighbors, the connectivity of an Sblock can also be specified using the same truth table. For example, if the input coming from the northern neighbor of an Sblock has no effect on the output of the truth table of the Sblock (is a "don't care") *and* the output of the Sblock has no influence on the output of the truth table of its northern neighbor (is a "don't care"), then that particular Sblock and its northern neighbor are effectively unconnected (Figure 10.3).

Hence, in order to specify the functionality of a grid of say N × N Sblocks with M-variable truth tables, each one requires $N^2 \times 2^M$ bits of information to

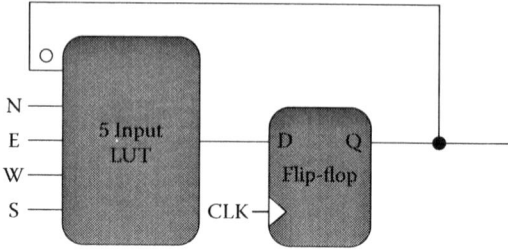

FIGURE 10.2
A block diagram of an Sblock. (Haddow, P.C. and Tufte, G. [2001] Bridging the genotype-phenotype mapping for digital FPGAs. In proceedings of the *Third NASA/DoD Workshop on Evolvable Hardware*, IEEE Computer Society.)

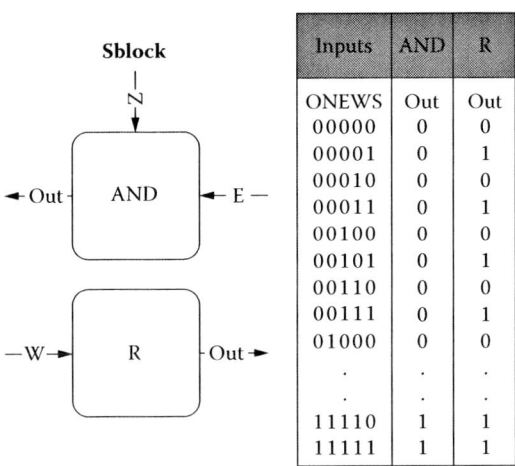

FIGURE 10.3
The truth tables of two Sblocks. (Haddow, P.C. and Tufte, G. [2001] Bridging the genotype-phenotype mapping for digital FPGAs. In proceedings of the *Third NASA/DoD Workshop on Evolvable Hardware*, IEEE Computer Society.)

be specified. If a traditional GA with a binary linear chromosome were used to evolve acceptable values for all the truth tables defining a grid of Sblocks, then the length of the chromosome representing a potential configuration would equal $N^2 \times 2^M$. For large N values, the length of the chromosome would probably render the search for acceptable configurations ineffective or at least highly inefficient.

One way to tackle this problem is to develop or grow a network of Sblocks starting from a single initial Sblock.

10.3.3 Developmental Procedure

If one views a programmed Sblock as a living cell, then a set of connected and programmed Sblocks may be viewed as a multicellular organism. Such an organism would have to grow (or develop) from an initial state, usually a single cell. Development within this environment means the programming of an Sblock by a neighborly cell. Obviously, cells described by different truth tables have different functionalities. Altering the connections between existing cells can also be done by changing values with the truth tables of the concerned cells. Cell "death" occurs when a cell's truth table is returned to its unprogrammed state.

The development program must decide which existing cells sprout (or not) new cells and which existing cells die (or not). The program must also decide the functionality and connectivity of every new cell. Since growth occurs on a cell-wise basis, with no central controller, this program must be part of every cell and can only utilize local information available to a cell.

Haddow et al.[3] adopted L-systems as the basis of their approach to development. L-systems are a form of production rules first proposed by Lindenmayer.[5] L-systems were first developed for modeling the way in which various plant morphologies come to be. One can devise an L-like system to define exact growth rules for more than one type of system. In an Sblock system, one can define two types of rules: (1) change rules and (2) growth rules. An example of the set of rules is shown in Figure 10.4 — an asterisk in a rule stands for a "don't care." The length of the right-hand side (or RHS) of a *change rule* is equal to the length of the rule's left-hand side (or LHS). As such, the application of a change rule to a cell (or Sblock) may lead to a change of the functionality and/or connectivity of that same cell, but has no effect on its neighbors (rules 1 through 4 below). In contrast, the application of a *growth rule* to a cell leads to the introduction of new cells with new functionalities.

The rules are ranked according to their specificity — the more specific, the higher the rank (with 1 being the highest rank). Since growth rules have equal specificity, they are assigned ranks on a random basis. It is worth noting here that this set of rules is initially randomly generated and is subject to artificial evolution in hope of improvement.

The output columns of the truth tables of all the Sblocks in the grid are placed in an array (the output columns are referred to simply as Sblocks). Through the application of a change rule to that array, a number of Sblocks may be changed. This occurs whenever the LHS of a change rule matches a substring (or more) of the array at certain location(s) — a substring might lie in one or two adjacent Sblocks. This leads to the replacement of the matching contents (in the array) with the RHS of the applied rule. Obviously, this results in a change of the functionality and/or connectivity of one or two Sblocks, but does not result in the introduction of any new Sblock (i.e., cell growth).

In order for a new cell to appear, two conditions must be satisfied: (1) the LHS of a growth rule must match a substring (or more) of the array and (2) at least one of the four neighbors of the matched Sblock(s) should be free. If and

```
01   *1*00110101*1*  ─────────────►  10110100001110
02   00001*110111  ───────────────►  100001110011
03   0101*11*1***  ─────────────────►  100100010100
04   ***00*10**  ───────────────────►  0111001110
05   01001 ──► 00101110011101100110000000100111
06   10111 ──► 100101110001110101010001100101101 10
07   01011 ──► 10101100111111011101100100011100
08   10101 ──► 01011110111001100100001011101010
09   00011 ──► 11111011001000011001011001101010
10   01011 ──► 10111101001010010011011110000101
11   01110 ──► 11010001001000110101011001000001
12   01100 ──► 11111100101100111000101010111001
```

FIGURE 10.4

L-systems type rules of change and growth. (Haddow, P.C. and Tufte, G. [2001] Bridging the genotype-phenotype mapping for digital FPGAs. In proceedings of the *Third NASA/DoD Workshop on Evolvable Hardware*, IEEE Computer Society.)

when these conditions are satisfied, the RHS of the growth rule (= 32 bits) is placed in one of the available free neighbors; which one depends on a preset priority scheme. North has first priority, followed by south, west, and east. As a result of the successful application of a growth rule, an Sblock, which was free but of an initial seed, gains a fully specified truth table and becomes a functioning connected cell.

As stated, rules are applied in order of their priority. Additionally, when a rule is matched and applied to a certain location, that location is protected from any further change until every other rule (in the rules' set) is tested for application. Once all the rules have been tested, all protection is removed from the array, and the whole cycle of selection, application, and protection of affected locations is repeated. In some, this iterative application of rules is repeated for a predetermined number of cycles. It is, however, conceivable to use a different criterion for termination of growth, such as lack of change.

10.3.4 Fitness Evaluation

The fitness of an evolutionary algorithm is always based firmly on the objective of the target application. If the objective is the design of a router (with a certain direction), then the fitness function must reflect the qualities that a designer wants in a router. Obviously, a router circuit made of Sblocks can consist of a string of routing Sblocks, where the output of one Sblock is fed into the input of the next Sblock, without change.

It is also true that fitness functions are sometimes augmented in ways that have an influence not only on the final product (i.e., a design with a certain functionality), but also on the manner in which the GA searches the space of all possible designs.

In fact, the fitness function used by Haddow et al.[3] is

$$F = 15*R + [6*(C\text{-}R) - (6*6in + 5*5in + 4*4in + 3*3in + 2*2in + 1*N1in)] \quad (10.1)$$

R is the number of routing blocks. C is the number of configured Sblocks (or cells), both routing and nonrouting. As such, the term (C-R) provides information about how close nonrouting blocks are to routing blocks. Xin is the number of blocks with X inputs where X is an integer (e.g., 4). N1in is the number of inverter SBlocks. As a perfect routing, Sblock has exactly one input and it does not invert the signal, low-input Sblocks are given a higher weight than high-input Sblocks. Inversion is associated with a small negative weight as well.

10.3.5 Overall Evolution

The purpose of evolution is finding a set of change and growth rules that, when iteratively applied to a grid of unconfigured Sblocks, would result in a group of configured and interconnected cells (i.e., an "organism"), satisfying a certain predetermined functionality. The degree of satisfaction is measured by the fitness function.

A standard GA is used in the experiments and the whole set of experiments was done on-computer (i.e., the actual programmable chip was not used during evolution). The GA parameters follow. Population size is three hundred; selection method is fitness proportional with elitism; crossover probability equals 0.8 (then 0 in successful runs); mutation probability is 0.1; the maximum number of generations was set to one hundred fifty.

10.3.6 Experimental Results

Figure 10.5 illustrates three snapshots of a typical run achieved by Haddow et al.[3] In the figure, a square grid of 16 × 16 Sblocks is shown. A black box represents a 0-input box or an unconfigured (or free) Sblock. The decrease in the number of inputs to Sblocks necessary for routing to occur is expressed by a change in the color from dark grey to white. White Sblocks stand for routing blocks. The experiments evolved 30 to 50 percent routers in less than one hundred evolutionary generations.

The trend shown toward increasing routing blocks demonstrates that the application of artificial evolution (i.e., a GA) to the problem of finding rules that can grow effective (though not optimal) digital circuits is a real technological possibility.

10.3.7 Summary and Challenges

In this section, we presented a brief introduction to the use of a genetic algorithm to search for a set of rules that can control a developmental process leading to a digital circuit with a particular functionality. Or, in short, evolution was used to configure development. The GA did not work as expected, as crossover had to be taken out before the evolutionary run started to stabilize. In addition, the resulting routing circuits, though functional, were far from optimal. There were other problems as well, but nevertheless, the experiment did produce the goods: Evolution did find a set of rules that grew an acceptable routing circuit on a simulation of a programmable logic device.

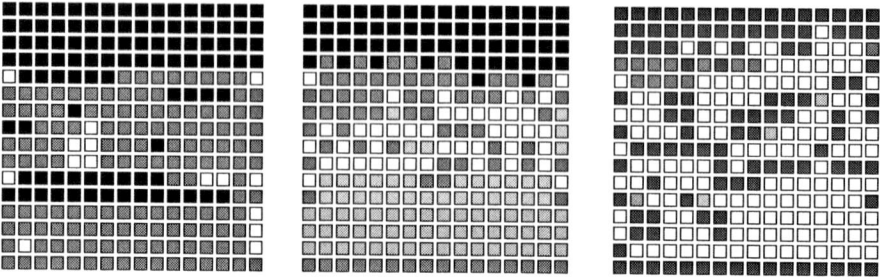

FIGURE 10.5
A grid of developing Sblocks after 3, 23, and 57 generations. (Haddow, P.C. and Tufte, G. [2001] Bridging the genotype-phenotype mapping for digital FPGAs. In proceedings of the *Third NASA/DoDWorkshop on Evolvable Hardware*, IEEE Computer Society.)

10.4 Cellular Automata-Like Systems and Evolution of Plane Trusses

Deval by Kowaliw, Grogono, and Kharma[6] is a general means of mapping between genotype and phenotype, where phenotype is realized in a dynamical system guided by the genome. The Deval algorithm was tested in the context of plane truss design, guiding evolution through the choice of fitness function — successful truss designs are found for each set of imposed constraints.

Plane Trusses are two-dimensional constructs consisting of (for this work) joints, beams, and grounds. A truss is any connected collection of these three components, regardless of usefulness or triviality. All beams are connected via joints, which may be connected to grounds. The typical purpose of a truss is to support other structures and to redistribute any external forces so as to retain its original form. Hence, one typically talks about the stability of a truss and the stress on any particular component. Given a truss, the obvious first question is whether or not it is stable, i.e., will it (approximately) retain its shape. The second question involves the stress placed on the members under some external force. If the maximum stress exceeds the yield strength of any particular beam, the truss may quickly become unstable. Another important issue involves the deformation of the truss members under strain. Given a beam and an external force, a beam will either compress or stretch, which in turn will cause the trusses' joints to dislocate. Some dislocation is to be expected, indeed, some dislocation is precisely the advantage to the use of joints rather than rigid connections. Too much, of course, will compromise the design. Figure 10.6 shows two trusses. The first is stable, but the second is not. Any external force would cause the second to deform drastically. In designing a plane truss, one typically operates under a set of criteria and

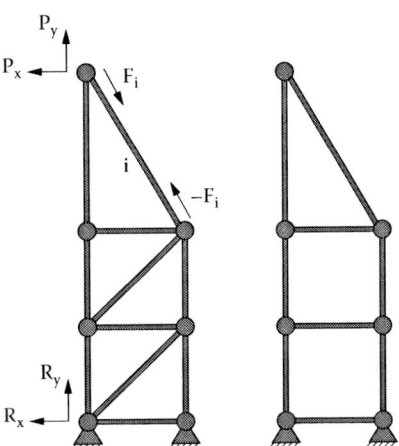

FIGURE 10.6
Two plane trusses; the left is stable, the right unstable.

attempts to design a truss that can maximize those criteria while being both stable and as resistant to external forces as possible. It will be assumed, for all future discussions, that the trusses are topologically connected, pin-connected, friction free, and that force is applied only at joints.

Truss Stress Analysis: The computation of member forces in an arbitrary plane truss is now examined. There exist some simple counting tests that may determine if a given truss is unstable. Failing that, one must attempt to compute the equilibrium state given some external forces; in the process, one obtains values for all member forces. In this example, all truss members are identical in terms of material and area, grown in a developmental space where units are measured in meters; EA is set to 1.57×10^4 N, corresponding to a modulus of elasticity for steel and a cylindrical member of diameter 1 cm. Consider a general truss with n joints and m beams; external forces are applied at joints and the member forces are computed. Let the structure forces be

$$\{P\} = \left\{P^1, \ldots, P^n\right\}^T,$$

structure displacements be

$$\{\Delta\} = \left\{\Delta^1, \ldots, \Delta^n\right\}^T,$$

and member forces be

$$\{F\} = \left\{F^1, \ldots, F^m\right\}^T.$$

One may relate the individual member forces to displacement and structure forces as follows:

$$\{F\}^i = \left[k\right]_a^i \left[\beta\right]^i \{\Delta\}$$

where $[\beta]^i$ is the connectivity matrix for the ith member beam and $[k]^i$ is its stiffness matrix, relating the deformation of the beam under a given force to the displacement at the joint. Hence, to solve for forces, it suffices to compute the displacements. The displacements may be computed through a truss stiffness matrix, a combination of the individual member stiffness matrices:

$$\{\Delta\} = \left[K\right]^{-1} \{P\}$$

Therefore, given a plane truss, one may first compute the stiffness matrix, then compute the displacements, then the individual member forces. The entire process is bounded by the calculation of a matrix inversion (or LU-Decomposition), and, hence, has running time $O(m^3)$.

10.4.1 Developmental Procedure

Deva1: Let us consider a model that consists of a developmental space, D; a collection of cell types (or colors), C; a set of actions, A; and a transition function, φ. The developmental space, simply $D = Z^2$, is a space in which one may grow an organism endowed with a discrete time. Each point in the lattice is a cell, possibly the empty cell; each nonempty cell may be viewed as an independent agent. Cells change in time by executing one of several actions. Which action is executed is determined by the cell's genome and the transition function.

The process of growth is described as follows. Developmental space is initialized empty everywhere, save the central point, which is initialized with a cell of type "1." At every time step, any nonempty cell examines its neighborhood and selects an action through the consultation of the transition function. If the cell has sufficient resources (measured via an internal counter, r_c) and has sufficient age, that action is executed. The set of possible actions is $A = \{nothing, divide, elongate, die, specialize(X)\}$, where $X \in C$. Through this process, the developmental space changes in time. Termination occurs when the space is identical to the space that preceded it (guaranteed to occur due to a finite maximum value of r_c). This process may be written more explicitly as:

```
Time t ← 0
Initialize developmental space Dₜ
while Dₜ ≠ Dₜ₋₁ do
        t ← t + 1
        Dₜ ← Dₜ₋₁
        for all Cell c ∈ Dₜ₋₁ do
                if c has sufficient age and cᵣ𝒸 then
                        Action a ← φ(μ𝒸)
                        Decrement cᵣ𝒸 appropriately for a
                        Execute a in Dₜ
                end if
        end for
end while
```

A *Deva1* transition function is a listing of descriptions of possible neighborhoods of a specified length, $|\phi|$. These rules are tuples of the form:

$$\left(c, h_1, \ldots, h_{n_c}, a\right)$$

where c is a color, $n_c = |C|$ is the number of cell types, a is an action, and h_i is a count of the number of neighbors of cell type i, or a *hormone level*. Hence, the size of the representation of such a transition function is $O(|\phi| \cdot n_c)$, and the number of possible transition functions is $n_c \cdot |\mu|^{n_c} \cdot |A|$, where A is the set of all actions and $|\mu| = 12$, the size of a neighborhood.

10.4.2 Representation of Potential Solution

First, a representation of general plane trusses constructed on a lattice is proposed. The purpose is to be able to map a specialized lattice of integers to some (possibly trivial or useless) plane truss.

Then, a set of cell types are defined; each nonempty cell will contain a joint and between zero and five beams. The beams will extend in directions π, $3\pi/4$, $\pi/2$, $\pi/4$, and 0, labeled g_0 through g_4, respectively. Conversion from Boolean gene values to an integer is accomplished through the following equation:

$$color = 2^4 g_4 + 2^3 g_3 + 2^2 g_2 + 2^1 g_1 + 2^0 g_0 + 1$$

The zero cell type is reserved for the empty cell, the one value is for a joint with no beams, and all other combinations exist in the set $\{2,\ldots,32\}$.

Finally, one may allow cells to be elongated in one direction, by an arbitrary number of cell lengths. For example, a cell of type 9 has an angle of $3\pi/4$ with the x-axis, and a length of $\sqrt{2}$. A single elongation in the y-direction would lead to a length of $\sqrt{5}$ and an angle of $7\pi/8$ with the x-axis. Hence, excluding elongation, any two-dimensional lattice of integers may be mapped to (some) truss. One such mapping, including elongations, is shown in Figure 10.7.

In Figure 10.8, the growth of an agent is shown (in grey), whereas the final organism (in black) is much smaller. This is the result of a trimming process applied to every organism following development. The trimming process serves to (1) remove obviously unstable sections, such as beams which do not

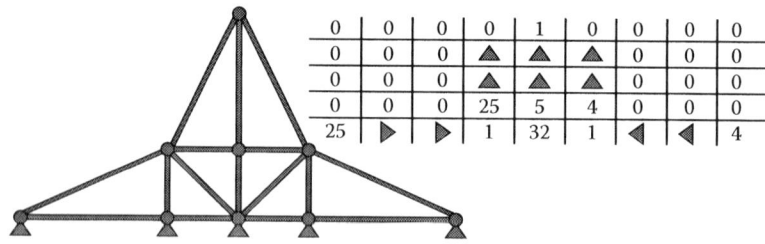

FIGURE 10.7
Example of a mapping between a lattice of integers and a plane truss.

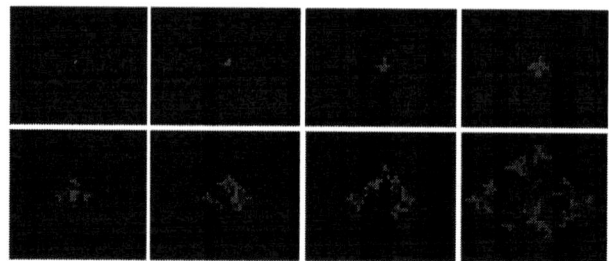

FIGURE 10.8
Example of growth achieved by a *Deva1* algorithm.

connect to joints at both ends; (2) to remove sections that are not connected to the base of the structure; and (3) to remove redundant joints, replacing them with longer beams. All three of these can be accomplished in a single pass of the untrimmed truss structure, allowing for processing in $O(n)$ time, where n is the number of beams.

The output of φ may be computed as follows, given a current cell c_0 and its neighborhood μ_{c_0}:

```
int  e_i  ← 0 for all  e_i  ∈  {e_0,...,e_{n_c}}
for all Cells  c ∈ μ_{c_0}  do
        int type ← cellType(c)
        e_{type} ← e_{type} + 1
end for
        Rule r_{min} ← φ
        int minDistance ← ∞
        for all Rules r ∈ φ do
            if r_{color} = color(c_0) then
                    float distance ← (e_0 - r_{h_0})² + ... +(e_{n_c} - r_{h_{n_c}})²
                    if distance < minDistance then
                        minDistance ← distance
                            r_{min} ← r
                    end if
                end if
            end for
return a_{min}
```

The null action is interpreted as "nothing." The running time of a transition function lookup is thus $O(|\varphi|)$. Cell actions are the sole means through which the developmental space changes in time. The possible actions are

- *Nothing*, the empty action.
- *Die*, which removes the cell, leaving an empty location.
- *Divide*, which creates a clone of the cell in the best free location.
- *Specialize(X)*, which changes the cell's specialization from one cell type to another, $X \in C$.
- *Elongate*, which causes the cell to elongate in the direction of previous elongation, or, if unelongated, in the best free location.

The best free location is defined as the empty adjacent location (in the von Neumann neighborhood surrounding the cell), which lies opposite to the greatest mass of nonempty cells (in the Moore neighborhood). Most cell actions come with a cost, decrementing a cell's r_c; this is meant to incorporate the notion of finite resources. If a cell cannot execute an action (no best free location, insufficient resources), it does nothing. A *Deval* growth is controlled then through a genome (transition function), and several system parameters

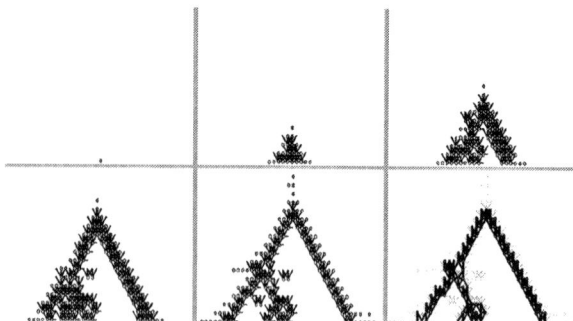

FIGURE 10.9
Another example of growth in developmental space.

(number of cell types, n_c, initial setting of resource counter, r_c). Figure 10.9 illustrated one possible example of *Deva1* growth.

10.4.3 Overall Evolution

The use of *Deva1* for the generation of designs is controlled overall via evolutionary computation (EC). That is, genomes are mapped to organisms via the *Deva1* algorithm, and the organisms are assigned fitness through the truss interpretation. The fitness serves to select a set of genomes for the next generation, and the actual selection and recombination is controlled through a GA.

A typical GA is used, as described by Eiben and Smith.[7] The GA uses elitism as well as crossover and mutation as defined above. Selection is accomplished through a tournament of five population members, using a tournament probability of $p = 0.7$. As will soon become clear, convergence is difficult to recognize, so trials were run for a fixed number of generations. Additionally, the initial population size was larger than the population size for successive generations, by a factor of *initMult*.

The evolution of plane trusses may be viewed as a multiobjective evaluation; there are many factors involved in the computation of fitness. These factors, defined for a general truss T, include:

- Selection for nontriviality, where T is trivial if it contains fewer than five cells, or fewer than three cell types: $t(T) = 1/2$ if T is trivial, $t(T) = 1$ otherwise.
- Selection for height, $h(T) = h/(r_c + 1)$, where h is the raw height of T.
- Selection for minimal material use, where $m \in [0, 1]$ varies linearly between 0 for maximal use of materials ($2(r_c +1)^2$ in beam length), and 1 for no use of materials.
- Selection for a small base, $b \in [1/2, 1]$, minimized when a minimal number of joints support the structure at the center of the space. The base count, b' is defined to be the sum of the distance of the existing base joints from the center.

$$b' = \sum_{i=-r_c}^{r_c} \begin{cases} |i| & \text{there exists a joint at location } i \\ 0 & \textit{otherwise.} \end{cases}$$

and define the base factor to be

$$b = \frac{1}{2} + \frac{1}{2}\left(\frac{\sum_{i=-r_c}^{r_c}|i| - b'}{\sum_{i=-r_c}^{r_c}|i|}\right)$$

- Selection for stability, where T is considered stable if the inverse stiffness matrix is nonsingular and, if under external force, there are no absurd deformations. The stability criterion is then defined as $s(T) = 1$ if T is stable, $s(T) = 1/4$ otherwise.
- Selection for distribution of pressure, $p \in [1/2, 1]$. Having applied some external force, the maximum absolute beam pressure in the truss, M, is measured. If pressure has exceeded the yield limit of 165 MPa, a p of ½ is returned, otherwise:

$$p = \frac{1}{2} + \frac{1}{2}\left(\frac{165MPa - |M|}{165MPa}\right)$$

At every joint, a force of 50 N is applied down and 50 N left, simulating gravity and a mild horizontal force. Additional forces are defined by the fitness function in question.

10.4.4 Fitness Evaluation

The first fitness function, f_{mat}, is designed to maximize height while minimizing overall material. Pressure is evaluated by applying 20 kN down and 5 kN right at the highest joint. In the case of several joints, the force is divided evenly between them. Hence, the aim is a tall, minimal structure, capable of supporting a large mass at the top, much like a tower supporting some additional structure at the peak. The fitness of a truss T is defined as:

$$f_{mat}(T) = t(T) \cdot h(T) \cdot m(T) \cdot s(T) \cdot p(T)$$

The second fitness function, f_{stoch}, is similar to f_{mat} in all ways except that rather than apply external forces at the highest joint, the forces are applied randomly. Hence, three nonbase joints are selected at random, and at each a force of 5 kN is applied down and 500 N either right or left (with equal probability at each joint).

Our final fitness function, f_{base}, is again similar to f_{mat} in that a force of 20 kN is applied down and 5 kN right divided between the highest joints. However, the length minimization factor is removed and instead the base minimization factor is included. Therefore, f_{base} selects for tall trusses capable of supporting a load at the peak occupying as little ground space as possible:

$$f_{base}(T) = t(T) \cdot h(T) \cdot b(T) \cdot s(T) \cdot p(T)$$

10.4.5 Experimental Results

For the fitness trials, the population size is two hundred; *initMult* is twenty; the probability of crossover is 0.8; the rate of elitism is 0.01; the probability of copy mutation is 0.05; the probability of power mutation is 0.05; *useSeed* is set to true; φ is one hundred.

There were ten runs of the fitness trials for each fitness and r_c setting, r_c = 16, 24. These runs are referred to as *fit.x.y.z*, where *x* is a fitness function, *y* is an r_c value, and $z \in \{0, .., 9\}$ is an index. Hence, *fit.stoch.24.3* is the third run of the r_c = 24 trial using the fitness function f_{stoch}.

In all trials, successful trusses were evolved; all runs found stable trusses, and fifty-six of the sixty found trusses capable of supporting the applied external forces. In general, heights of approximately 9 m were found in the r_c = 16 trials, and heights of 18 m when r_c = 24. There were several general trends in evolved solutions for each fitness function. For the f_{mat} function, all high fitness population members somewhat resembled the exemplar, a simple triangle-shaped truss. Organisms varied greatly, however, in the f_{stoch} and f_{base} runs. For the f_{stoch} function, sparse pyramids were common. Also, there were many agents with thin, tall upper sections and large bases. For the f_{base} function, some tall trusses with small, central bases were found. Additionally, large pyramid trusses with sparse bases were also common. Figure 10.10 shows exemplar population members illustrating these phenomena.

The maximum fitness of agents in the r_c = 16 fitness trials are graphed in Figure 10.11. Note that frequent plateaus are found in each run, also present in the r_c = 24 runs. This suggests that the genetic operators are more frequently impotent or destructive, relative to more typical experiments in GAs. It is also reminiscent of theories of evolution via punctuated equilibria, where (real world) evolution is believed to work via infrequent jumps.

As illustrated in Figure 10.12, a visual continuity between the phenotypes of members could typically be seen. In the example, agents show many similar qualities, including the presence of single-unit length crossbeams, hollowed-out centers, and elongated base supports.

10.4.6 Summary and Challenges

Kowaliw et al.[6] presented a new model of AE, *Deval*. *Deval* has been shown to be capable of evolving plane trusses, that is, evolving designs of structure that are stable, capable of effectively distributing external forces, and also optimizing other constraints imposed by a fitness function.

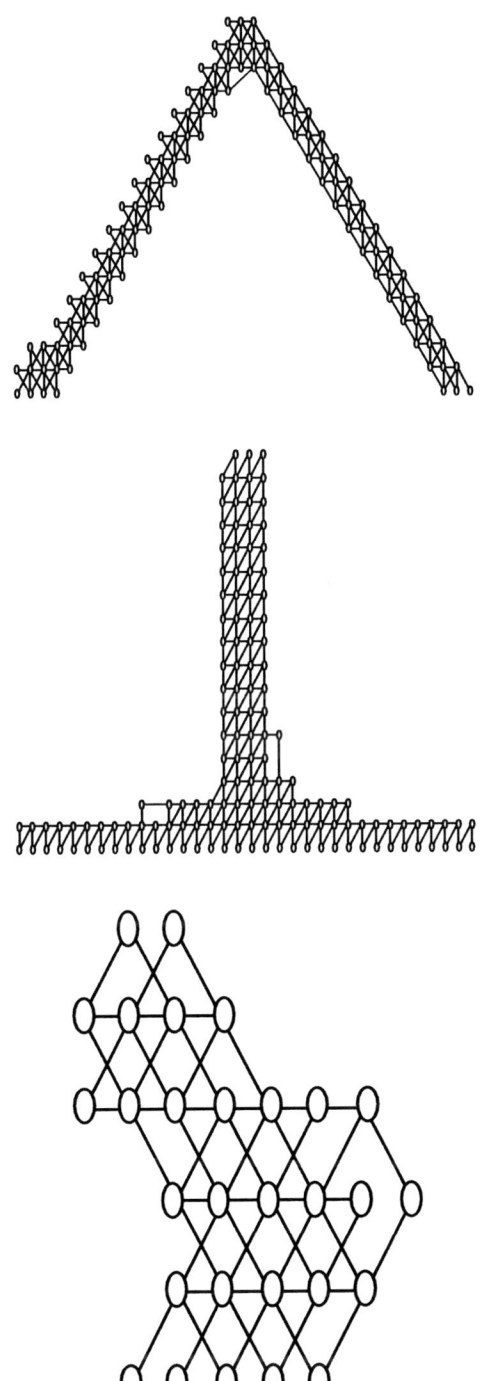

FIGURE 10.10
Exemplar organisms from the fitness trials.

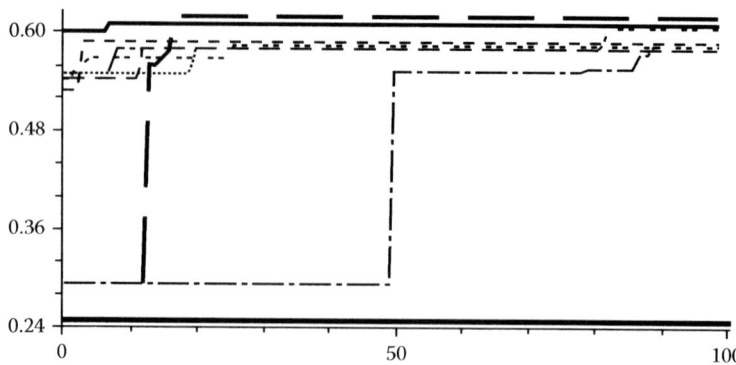

FIGURE 10.11
Plot of fitness (y) against generation number (x).

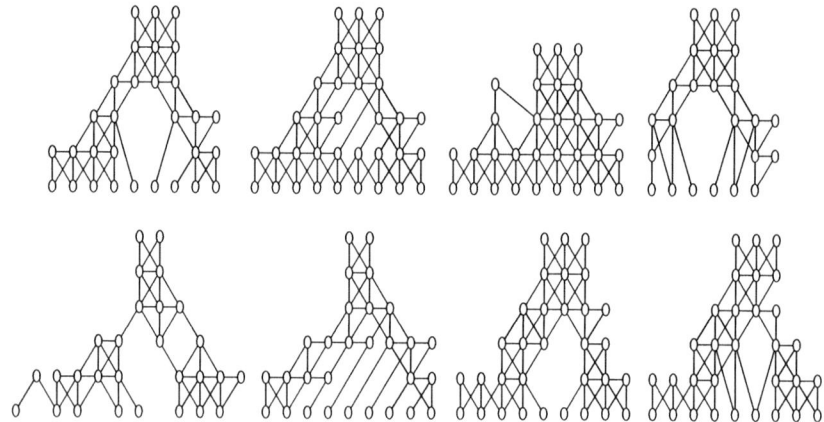

FIGURE 10.12
The fittest ten individuals (excluding repeats) in the hundredth generation; all are stable.

Given the nonlinearity of recursive growth, predicting the aggregate differences (or lack thereof) resulting from model-level decisions is very difficult, perhaps impossible. Considering the breadth of models currently existing in AE, empirical comparison on an accepted test problem should prove valuable.

10.5 Cellular Growth and the Evolution of Functions

10.5.1 Description of Sample Problem

The purpose of the exercise is the evolution of a "developmental program" that can guide the growth of one cell into an interconnected graph of cells with an overall property, such as a particular input/output relationship or complex shape. This work is done by Miller et al.[8] and builds on his previous work in Cartesian genetic programming (CGP).[9]

10.5.2 Representation of Potential Solution

In CGP (which does not have development), a single node has a function (e.g., addition or multiplication), inputs (e.g., 2), and an output representing the result of the application of the node's function to its inputs. Each node has a number representing its position and each of its inputs has a number indicating its connectivity.

In CGP, the genome corresponding to a graph is an expansive description of (1) each node in the graph and (2) how it connects to other nodes. An example of such representation follows.

Genome (1 chromosome):

0	1	2	3	4	5	6	7
x	001	101	111	220	341	510	600

Phenome (directed graph) (Figure 10.13):

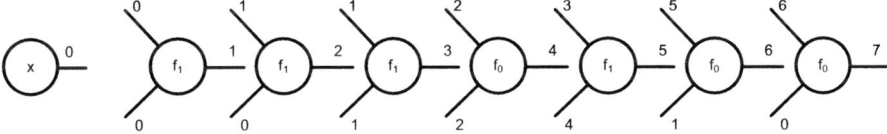

FIGURE 10.13
A genome and its corresponding phenome in Cartesian genetic programming (CGP). (Miller, J.F. and Thomson, P. [2003] A developmental method for growing graphs and circuits. In proceedings of the *Fifth International Conference on Evolvable Systems*, Springer-Verlag, Berlin, pp. 93–104.)

The first part of the genome represents the external input (x) that has position 0. Next, comes node 1 then 2 and so on up to node 7. The inputs of each node can only connect to outputs of nodes preceding them or external inputs. All nodes are described similarly and so only that part of the genome corresponding to node 5 will be described, the rest of the genome can be interpreted similarly. Node 5 has two inputs connected to the outputs of node 3 and 4, respectively, and it has function 1, which is multiplication (0 stands for addition).

10.5.3 Developmental Procedure

The core difference between Developmental CGP (or DCGP) and CGP is that the genome of a DCGP solution represents a single cell, which can develop into a graph of cells. To endow a cell with the ability to grow, the genome of a DCGP cell contains a "developmental program" (DP). This DP senses the

state of the cell, i.e., its current function, connectivity, and position. These are
the program's inputs, and its outputs are its next function connectivity and
whether it will divide or not. The position of a node does not change. Nodes
operate synchronously (i.e., they have a clock that control transitions from
one state to the next) (Figure 10.14).

At every clock tick, the developmental program of each cell (in the graph
of interconnected cells) computes its next state and whether it will produce a new cell or not. Division produces a new cell with exactly the same
(unchangeable) developmental program as the mother cell, but with a new
location = location of the mother cell + 1. Because CGP and by extension
DCGP only allow feed-forward graphs, the inputs of all cells will come from
external inputs and/or the outputs of other cells, which are directly/indi-
rectly connected to the external inputs. Hence, if the external inputs stay
stable, then so will the outputs of all the cells in the graph.

This stability means two things: (1) the functionality and connectivity of
the cell will stay constant (positions, as we said, never change) and (2) the
cell will either always divide or always not divide. This means that there
is a need for an external mechanism that determines when division (of all
dividing cells) is to halt. At present, this is a parameter value determined
beforehand by the human experimenter. All in all, one cell with a DP and
external inputs to that DP may divide into a graph of cells, all with the same
DP, but with potentially different functionalities and connectivities. After
a predetermined number of iterations, all dividing cells will stop dividing,
and the result of development will be a fixed graph of interconnected cells
with one overall functionality. This functionality is realized via a multicellular "organism" lying between the external inputs and the output of the
final node.

One issue remains. How is the DP actually implemented? We know that
the DP is basically a mapping that accepts four integers as inputs: connec-

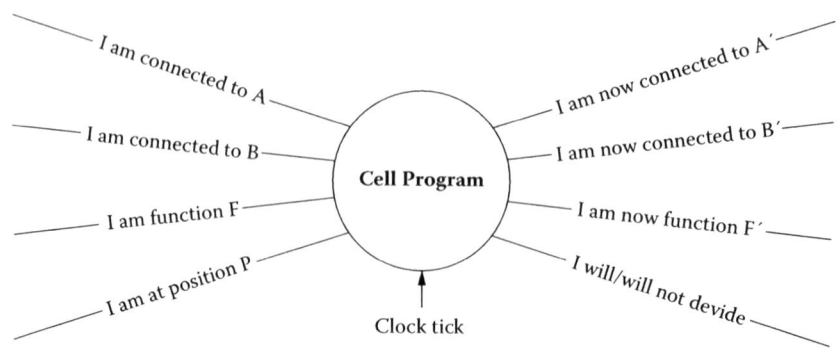

FIGURE 10.14
The general model of a node in developmental Cartesian genetic programming (CGP). (Miller,
J.F. and Thomson, P. [2003] A developmental method for growing graphs and circuits. In proceedings of the *Fifth International Conference on Evolvable Systems*, Springer-Verlag, Berlin, pp.
93–104.)

tivity of first input A, connectivity of second input B, function type F, and position P. The mapping produces three integers, Z, Y, and X, as well as a single bit W as outputs. These outputs are used to compute the next state and divide/not divide bit of the cell. Such a mapping can be implemented in various ways, including truth tables. However, since CGP allows us to define functions (with multiplication and addition operators), then there is no harm in defining the DP mapping in terms of four CGP chromosomes. A handcrafted set of four chromosomes, translated into their equivalent phenomes (functions), follow below (Figure 10.15).

However, these functions on their own may produce erroneous outputs to valid inputs; this is because there are no upper bounds placed on their outputs. A possible solution is to modulate the raw outputs of Z by P and Y by P because inputs of a cell can only connect to the outputs of cells at preceding positions, modulate X by N_f (the number of functions, i.e., 2), and modulate W by 2 (resulting in the divide/not divide bit value).

The modulated functions *are* the developmental program of the first (and every other cell) of the organism. Growing the entire organism requires initial values for the inputs to the DP of the first cell (say, A = 0, B = 1, F = 0 and P = 2) and the computation of the outputs of the DP by applying the phenomes (i.e., the modulated functions of Figure 10.16) to the first cell, and iteratively to any and every new cell, until the last predetermined cycle of growth. This will result in the graph shown in Figure 10.17.

Since the genomes representing the DP are no more than four binary strings, it is possible to use a GA to search for a set of four chromosomes (a genome) for a DP, instead of handcrafting the chromosomes as had been done so far.

$$Z = 2A+B$$
$$Y = Z+F$$
$$X = Y + P$$
$$W = ZX$$

FIGURE 10.15
The unmodulated functions of the developmental program of a developmental Cartesian genetic programming (DCGP) node.

$$Z = (2A+B) \mod P$$
$$Y = (Z+F) \mod P$$
$$X = (Y + P) \mod N_f$$
$$W = (ZX) \mod 2$$

FIGURE 10.16
The modulated functions of the developmental program of a DCGP node.

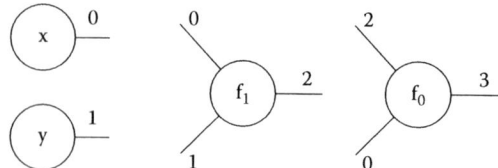

FIGURE 10.17
The final phenome resulting from the application of the developmental program (DP) in every node. (Miller, J.F. and Thomson, P. [2003] A developmental method for growing graphs and circuits. In proceedings of the *Fifth International Conference on Evolvable Systems*, Springer-Verlag, Berlin, pp. 93–104.)

10.5.4 Fitness Evaluation

In our example, the object of evolution is a developmental program for a cell. This DP controls the growth of the cell into a graph of cells representing a mathematical function. Why would one want to effectively evolve a function? A possible application is curve fitting.

In curve fitting, a set of points is given as input, and the required output is a function that passes through or close to the given points and possibly predicts the locations of future points as yet unknown. Complexity of the evolved function, in terms of number of operators used or modality, can also be taken into consideration when fashioning a fitness function of this kind.

10.5.5 Overall Evolution

The following simple evolutionary algorithms can be used for evolution:

```
Generate N chromosomes randomly
Repeat
   Evaluate the fitness of each genome.
   Determine the fittest chromosome CurrentBest.
   Generate N more genomes by mutating CurrentBest.
   Compile the next population from CurrentBest and its N mutants.
Until stopping criteria is satisfied.
```

Termination criteria usually combine two terms: one relating to the actual ultimate objective of the exercise, such as a perfectly fitting function and another term setting an upper limit on the number of generations that the evolutionary algorithm can run.

10.5.6 Experimental Results

There are a few real world experimental results for the DCGP approach to function synthesis. However, this approach was modified in order to apply it

FIGURE 10.18
Snapshots of the development of a cell program into a French flag.

to the problem of evolving two-dimensional colored patterns, specifically a French flag. That attempt involved a much larger number of inputs and outputs to the developmental program (of the growing cell). In fact, the genome was four hundred integers long. The evolutionary algorithm used was similar to the one described above. The fitness function measured the closeness of the emerging multicellular organism to a French flag. Those interested in this work are referred to Miller et al.[8] Snapshots of the unfolding of one developmental program produced in that study are shown above. They show a steady progression from a single (white) cell to a fairly well-defined red, white, and blue flag by the ninth iteration (Figure 10.18).

10.5.7 Summary and Challenges

It is obvious that developmental encodings, i.e., genomes that encode developmental programs rather than direct mappings, are difficult to design. In the most abstract terms, developmental encodings shrink the space necessary for describing a phenome. This is a good thing when the phenome is a very large design. However, this reduction in space is achieved at the expense of an expansion in the time necessary to generate the phenome from the genome, usually via an iterative or recursive process. There is no free lunch awaiting at the end of the developmental rainbow. Hence, the additional difficulty in devising developmental programs and the additional time that a developmental program requires to grow a phenome, makes it necessary for scientists to establish clear guidelines as to the exact real and potential benefits of using a DP (e.g., scalability) and the conditions that problems should meet in order for them to make good candidates for DP-based approaches.

10.6 Genomic Regulatory Networks and Modeling Development

10.6.1 Description of Sample Problem

The purpose of the work by Dellaert[10] is the simulation of a biologically defensible model of biological development, and the use of this model to evolve functionally autonomous agents.

10.6.2 Representations of Potential Solutions

Dellaert proposed two models of development, a complex model and a simplified model. Both models assume that agents are made of cells controlled by genomes. As such, this section presents the model of genomic regulation in both models. This is followed in the next section by an explanation of how the simple model can guide a process of development, which starts with a single cell and concludes with a complete organism.

10.6.2.1 Complex Model of Development

The complex model is called the genome–cytoplasm model. In this model, an artificial cell contains a genome consisting of a set of *operons*; all cells have the same genome. An operon is made of a small number of input tags, a Boolean logic function of the tags, and product tags, representing different

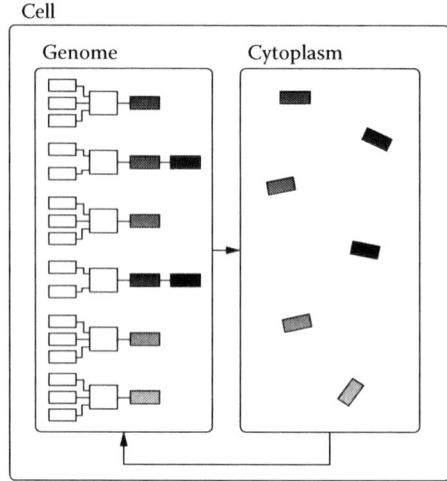

FIGURE 10.19
A graphical representation of genomic regulation in the complex model of development.

proteins. As shown in Figure 10.19, the product(s) are *produced* and injected into the cytoplasm if the Boolean function is satisfied or true (i.e., the operon is activated), otherwise, the product(s) are not produced.

The cytoplasm represents the rest of the cell and it contains the products produced via the activation of any and all of the operons (in the genome). The state of the cell (or its *type*) is determined at any time during development by the mix of protein products available within its cytoplasm. In addition, the proteins within the cytoplasm of a cell determine the next set of operon activations in the genome.

10.6.2.2 Simple Model of Development

The simple model may be called the random Boolean network (RBN) model (Figure 10.20). RBNs were first introduced by Kauffman[11] as an abstraction of real genetic regulatory networks. An RBN is a directed graph with N nodes, each with K incoming edges, which may be recurrent. Each node has its own Boolean logic function and is updated synchronously (i.e., on a global clock). The state of a node, as defined by its logic function, represents the presence (if = 1) or absence (if = 0) of a *protein*. Hence, if at a point in time, nodes (1, 2, 3) have the state vector (1, 0, 1), then that symbolizes the presence of proteins 1 and 3 in a cell represented by the RBN.

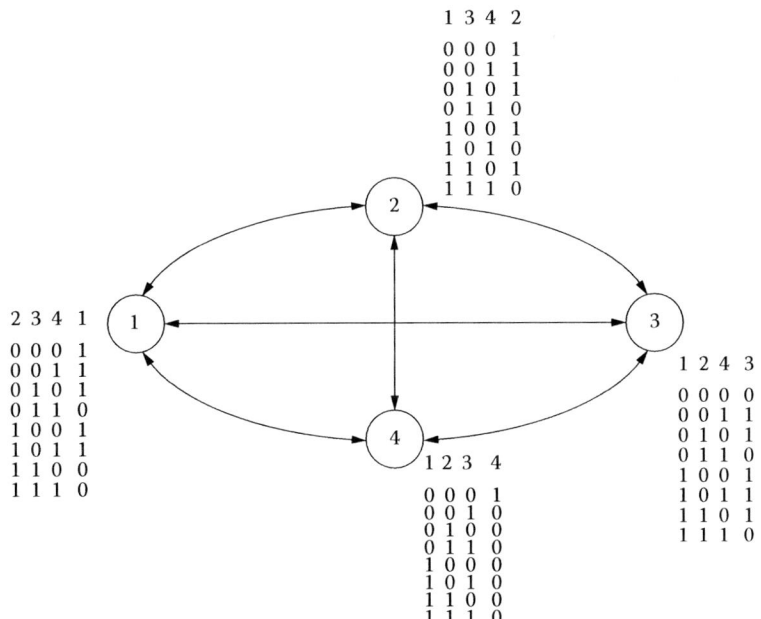

FIGURE 10.20
Genomic regulation in the simple model development.

Naturally, the state of an RBN will change with time. If there are no external influences, then it has been shown that the state of any RBN will settle into either a point attractor (i.e., steady state) or a cycle.

10.6.3 Developmental Procedures

The process of development described here concerns the simple (RBN-based) model, though similar operations do take place in a more complex model. Development occurs at three hierarchical levels: the genomic level, the cellular level and the organism (or agent) level.

At the genomic level, an RBN models genomic regulation. The state of a cell is described via its state vector. A state vector may be viewed as a set of the proteins that exist in a cell at some moment in time. However, even in this minimal model, two other properties of real cells must be included: cell division and spatial patterning.

To simulate cell division, a cell cycle made of two phases is used: interphase and mitosis. During interphase, the synchronous updating of the random Boolean network is carried out. This is done repeatedly for a (prespecified) fixed period of time until the RBN reaches a point attractor. Any organism with an RBN that settles into a cycle or takes too long to settle is discarded. Once the RBN has settled down, the state vector will decide the nature of the next phase. If a particular gene (i.e., a prespecified node in the RBN) is switched off, then the cell will not divide, but simply wait for the next interphase to start. But, if that gene is on, then this would signify mitosis and the cell will divide. The new cell will by definition have the same RBN as the mother cell; it will also inherit its state vector. This means that unless an input (from another cell or the external environment) changes, the new cell will have exactly the same state and, hence type as the mother cell (Figure 10.21).

To simulate spatial patterning, the organism is depicted as a dividing square of fixed area. Only the number of cells changes and it does so by adding a vertical or a horizontal division to an existing cell, i.e., the size of the cell is unimportant. Each existing cell is either a square or a rectangle, and it has a color that reflects its type. This simple scheme of representing spatial organization of cells is obviously not realistic; it does, however, reduce the computational needs of the simulation.

FIGURE 10.21
The initial cell dividing twice into three cells.

At the organism level, there are three issues that must be tackled, besides that of visual representation: the first division, breaking of symmetry, and intercellular communication. The simulation forces the first cell to divide by initializing the node (gene) signifying division to 1. This ensures that the organism divides at least once.

In order to ensure long-term asymmetric divisions, each cell is provided with information about its spatial positioning with respect to the external boundary and the midline of the organism. This information is provided to the growing organism via a special mechanism, which also deals with intercellular communication, i.e., information signals coming from neighboring cells.

A great deal of thought has gone into ways of simulating intercellular communication (or induction). Induction is concerned with the means by which one cell or group of cells can affect the developmental fate (or settled cell type) of another cell or group of cells via passing signals. A direct way to simulate this is to implement a modification, which would allow for some of the edges of the random Boolean network of a cell to come from a *neighborhood vector*. The neighborhood vector is the logical OR of the state vectors of all neighboring cells. To differentiate between an edge coming from the cell's own state vector and an edge coming from the neighborhood vector, a minus sign is attached to every input coming from the neighborhood vector. Finally, specific bits in the neighborhood vector of a cell are used as binary (ON/OFF) indicators of midline and externality. This whole arrangement is shown in Figure 10.22.

10.6.4 Fitness Evaluation

To explore the operation of the simple model of development, an artificial organism was evolved, one that simulates a chemotactic agent; an agent that

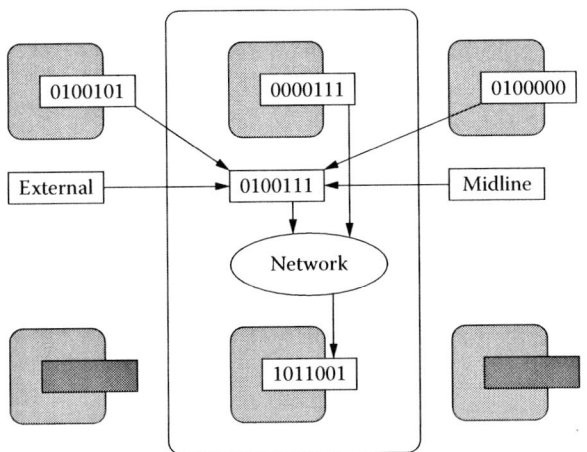

FIGURE 10.22
A simplified graphical representation of intercellular communication.

moves toward a certain external chemical signal. For chemotaxis, a possible straightforward design would involve a bilaterally symmetric agent with sensors placed at the front, actuators placed at the back, and a control structure using the information coming from the sensors to properly direct the actuators. An instance of such a design is shown in Figure 10.23.

In Figure 10.23, 1 stands for control cells, 2 stands for sensors, and 4 stands for actuators. Cells of type 0 and 6 are unwanted types. The organism on the left of the figure is the "ideal" organism used to compute fitness; an evolved organism matching it perfectly would receive top fitness numbers. The organism on the right of the figure (called *seeker*) is an example of an actual organism evolved by Dellaert. The fitness function used to calculate the fitness of an evolved organism relative to the ideal organism is presented below (in two parts):

$$\text{Total score} = \text{number of cells} * 10 + \text{sum over colors } (s_c) \qquad (10.2)$$

where for each requested cell type or color c

$$s_c = [\text{anywhere}]*300 + [\text{right place}]*10000 + \text{sqrt (match percentage)}$$

The fact that the sheer number of cells is rewarded positively encourages division. If cells of a right type grow in the wrong place (i.e., anywhere), then that also is rewarded. However, if cells of the right type grow in the right place then the reward is highest. Finally, the square root term is meant to balance matching percentages when all wanted cell types are found. The ideal organism has the highest fitness value of thirty, and only an organism with two cells (as one is not possible) of unwanted types would receive the lowest fitness value.

4	4	1	1	1	1	2	2
4	4	1	1	1	1	2	2
1	1	1	1	1	1	1	1
1	1	1	1	1	1	1	1
1	1	1	1	1	1	1	1
1	1	1	1	1	1	1	1
4	4	1	1	1	1	2	2
4	4	1	1	1	1	2	2

6	6	6	6	6	6	2	2
4	0	0	0	0	0	0	2
4	0	0	0	0	0	0	4
1	1	1	1	1	1	1	1
1	1	1	1	1	1	1	1
4	0	0	0	0	0	0	4
4	0	0	0	0	0	0	2
6	6	6	6	6	6	2	2

FIGURE 10.23
An ideal (left) and an actual evolved organism (right).

10.6.5 Overall Evolution

A *steady state* GA is used. It proceeds on an individual per individual basis. As is illustrated in Figure 10.24, at each generation two of the fittest individuals in the population are selected using *tournament selection*. They are then used to generate one offspring. This is followed by the application of crossover then mutation. The single offspring replaces one of the less fit individuals in the population, with a probability dependent on the fitness of the individual considered for replacement. In addition, there is a certain probability that the newly generated individual is not used, but that a completely new and random genome is used instead. The specific details of the implementation are dependent on the encoding used. Also, for each run the following parameters must be chosen: population size, tournament size, maximum number of generations, random fraction, mutation rate, and probability of crossover.

Each individual results in a genome capable of developing into a fully grown individual. After development, the fitness can be calculated and the evolutionary process gets its feedback. The following example explains the development process of a "good" individual evolved by Dellaert.[10]

10.6.6 Extended Example

The genome of *seeker* represents the core control circuitry, so to speak, of each cell of the organism, not just at the start of its development, when it is made of exactly one cell, but also after development has concluded, resulting in a

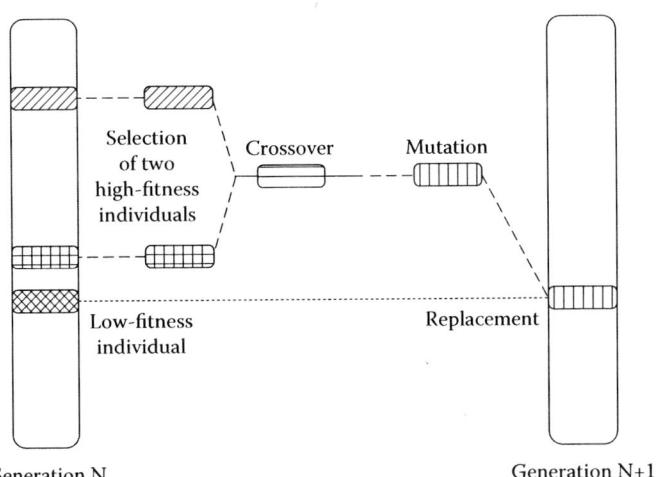

FIGURE 10.24
The steady state genetic algorithm.

functional multicellular organism. The genome is the only part of the organism that undergoes evolution; all other mechanisms (e.g., the specific way in which cells are allowed to communicate via a neighborhood vector) are fixed throughout the simulation. The genome describes the RBN of a cell: the number of nodes it has, the Boolean logical function of each node, and how the nodes connect to each other (as well as to external signal sources).

The genome in Table 10.1 describes an RBN with N (number of nodes) = 6 and K (node connectivity) = 2. Each node is fully described by its inputs and the logic function applied to them. Because each node has 2 inputs, it requires a $2^2 (= 4)$ row truth table to fully specify its output (state). Since the truth table has a fixed format (00/10/01/11), it is sufficient to list the output column of the truth table of a node to describe its functionality. For example, node # 3 has "0001" as its logic function. This is read: Input values of 00 return an output of 0, 10 return a 0, 01 return a 0, and 11 return a 1. This is equivalent to the AND function, applied in this case to inputs (−5) AND (5). The rest of the rows, representing nodal functions, are read similarly. It is important to note, however, that the logic function at node #4 will always result in a value of 1 (or TRUE) making this genome one that favors cell division, something that was favored in the fitness function seen in equation (10.2) earlier.

It is essential to note that input (5) comes from node #5 of the RBN of the cell itself. In contrast, the minus sign of input (−5) indicates an external input coming from the neighborhood vector, where the (−5) and (−6) inputs are reserved inputs. The value of (−5) equals 1 if the cell is on the boundary of the organism; it is 0 otherwise. Another external input, (−6) equals 1 if the cell borders the midline of the organism; it is 0 otherwise. If the information in Table 10.1 is used to construct the RBN of a seeker cell, then the graphical representation in Figure 10.25 would emerge.

In order to understand the mechanism that results in a fully developed organism, let us look at the start of the process after having — through

TABLE 10.1

Node #	Logic Function 0101 0011	Inputs	
1	0010	3	−6
2	1100	−2	−1
3	0001	−5	5
4	1101	4	4
5	0110	6	−6
6	0111	6	−1

Source: Dellaert, F. (1995) Towards a Biologically Defensible Model of Development. Master's thesis, Department of Computer Engineering and Science, Case Western Reserve University, Cleveland, OH.

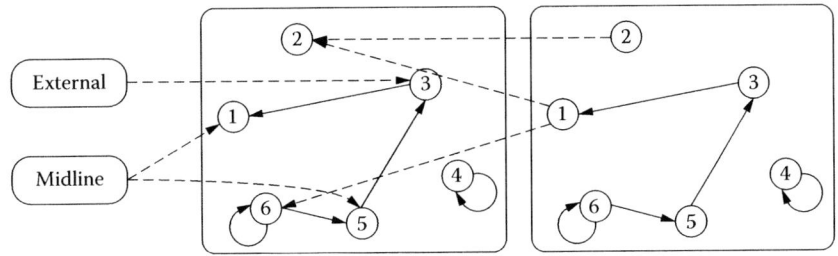

FIGURE 10.25
The random Boolean network of a seeker cell showing internal and external connections (via edges).

evolution — settled on a genome. The genome explains what to do with an input (be it internal or external), but does not specify the value of that input. Therefore, to kickstart the process, all nodes (inputs) are set to 0 except node #4 responsible for cell mitosis (division). This will ensure a first division.

By definition, cell mitosis results in two daughter cells, identical to each other and to the parent cell. So in order to break symmetry, bit 1 in the state vector is set to 1 in one cell, while it is set to the reverse (0) in the other cell. The first 3 bits of the state vector are prespecified as the ones indicating cell type — in reverse. For example, a cell with state vector = 001110 has type 001-reversed, which is 100 or 4 (in decimal). This entails that the first division inevitably results in two cells of different types with state vectors 000100 and 100100, respectively.

From this stage onward, interphase and mitosis (or division) alternate until the final organization of the agent is reached. Again, interphase is the phase in which each and every cell (1) computes its neighborhood vector, (2) allows its RBN to settle into a fixed state, and (3) decides whether to divide or not based on the state of bit 4 of its (settled) state vector. One more cycle of development will be described in detail. All following cycles, which are too many to be described here in detail, follow the same pattern.

Applying the procedure outlined above implies the application of the update rules of Table 10.1, with an eye to Figure 10.24, to the two cells resulting from the first division, giving us:

Cell L	Cell R
Neighborhood vector = 100110	Neighborhood vector = 000110
State Vector = 000100→000101→001111	State Vector = 100100→010100
Cell Type = 4	Cell Type = 2
State Vector Bit [4] = 1 (Divide)	State Vector Bit[4] = 1 (Divide)

The tabulated form above accurately describes the process all cells go through. What follows is a crude step-by-step explanation for one of the daughter cells.

First, the *neighborhood vector* is computed — by ORing the neighboring state vectors (in this case only one), and by setting bits (5) and (6) accordingly. Second, the *state vector* describes the stages that the RBN goes through until it settles into a fixed state and a cell type is generated. Iteration –1– involves setting bit (6) by following the rule at node #6, i.e., bit (6) = (6) OR (–1). Iteration –2– follows similarly and the RBN settles resulting in cell type #4. Third, a decision on dividing is made.

After cells L and R divide, we get four cells: LT, LB, RT, and RB (where L is left, R is right, T is top, and B is bottom). All cells have the same genome (and, thus, update rules), but they also inherit the steady state vectors of their parents. The steady states of new cells RT and RB are not perturbed by any external influence, in accordance with the update rules of Table 10.1. However, for new cells LT and LB, the resetting of bit 1 in the neighborhood vector results in bit 2 of their state vectors switching ON. Since this bit is part of the 3-bit cell type bits, the type (and, thus, color) of both resulting cells change. This process is summarized below.

Cells LT and LB
Neighborhood vector = 011110
State Vector = 001111 → 011111
Cell Type = 6
State Vector Bit[4]= 1 → Divide

Gradually, and after a large number of divisions and differentiations, the final organism emerges; it is the one shown earlier in Figure 10.23.

10.6.7 Summary and Challenges

In this section, two models of development were presented, a complex model consisting of a multioperon genome and a cytoplasm, and a simple model based on random Boolean networks. The simpler model was explained in more detail, as it is the basis for the extended example described here. This model utilizes both development and evolution to get to a cell that can develop into a multicellular organism able to seek a chemical trace.

This process of development, which starts with a single cell, is controlled by the genome, which describes an RBN. At any moment, all the cells of the organism will have the same genome (and, thus, RBNs), but may well have different inputs to these RBNs. Inputs to a cell's RBN come from the RBN itself as well as an external neighborhood vector. All cells in the organism follow the same three-step update procedure in order to arrive at a steady state. This steady state determines the type of the cell (which is interpreted as a color) as well as whether the cell will divide or not. This process eventually leads to a complete functioning organism, such as the one shown in Figure 10.23.

A steady state GA is used for evolution. Evolution is applied to the genome of the organism. An example is shown in Table 10.1. The purpose of evolution is to devise a genome (= RBN) that can successfully guide a process of development, starting with a single cell and ending with a satisfactorily functioning multicellular organism. Fitness is only measured after the organism is fully developed, but is assigned to the genome. The cycle of evolution continues until a good genome is found.

10.7 Summary

This chapter began with an introduction to the often-neglected *mapping* between the genome and the phenome of an individual within the context of evolutionary computation. In most evolutionary algorithms (EAs), this mapping is seen simply as a way of converting the genome of an individual into an entity that can be evaluated for fitness. However, mappings can be very elaborate, and in natural organisms such mappings take a considerable amount of time and resources to unfold. It is the time necessary for a seed to develop into a tree and an embryo into an adult. In evolvable developmental systems (EDS), development is simulated using mappings and evolution is used to find a satisfactory mapping or an appropriate configuration for a generic mapping.

The field of EDS is relatively new and expanding. There are not yet standard techniques or standard methodologies to apply, not even in specialized fields of application (e.g., digital chip design). Nevertheless, there are a number of approaches that seem to be attracting a fair bit of attention by EA practitioners and researchers. These include production rules (e.g., L-systems), cellular automata (e.g., *Deva1*), cellular growth mechanisms (e.g., developmental CGA), genomic regulatory networks (e.g. RBNs), and others. Sections 10.3 to 10.6 of this chapter present examples of each one of these approaches. They have different areas of "application," but since real-world applications are not yet the main aim, the emphasis in our explanations is focused on methodology.

In section 10.3, Haddow et al. show how they successfully evolved a set of production rules, which grew a digital circuit that operated as a router. It was not an optimal router, but it showed the possibility of using evolution and development to design circuits. In section 10.4, Kowaliw et al. evolved single cells with the ability to grow into different types of plane trusses. None of the evolved designs are optimal by civil engineering standards, but they all appeared to substantially satisfy whatever fitness function was used in evolving them. Section 10.5 presents an incremental growth (= developmental) version of Cartesian genetic programming, as proposed by Miller at al. This approach was used to evolve both mathematical functions and colored two-dimensional images (flags). In the case of functions, development starts with a single node (which contains the developmental program) and ends with a completely specified and interconnected network of nodes, representing a mathematical function. Finally, section 10.6 briefly describes two computational models of genomic regulation developed by Dellaert. The simpler one (utilizing random Boolean networks or RBNs) is used to develop, via division and specialization, a multicellular simulated robot with the ability to steer itself toward a (simulated) chemical trace.

This chapter concludes with a fairly abstract introduction to future challenges, as we see them, and a brief science fiction story, which also functions as an epilogue.

10.8 Future Challenges and Epilogue

There is a general malaise in the evolutionary computation (EC) community. It is that EC has been reduced to genetic algorithms (GA) and genetic programming (GP), that GA and its close relatives (e.g., evolutionary strategies) are essentially optimization techniques and, that despite valiant attempts by John Koza and colleagues, GP has not been able to deliver automatic general-purpose programming tools. In both cases, representation and fitness evaluation have been at the heart of endless problems as well as some opportunities. Our own experience has been with GA and GA-like computational systems, thus our focus will henceforth be on GA.

Representation requires that the designer of a typical evolutionary computation algorithm (EA) formulates one inadaptable blueprint for the solution of some problem, then present the variables of that blueprint in a form that is amenable to manipulation by the genetic operators of the EA. Fitness evaluation, on the other hand, has limited GA in two distinct ways: (1) it has limited environmental feedback to the confines of a formula or algorithm, which reflects accurately and exclusively the quality of the complete candidate solution from the perspective of the human designer. In addition, (2) fitness evaluation has proven to be the most computationally costly part of a typical EA. Note that elaborate developmental mappings actually increase that computational cost. However, our interest here lies in the limiting effects of representation.

Representation, in GA, means three entities: (1) the genome, which is the (initial) encoding of the general form of a solution; (2) the phenome, which is the candidate solution; and (3) the mapping, which takes us from the genome to the phenome. In the great majority of GA applications, there is a direct mapping between genome and phenome. There are examples of mappings that are more elaborate, but it was only recently that serious attempts were made to design multistage mappings, and to assess their value, using empirical and theoretical means. These attempts are now coalescing under a set of related names, including development, embryology, and generative representations. What they all have in common is an elaborative mapping procedure that starts with a simple genome and produces a significantly more complex phenome. In some cases, this mapping is explicitly described and in others it is implicit.

Our academic purpose in presenting various examples of evolvable developmental systems is to raise people's attention to the need for computational techniques that are

- *Adaptable*: Go through an embryonic stage, which allows the developing phenome to adapt, in terms of size, complexity, and possibly other attributes to certain fixed attributes of the environment (or problem). The specific settings of these attributes are considered as inputs to the developmental program, which controls the embryo from conception to adulthood. Hence, the adult will reflect in some of its unchanging "structural" attributes, relevant long-term features of the environment, in which an adult will have to function.

- *Adaptive*: The resulting adult will need to have the ability to adapt to short-term attributes of the environment, via physical and cognitive *learning*, in order to function as optimally as possible, within a dynamically changing environment. This requires that the resulting adult have prespecified machine learning mechanisms, such as neural nets or classifier systems. Also, the organism must be *robust*, in that it must be resistant to temporary and lasting changes to its form and function. In terms of form, the ability to regenerate or use redundant components comes to mind; as to functionality, the ability to reestablish past functionality or switch to an acceptable alternative functionality after a brief readjustment period comes to mind.

- *Evolvable*: Besides life-time adaptation, which takes place during embryonic development and adult learning, intergenerational adaptation of heritable patterns (e.g., genome) will have to take place, via selection-and-diversification procedures acting at both individual and population levels, in order to ensure long-term survival of the species in a slowly changing environment. This adaptation can be carried out via GA, ES, GP, or other flavors of EA.

Therefore, a key goal for future research is to establish a clear methodology by which self-adapting computational organisms can adapt, both short-term and long-term, in order to enhance their survivability, in a changing competitive world.

In addition, definitive quantitative measures need to be defined to allow us to measure the potential survival advantages of adaptation at all three stages. This requires that we set out a fair comparison to nonadaptive versions of the same organisms, and to run the whole simulation in an artificial world, which is an accurate reflection of the actual environments that these organisms are expected to function in.

10.8.1 Epilogue

It was a beautiful summer day of the year 16042 when the expansion of the universe halted and, for a moment, time stood still. It was well known that the spatial expansion of the universe was slowing down, and at an increasing rate. Scientists were even able to narrow down the window, within which the great reversal would start, to within five years. Only a small minority, however, argued that time and not just space will reverse direction. They had warned of the dangers of completely neglecting an improbable, but highly significant, scenario. Then it came. Rivers flowed backward from the sea to their source and plants grew back into seeds. Nevertheless, the animals did not seem to mind this unusual change of order. Only humans fretted depressively about their impending doom, as their minds grew dimmer and their bodies younger. The governments of the richest nations came together in an urgent World Summit. For days on end, they produced and debated possible solutions to their unusual condition. Their main concern was the survival of the species — especially the human one. They knew that they had a fair bit of time to agree on a realizable solution, but not infinity. For all humanity was degenerating, deevolving, in slow motion, and at some point in the future-past, Man himself is bound to go extinct. Twenty-one and a half days later, the Subcommittee for Biocomputing Alternatives proposed a detailed remedy. Match the great reversal with a reversal of development. Gather all the knowledge that humanity has been able to generate about biological development— especially human development. Engineer a virtual machine — not a living creature, but a software entity, an entity with a thousand times the intellect of a genetically enhanced human. But this would be no ordinary entity; it would have the ability to grow, to develop, as humans do or rather, used to. The entity would "live" and develop in a virtual world of prereversal laws, where flowers gave rise to fruits and the sun rose from the east. The entity would also have access to an unlimited supply of energy and would be linked to the online crucibles (libraries) of the advanced world. Critically, however, this entity would develop the old-fashioned way: From a single cell to a fully formed and superintelligent adult. A team of nine top scientists, one hundred and twenty-one engineers and an army of technicians and support staff were put to work. No cost was spared, and all distracting voices were kept at bay. The raging discussions and even confrontations about the possible futility of this expensive technology were kept off the team's airwaves. Indeed, the team was totally preoccupied with the success of the first simulation. So, it was no surprise that the sudden and inexplicable "death" of EV1 (what the entity came to be called) almost killed the whole project. Seven of the nine original scientists persisted, developed software modules carefully and reverted to safer means of implementation of increasingly biological simulations. They launched one trial after another, mostly in secret, using: e-GRNs, PBNs, even G-Nets. Five days shy of two years of development, on an equally gorgeous summer day of the year 16039, EV5 was unleashed — its (hardware) blue, green, and red lights flickering in seemingly random patterns. The entity grew into a bulb of triangular then hexagonal "cells." The bulb morphed into a slithering snake-like "worm" with hundreds of pulsating cells, emitting high-frequency shrieks. The cells divided again and again, always changing shape, color, and behavior. This continued for about 32 weeks; by then EV5 had developed into a beautiful and hyperintelligent baby girl. Everyone was anxiously awaiting "her" first words. What would they be? Would they be "milk" or perhaps "mama" or a word in some ancient language, which none of them understood? The days went by, and suddenly, a completely formed paragraph came out of the girl's virtual mouth:

> Your eyes of wonder
> are stained with blood.
> Your brains a sunder
> with fears of mud.
> Why not let go,
> let time run raw,
> and make your dreams
> of nature grow?

Acknowledgments

In order to complete this chapter, I had to call on the help of three friends, who I would like to acknowledge. First, I thank Dr. Luc Varin for providing me with unfettered access to his office at the Loyola campus of Concordia, away from the distractions of Montreal's downtown. I also am thankful to Imad Hoteit for reading and providing detailed feedback on the first draft of the chapter. Last, but not least, I would like to thank Mohammad Ebne-Alian for the excellent technical drawings and recreations used to illustrate parts of this chapter.

References

1. Kumar, S. and Bentley, P.J. (1999) Computational embryology: Past, present and future. In Ghosh, A. and Tsutsui, S. (eds.) *Theory and Application of Evolutionary Computation: Recent Trends*, Springer-Verlag, Berlin.
2. Stanley, K.O. and Miikkulainen, R. (2003) A taxonomy for artificial embryogeny. *Artif. Life*, 9:93–130, MIT Press, Cambridge, MA.

3. Haddow, P.C., Tufte, G., and Van Remortel, P. (2003) Evolving hardware: Pumping life into dead silicon. In Kumar, S. and Bentley, P.J. (eds.) *On Growth, Form and Computers*, Elsevier Academic Press, Amsterdam, pp. 405–423.

4. Haddow, P.C. and Tufte, G. (2001) Bridging the genotype-phenotype mapping for digital FPGAs. In proceedings of the *Third NASA/DoD Workshop on Evolvable Hardware*, IEEE Computer Society, Washington, D.C., pp. 109–115.

5. Lindenmayer, A. (1975) Developmental algorithms for multicellular organisms: A survey of L-systems. *J. Theoret. Biol.*, 54:3–22, Elsevier, Amsterdam.

6. Kowaliw, T., Grogono, P., and Kharma, N. (2007) The evolution of structural design through artificial embryogeny. In proceedings of the *IEEE Symposium on Artificial Life* (ALIFE'07), IEEE Computer Society, Washington, D.C., pp. 425–432.

7. Eiben, A.E. and Smith, J.E. (2007) *Introduction to Evolutionary Computing*, Springer-Verlag, Berlin.

8. Miller, J.F. and Thomson, P. (2003) A developmental method for growing graphs and circuits. In proceedings of the *Fifth International Conference on Evolvable Systems*, Springer-Verlag, Berlin, pp. 93–104.

9. Miller, J.F. and Thomson, P. (2000) Cartesian genetic programming. In proceedings of the *Third European Conference on Genetic Programming*. LNCS, 1802:121–132, Springer-Verlag, Berlin.

10. Dellaert, F. (1995) Towards a Biologically Defensible Model of Development. Master's thesis, Department of Computer Engineering and Science, Case Western University, Cleveland, OH.

11. Kauffman, S. (1969) Metabolic stability and epigenesis in randomly constructed genetic nets. *J. Theoret. Biol.*, 22:437–467, Elsevier, Amsterdam.

Index

D

E